Results and Problems in Cell Differentiation

Series Editor:
W. Hennig

37

Springer

Berlin
Heidelberg
New York
Barcelona
Hong Kong
London
Milan
Paris
Tokyo

Kevin Moses (Ed.)

Drosophila Eye Development

With 58 Figures

Springer

Dr. KEVIN MOSES
Department of Cell Biology
Emory University School of Medicine
1648 Pierce Drive NE
Atlanta, Ga 30322-3030
USA

ISSN 0080-1844
ISBN 3-540-42590-X Springer-Verlag Berlin Heidelberg New York

Library of Congress Cataloging-in-Publication Data

Drosophila eye development / Kevin Moses (ed.).
p. cm. – (Results and problems in cell differentiation ; 37)
Includes bibliographical references.
ISBN 354042590X (hardcover)
1. Drosophila melanogaster – Sense organs. 2. Eye – Differentiation. I. Moses, Kevin, 1959– II. Series.

QH607 .R4 vol. 37
573.8′815774 – dc21

Springer-Verlag Berlin Heidelberg New York
a member of BertelsmannSpringer Science+Business Media GmbH
http://www.springer.de

Printed in Germany

Cover design: Meta Design, Berlin
Typesetting: SNP Best-set Typesetter Ltd., Hong Kong
SPIN 10780814 39/3130 – 5 4 3 2 1 0 – Printed on acid-free paper

Contents

NOTCH and the Patterning of Ommatidial Founder Cells in the Developing *Drosophila* Eye
N.E. Baker

The Epidermal Growth Factor Receptor in *Drosophila* Eye Development
J. Kumar

Cell Fate Specification in the *Drosophila* Eye
R. Nagaraj, J. Canon, and U. Banerjee

Evolution of Color Vision
F. Pichaud and C. Desplan

Developmental Regulation Through Protein Stability
J.A. Fischer

The Establishment of Retinal Connectivity
U. Gaul

Homologies Between Vertebrate and Invertebrate Eyes
V. Hartenstein and T.A. Reh

Applications of the *Drosophila* Retina to Human Disease Modeling
N.M. Bonini and M.E. Fortini

Introduction

Kevin Moses[1]

It is now 25 years since the study of the development of the compound eye in *Drosophila* really began with a classic paper (Ready et al. 1976). In 1864, August Weismann published a monograph on the development of Diptera and included some beautiful drawings of the developing imaginal discs (Weismann 1864). One of these is the first description of the third instar eye disc in which Weismann drew a vertical line separating a posterior domain that included a regular pattern of clustered cells from an anterior domain without such a pattern. Weismann suggested that these clusters were the precursors of the adult ommatidia and that the line marks the anterior edge of the eye. In his first suggestion he was absolutely correct – in his second he was wrong. The vertical line shown was not the anterior edge of the eye, but the anterior edge of a moving wave of patterning and cell type specification that 112 years later (1976) Ready, Hansen and Benzer would name the "morphogenetic furrow". While it is too late to hear from August Weismann, it is a particular pleasure to be able to include a chapter in this Volume from the first author of that 1976 paper: Don Ready!

These past 25 years have seen an astonishing explosion in the study of the fly eye (see Fig. 1) and this has been due to a happy coincidence of human and technical factors. The compound eye is a beautiful structure and its development illustrates many of the most significant questions in developmental biology: the regulation of cell growth, cell shape and cell division, the specification of cell type and differentiation, the ordering and connection of the nervous system and the exquisite control of biological pattern. But above all this, the genetic and molecular technology available in *Drosophila* has allowed an extensive and continuing genetic dissection of eye development. After the discovery of the morphogenetic furrow the cellular events that follow it were described at the EM level (Tomlinson 1985), and this knowledge base has allowed us unusual precision in studying normal and mutant development – in much the same way as lineage studies did for *C. elegans*. The first important molecular reagents were monoclonal antibodies that allow the detection of very early cell differentiation (Fujita et al. 1982; Zipursky et al. 1984; Tomlinson and Ready 1987). However, the real key has been the discovery, phenotypic and molecular analysis of mutants. Beginning with *sevenless* and com-

[1] Department of Cell Biology, Emory University School of Medicine, 1648 Pierce Drive NE, Atlanta, Georgia 30322-3030, USA

Results and Problems in Cell Differentiation, Vol. 37
K. Moses (Ed.): Drosophila Eye Development

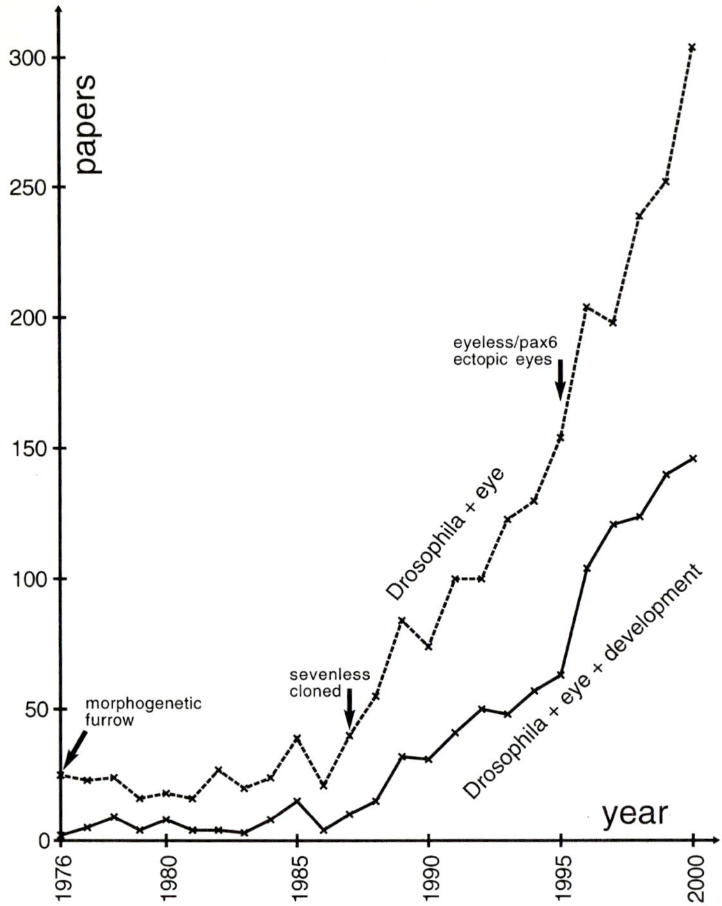

Fig. 1. Graph showing number of papers published each year since 1976. *Dotted line* indicates papers listed in MedLine for the search terms "*Drosophila* and eye". *Solid line* indicates papers listed for "*Drosophila* and eye and development"

binatorial models for cell-type specification, a series of molecular insights poured forth (Banerjee et al. 1987; Hafen et al. 1987): the control of the furrow by segmentation genes, the control of retinal tissue polarity by Notch and other signals, the control of ordered spacing of the ommatidia.

In the last few years the greatest surprise has been the outrageous discovery of homology between flies and ourselves, not only in the specification of the eye by Pax6/Eyeless, but even now in finer details of patterning via Hedgehog and the EGF receptor (Quiring et al. 1994; Neumann and Nüsslein-Volhard 2000). When I published my first paper in this field 12 years ago, I thought it hardly worth looking for a vertebrate homologue of the gene that I had just cloned. No one would be so foolish today. Finally, perhaps the biggest surprise to me is that the fly eye has become useful! It is now in use in at least a dozen labs and two companies as a model for human disease and a means to discover new disease genes.

These technical and scientific advances would never have come along but for the extraordinary collection of outstanding scientists who have come to work on the eye. It has been a genuine pleasure to work with many of them, and to come to know many more through conferences and other meetings. There is not space here to mention all of their names – some have contributed chapters to this book, but there are many more. Two of course stand out, and this book is dedicated to them.

Dedication

This book is dedicated to the two "fathers" of the fly eye: Seymour Benzer and Gerry Rubin. Two rather different but remarkable scientists without whom we would not have the eye as our companion and playground.

References

Banerjee U, Renfranz PJ, Pollock JA, Benzer S (1987) Molecular characterization and expression of *sevenless*, a gene involved in neuronal pattern formation in the *Drosophila* eye. Cell 49:281–291

Fujita SC, Zipursky SL, Benzer S, Ferrús A, Shotwell SL (1982) Monoclonal antibodies against the *Drosophila* nervous system. Proc Natl Acad Sci USA 79:7929–7933

Hafen E, Basler K, Edstroem JE, Rubin GM (1987) *sevenless*, a cell-specific homeotic gene of *Drosophila*, encodes a putative transmembrane receptor with a tyrosine kinase domain. Science 236:55–63

Neumann CJ, Nüsslein-Volhard C (2000) Patterning of the zebrafish retina by a wave of sonic hedgehog activity. Science 289:2137–2139

Quiring R, Walldorf U, Kloter U, Gehring WJ (1994) Homology of the *eyeless* gene of *Drosophila* to the *small eye* gene in mice and *Aniridia* in humans. Science 265:785–789

Ready DF, Hanson TE, Benzer S (1976) Development of the *Drosophila* retina, a neurocrystalline lattice. Dev Biol 53:217–240

Tomlinson A (1985) The cellular dynamics of pattern formation in the eye of *Drosophila*. J Embryol Exp Morphol 89:313–331

Tomlinson A, and Ready DF (1987) Neuronal differentiation in the *Drosophila* ommatidium. Dev Biol 120:366–376

Weismann A (1864) Die nachembryonale Entwicklung der Musciden nach Beobachtungen an *Musca vomitoria* und *Sarcophaga carnaria*. Zeit Wiss Zool 14:187–336

Zipursky SL, Venkatesh TR, Teplow DB, Benzer S (1984) Neuronal development in the *Drosophila* retina: monoclonal antibodies as molecular probes. Cell 36:15–26

Retinal Specification and Determination in *Drosophila*

Kartik Pappu[1] and Graeme Mardon[1,2]

1 Introduction

Dipteran insects such as *Drosophila* obtain visual information using compound eyes. In *Drosophila*, these compound eyes are composed of approximately 800 unit eyes called ommatidia. An ommatidium contains eight distinct photoreceptor cells, each of which projects an axon directly to the optic lobe of the brain. This structure contrasts sharply with the mammalian eye, which contains a single lens and a retina with multiple layers of neurons. In spite of these and other substantial differences in the morphological appearance of insect and vertebrate eyes, work in the last several years has revealed common underlying genetic pathways controlling retinal cell fate specification. This discovery is surprising since the eye was considered an extreme case of convergent evolution, evolving independently as many as 40 different times (reviewed in Land and Fernald 1992). Much of the flurry of molecular and genetic data that has accumulated in recent years challenges this notion and suggests divergence from a single, prototypical visual processing unit. Thus, *Drosophila* has proven to be an excellent model system for identifying new genes that are conserved in vertebrate retinal development. This chapter will mainly be concerned with describing the factors responsible for specification and determination of retinal cell fate in *Drosophila*.

2 Structure and Early Development of the *Drosophila* Eye

The compound eye of *Drosophila* contains between 750–800 individual light-sensing units, or ommatidia, that are arranged in a regular hexagonal array. Each adult ommatidium consists of 19 cells including 8 photoreceptors, 4 cone cells, 6 pigment cells, and a mechanosensory bristle at alternate ommatidial vertices (reviewed in Wolff and Ready 1993). The adult eye develops from an epithelial bilayer called the eye imaginal disk, which is derived from a group

[1] Program in Developmental Biology, Baylor College of Medicine, One Baylor Plaza, Houston, Texas 77030, USA

[2] Departments of Pathology, Neuroscience, Molecular and Human Genetics and Ophthalmology, Baylor College of Medicine, One Baylor Plaza, Houston, Texas 77030, USA

Results and Problems in Cell Differentiation, Vol. 37
K. Moses (Ed.): Drosophila Eye Development

Table 1. The Retinal Determination genes (see text for details. PR-Photoreceptor)

Gene	Loss of function	Gain of function
twin of eyeless (*toy*)	Not reported	Strong ectopic eye induction
eyeless (*ey*)	Reduced or no eyes	Strong ectopic eye induction
eyes absent (*eya*)	Reduced or no eyes	Weak ectopic eye induction
sine oculis (*so*)	Reduced or no eyes	No phenotype
dachshund (*dac*)	Reduced or no eyes	Weak ectopic eye induction
teashirt (*tea*)	No phenotype in the eye	Weak ectopic eye induction
optix (*optx*)	Not reported	Weak ectopic eye induction
eyegone (*eyg*)	Reduced or no eyes	Ectopic PRs in the eye disk
homothorax (*hth*)	Ectopic PRs in eye disk clones	Loss of PRs
extradenticle (*exd*)	Ectopic PRs in eye disk clones	Loss of PRs

of about 20 cells set aside during embryonic development (Garcia-Bellido and Merriam 1969). During the first two larval instar stages, cells of the eye disk remain undifferentiated but undergo repeated divisions to produce almost all of the cells that will form the adult eye. In the third and final larval instar, photoreceptor differentiation begins at the posterior margin of the eye disk and proceeds anteriorly following a dorsal-ventral groove termed the morphogenetic furrow (Ready et al. 1976). Thus, photoreceptor differentiation occurs progressively over a period of about 2 days. While cells anterior to the furrow do not express neural markers, they are already committed to become retinal tissue (Lebovitz and Ready 1986). During the last decade, molecular and genetic analyses have elucidated a pathway controlling specification and determination of retinal cell fate in *Drosophila*. These studies demonstrate that early eye development is regulated by a group of conserved, nuclear proteins that function in a network. Retinal specification is primarily under the control of two *Pax* family genes, *eyeless* (*ey*) and *twin of eyeless* (*toy*). Following specification, *eyes absent* (*eya*), *sine oculis* (*so*), and *dachshund* (*dac*) cooperate with *ey* to lock in or determine retinal cell fate. Together, these four genes constitute the core of the retinal determination (RD) network. In addition, several other genes have been identified that are likely to contribute to early eye development but whose function relative to the RD network have not been determined. Finally, the secreted patterning factors encoded by *hedgehog* (*hh*) and *decapentaplegic* (*dpp*) cooperate with Eyeless to regulate RD gene expression and have been incorporated into the RD network as well. Each of these genes will be discussed in detail in the following sections (see Table 1).

3 The Retinal Determination Network

The core RD genes (*ey*, *eya*, *so* and *dac*) are defined by a set of common features. Each RD gene encodes a conserved, nuclear protein that is required for normal retinal development (Fig. 1). In addition, each RD gene is expressed anterior to the morphogenetic furrow and prior to initiation of neural differ-

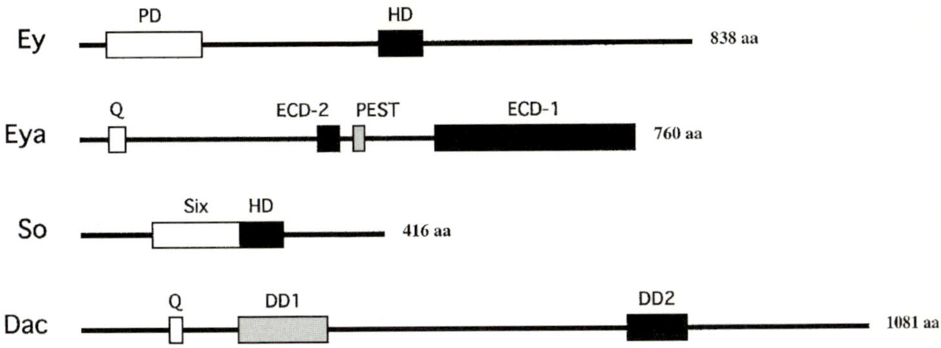

Fig. 1. Predicted structures of the proteins encoded by the core RD genes. The *PEST* sequence represents a domain rich in the amino acids proline, glutamic acid, serine and threonine and is predicted to function in protein turnover. Predicted proteins and their domains are drawn to scale

entiation, consistent with a role in early eye development. Moreover, ectopic expression of any one RD gene (except *so*) is sufficient to initiate the entire program of retinal development. Furthermore, coexpression of most combinations of the RD genes results in strong synergistic induction of ectopic eye formation. The genetic synergy observed among the RD genes is likely to be mediated by physical interaction of the encoded proteins through highly conserved domains. Finally, each member of the RD network is absolutely required for ectopic eye induction by any combination of other RD genes, suggesting that these genes do not function in a simple linear hierarchy. Taken together, these results have led to a model in which these genes function in a highly interactive network to regulate each other and to control the entire program of eye development in *Drosophila* (Fig. 2).

3.1 *twin of eyeless* and *eyeless*

toy and *ey* are *Drosophila* homologues of the vertebrate *Pax6* gene (Quiring et al. 1994; Czerny et al. 1999). Both are members of the *Pax* family of genes that encode sequence-specific DNA binding transcription factors (reviewed in Dahl et al. 1997). The characteristic DNA binding motif of the Pax proteins is a stretch of 130 amino acids called the Paired domain (PD, Fig. 1). In addition, both proteins have another putative DNA binding motif, the homeodomain (HD). Since both *toy* and *ey* are located near each other on the fourth chromosome and share splice sites that are not found in *Pax6* genes from other species, it seems likely that *toy* and *ey* arose as a result of a gene duplication event during arthropod evolution (Czerny et al. 1999). Both *toy* and *ey* transcripts are present in the optic primordia in the embryo. They are also expressed in the embryonic brain and the central nervous system (CNS), but

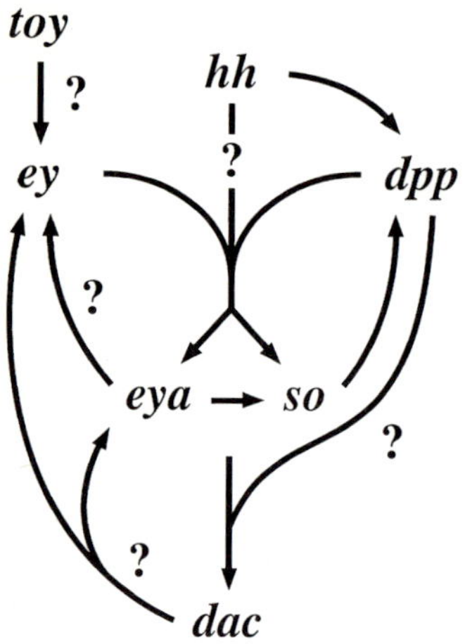

Fig. 2. The Retinal Determination network. A model for the genetic interactions that specify and determine retinal cell fate. Based upon loss-of-function studies, initiation of transcription of the core RD genes *ey*, *eya*, *so*, and *dac* can be placed in a primarily linear pathway. The patterning gene *dpp* interacts at multiple levels with the core RD genes while *toy* appears to act exclusively upstream of *ey*. The role of *hh* in the RD network is less clear. *hh* may act with *ey* alone or with *ey* and *dpp* together to regulate *eya* and *so* expression (our unpublished observations). Ectopic expression of *dac* induces the expression of *ey* in the antennal disk but whether this regulation is direct or mediated by *eya* is not known. *Question marks* represent transcriptional relationships suggested by ectopic gene expression studies but that have not been verified during normal eye development

the early expression of *toy* precedes expression of *ey* in the embryo (Quiring et al. 1994; Czerny et al. 1999). In first and second instar larval eye disks, *toy* and *ey* transcripts as well as Ey protein can be detected in the entire eye field (Halder et al. 1998; Czerny et al. 1999). Expression of *ey* and *toy* is downregulated in differentiating cells posterior to the morphogenetic furrow but remains high in undifferentiated cells anterior to the furrow in late third instar eye disks (Czerny et al. 1999). Hypomorphic mutations in the *ey* gene result in a partial or complete loss of compound eyes in adult flies while there are no reported mutant alleles of *toy* (Quiring et al. 1994; Czerny et al. 1999).

Major advances in our understanding of retinal development in *Drosophila* have come from ectopic expression studies using the GAL4-UAS system (Brand and Perrimon 1993). Targeted expression of either *toy* or *ey* in imaginal disks other than the eye disk results in strong induction of large ectopic eyes on all major appendages, including the legs, wings and antennae. Since *ey* and *toy* are expressed in the embryonic eye primordia prior to any of the other RD genes and are the most potent known inducers of ectopic eyes, *ey* and *toy* are likely to be the primary genes responsible for specifying retinal cell fate in *Drosophila*.

Molecular and genetic data suggest that *toy* functions upstream of *ey*. First, *toy* expression is unaffected in an *ey* mutant background. Second, Toy binds to an eye-specific enhancer of *ey* in vitro and this enhancer is required for *toy* activation of *ey* transcription in vivo. Third, misexpression of *ey* does not induce detectable *toy* expression. Finally, *toy* cannot induce ectopic eyes in an *ey* mutant background (Czerny et al. 1999). Taken together, these data suggest that *toy* acts upstream of *ey* in the RD network. However, it will not be possible to determine whether *ey* functions exclusively downstream of *toy* until loss-of-functions mutations in *toy* are available.

3.2 *eyes absent* and *sine oculis*

eya and *so* function downstream of *ey* in the genetic cascade that determines retinal fate and *so* is likely to be a direct target of *ey* activity. *eya* encodes a novel nuclear protein with two Eya conserved domains (ECD-1 and -2), a PEST domain, and a polyglutamine (Q) domain (Fig. 1; Bonini et al. 1997; Bui et al. 2000). *so* encodes a protein with a divergent homeodomain (HD) and a highly conserved Six domain(Six) immediately N-terminal to the homeodomain (Fig. 1; Cheyette et al. 1994). Prior to photoreceptor differentiation, both Eya and So are expressed in a gradient with highest levels in the posterior-lateral margins of the eye disk and tapering off toward the center (Bonini et al. 1993; Cheyette et al. 1994; Halder et al. 1998). During morphogenetic furrow progression, both proteins are expressed within as well as anterior and posterior to the furrow (Bonini et al. 1993; Cheyette et al. 1994). Although *eya* and *so* are both expressed in and required for normal larval visual system development (i.e., Bolwig's organ) and optic lobe development, neither gene is expressed specifically within the embryonic primordia of the adult eye (Cheyette et al. 1994; Bonini et al. 1998; Halder et al. 1998; Daniel et al. 1999; Suzuki and Saigo 2000). While null mutations in *eya* or *so* are embryonic lethal, deletions of eye-specific regulatory sequences of either gene cause a no-eye phenotype (Bonini et al. 1993; Cheyette et al. 1994; Niimi et al. 1999; Zimmerman et al. 2000). This phenotype reflects a requirement for both *eya* and *so* during morphogenetic furrow initiation. In addition, both genes are essential for furrow progression and photoreceptor differentiation (Pignoni et al. 1997).

Targeted misexpression of *eya* alone, but not *so*, is sufficient to produce ectopic retinal tissue (Bonini et al. 1997; Pignoni et al. 1997). However, *eya* is significantly less effective than *ey* in ectopic eye induction, both in frequency and magnitude. For example, using the *dpp-GAL4* driver, misexpression of *ey* results in large ectopic eyes (up to 50% of the size of a normal eye) on all major appendages in 100% of animals observed. In contrast, misexpression of *eya* using the same driver causes only small ectopic eyes (typically less than 5–10% of the normal eye) in about 10–20% of animals (Bonini et al. 1997; Pignoni et al. 1997; Seimiya and Gehring 2000). These observations suggest that *ey* may act upstream of *eya* during normal eye development. Indeed, *ey* is required for *eya* expression in the early larval eye disk while *ey* expression is unaffected in *eya* mutants (Halder et al. 1998). Furthermore, *eya* is required for ectopic eye induction driven by *ey* misexpression (Bonini et al. 1997). Since *ey* is a more potent inducer of retinal development than *eya*, *ey* must be able to regulate other genes that are important for retinal cell fate determination. In fact, *so* is a direct downstream target of *ey* function. Specifically, *ey* is required for *so* expression, Ey protein can bind to an eye-specific regulatory domain in the terminal intron of the *so* gene, and this intron is required for *ey* induction of *so* expression (Cheyette et al. 1994; Halder et al. 1998; Niimi et al. 1999). Moreover, *so* is required for ectopic eye induction by *ey* and is neither necessary nor sufficient for induction or maintenance of *ey* expression (Halder et al. 1998). *eya* is likely to function upstream of *so* since *so* expression is abolished in an *eya* mutant background while *so* is not required for *eya* expression (Halder et al. 1998). It is not known if *eya* and *so* are direct targets of *toy*.

The data described thus far suggest a linear transcriptional hierarchy such that *ey* acts upstream of *eya* which in turn is upstream of *so*. However, targeted expression of *eya* alone can induce expression of *ey* and *ey* is required for ectopic eye induction by *eya* (Bonini et al. 1997; Pignoni et al. 1997). These data suggest that *ey* and *eya* regulate each other during retinal determination and places *eya* both upstream and downstream of *ey* in a genetic regulatory pathway (Fig. 2). Moreover, coexpression of *ey* and *eya* is more effective in generating ectopic retinal tissue than either gene alone, indicating a genetic synergy between these two genes (Bonini et al. 1997). Similarly, strong synergistic effects are also observed upon coexpression of *eya* and *so*. This synergy is likely to be mediated by the physical interaction of the Eya and So proteins via their conserved domains (Pignoni et al. 1997). Thus, while *ey* is required to induce the expression of *eya* and *so*, these target genes are likely to, in turn, cooperate with each other to maintain the expression of each other and *ey*. Nevertheless, the range of tissues in which ectopic eyes can be generated by co-expression of *eya* and *so* is only a subset of those where *ey* alone is effective, suggesting that other important targets of *ey* function exist. Indeed, ectopic expression of *ey* or *eya* induces expression of another downstream gene required for normal eye development, *dachshund* (Chen et al. 1997).

3.3 *dachshund*

Drosophila dachshund (*dac*) encodes a novel nuclear protein that has weak similarity to the vertebrate Ski/SnoN proto-oncoproteins (Mardon et al. 1994; Hammond et al. 1998; Caubit et al. 1999; Kozmik et al. 1999). Comparison of *Drosophila* and mammalian *dac* genes reveals three conserved structural features: two protein domains called Dachshund Domain 1 (DD1) and Dachshund Domain 2 (DD2), and a trinucleotide repeat (Fig. 1; Hammond et al. 1998; Davis et al. 1999). Of these three domains, DD1 is particularly interesting as it is very highly conserved (78% identity between fly and human) and is similar to a domain contained in the proto-oncoproteins Ski and SnoN. In Ski and SnoN, this domain is required for transcriptional regulation and cellular transformation (Zheng et al. 1997a,b; Cohen et al. 1998). In Dac, the DD1 domain may be associated with transcriptional activation (Chen et al. 1997). *dac* is expressed at the posterior margin of the eye disk in the second and early third instar larval stages. During progression of the morphogenetic furrow, *dac* is expressed within as well as anterior and posterior to the furrow (Mardon et al. 1994). Null mutations in *dac* result in animals with no eyes and, similar to *eya* and *so*, this is caused by a failure of morphogenetic furrow initiation during larval development. However, unlike *eya* and *so*, *dac* is not required for furrow progression or photoreceptor differentiation but plays a role in normal ommatidial organization (Mardon et al. 1994).

Loss-of-function studies place *dac* downstream of *eya* and *so* during normal eye development (Fig. 2). Specifically, *dac* is not required for expression of any of the other core RD genes (*ey*, *eya*, or *so*) while each of these genes is required for *dac* expression. In addition, *dac* is required for ectopic eye development driven by any combination of *ey*, *eya* and *so* misexpression (Chen et al. 1997; Pignoni et al. 1997; Shen and Mardon 1997). Finally, misexpression of either *ey* or *eya* alone or a combination of *eya* and *so* induces ectopic *dac* expression (Chen et al. 1997; Pignoni et al. 1997; Shen and Mardon 1997). Targeted expression of *dac* in imaginal tissues other than the eye disk induces ectopic photoreceptor formation in a manner very similar to that caused by *eya* (Shen and Mardon 1997). Moreover, *dac* and *eya* act synergistically to induce ectopic eye development and the proteins they encode physically interact, again through conserved domains. Although loss-of-function studies place *dac* downstream of the other core RD genes, *dac* misexpression is sufficient to induce expression of *ey*, *eya* and *so*. Finally, ectopic eye induction driven by *dac* misexpression requires the function of each of the other RD genes (Chen et al. 1997). Taken together, these data strongly support a network model of retinal cell fate determination where the RD genes extensively cross-regulate each other and act synergistically to induce ectopic eye development (Fig. 2). An alternate model has also been proposed in which a linear hierarchy of RD genes is used reiteratively during several phases of eye development (Desplan 1997).

In summary, specification, determination, and differentiation of the eye represent temporally separable but molecularly overlapping stages of development (Fig. 3). *toy* and *ey* are expressed in the embryonic eye primordia prior to the other RD genes and act to specify retinal cell fate. Later, *ey* induces the expression of *eya*, *so*, and *dac* which function with *ey* to irreversibly determine retinal fate. Specification and determination set the stage for differentiation, which occurs posterior to the morphogenetic furrow and is likely to no longer involve the specification genes *toy* and *ey*. In contrast, the retinal determination genes *eya*, *so*, and *dac* are also required for normal photoreceptor differentiation and ommatidial assembly posterior to the furrow. Thus, these RD genes are expected to regulate the expression of other genes governing differentiation. For example, expression of the proneural gene *atonal* (*ato*) is lost in *eya* or *so* mutant eye disks (Jarman et al. 1995). *ato* is expressed in the furrow and is the earliest known gene required specifically for photoreceptor differentiation (Jarman et al. 1994). However, whether regulation of *ato* by the RD genes is direct or indirect remains to be determined. Thus, retinal development in *Drosophila* is controlled by the sequential, partially overlapping expression of a conserved network of regulatory factors.

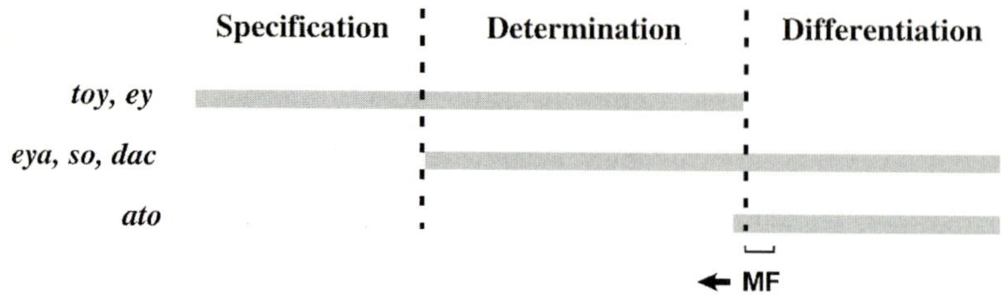

Fig. 3. A model for retinal development in *Drosophila*. See text for details. Anterior is to the *left*; *MF* morphogenetic furrow

4 Retinal Determination Genes: New Members?

In addition to the core RD genes, ectopic expression of *teashirt* (*tsh*), *optix* (*optx*) and *eyegone* (*eyg*) are also sufficient to induce retinal tissue development. Furthermore, loss-of-function clones in *homothorax* (*hth*) or *extradenticle* (*exd*) cause ectopic eye development in the head cuticle region. However, whether these genes act within the RD network has not been determined. We describe each of these genes in the following sections.

4.1 *teashirt*

tsh encodes a zinc finger transcription factor that is expressed anterior to the morphogenetic furrow in the developing eye disk (Fasano et al. 1991; Pan and Rubin 1998). Misexpression of *tsh* in the antennal disk induces ectopic expression of *ey*, *dac*, and *so* and produces ectopic eyes on adult antennae (Pan and Rubin 1998). Conversely, *ey* induces ectopic *tsh* when misexpressed in the same region of the antennal disk. Furthermore, the ability of *tsh* to form ectopic eyes is lost in an *eya* or *so* mutant background (Pan and Rubin 1998). These data suggest that *tsh* acts both upstream and downstream of *ey* and upstream of *eya* and *so* in the retinal determination network. Given these data, it is perhaps surprising that patches or clones of cells that are homozygous for a null mutation in *tsh* develop normally (Pan and Rubin 1998). Thus, it is possible that either *tsh* plays no role in eye development or that *tsh* function in the eye is redundant with another unknown factor. Consistent with the latter model, it has been proposed that *tsh* and *cubitus interruptus* (*ci*), a downstream effector of *hedgehog*, may act redundantly to regulate Wingless (Wg) signaling in the *Drosophila* embryo (Gallet et al. 2000). Further refinement of the role that *tsh* plays during normal eye development and with respect to the core RD gene network will require additional molecular and genetic data.

4.2 *optix*

optix (*optx*), another *Drosophila* member of the *so/Six* family of genes, has recently been isolated (Toy et al. 1998). Sequence analysis has revealed that *optx* is the true ortholog of the vertebrate *Six3* gene while *so* is most similar to *Six2* (Seimiya and Gehring 2000). Moreover, *so* and *optx* are closely linked in the *Drosophila* genome as are the mammalian orthologs *Six2* and *Six3*, suggesting that a gene duplication event involving these loci preceded the divergence of insects and vertebrates. Unlike *so*, *optx* is not expressed in the embryonic or larval visual systems and is expressed anterior to the morphogenetic furrow in a pattern that is highly similar to *ey* and *tsh* (Seimiya and Gehring 2000). While there are no known mutant alleles of *optx*, a recent study has reported that targeted misexpression of *optx* in the antennal disk can induce ectopic eye development (Seimiya and Gehring 2000). This stands in sharp contrast to *so*, which alone has no eye-inducing capability. However, the mechanism of ectopic eye induction by *optx* is distinct from that of the other RD genes: although *optx* does require *eya* and *so* for ectopic retinal development, it is *ey*-independent and does not act synergistically with *eya* (Seimiya and Gehring 2000). Since there are no known loss-of-function mutants in *optx*, further genetic analysis has been hampered and the exact relationship between *optx* and the core RD network remains to be determined.

4.3 *eyegone*

Another *Pax* family gene encoding a truncated Paired domain implicated in eye development is *eyegone* (*eyg*). Although the expression pattern of *eyg* has not been reported, loss-of-function mutations in *eyg* cause a reduced or no eye phenotype and eye disks prepared from *eyg* mutants do not show any photoreceptor differentiation (Hazelett et al. 1998). The mechanism of *eyg* function is not clear but is likely to involve regulation of the *wingless* (*wg*) pathway. Normally, *wingless* acts as a negative regulator of morphogenetic furrow initiation (Ma and Moses 1995; Treisman and Rubin 1995). In *eyg* mutant eye disks, *wingless* is ectopically expressed at the posterior margin. Furthermore, repression of *wingless* signaling at the posterior margin of *eyg* mutant eye disks is sufficient to restore initiation of photoreceptor differentiation. These and other data lead to the hypothesis that *eyg* may promote eye development by negatively regulating *wingless* signaling (Hazelett et al. 1998). Ectopic expression of *eyg* using the *dpp*-GAL4 driver produces eyes on the ventral regions of the adult head (H. Sun, personal communication). However, *eyg* is not sufficient to induce ectopic retinal development in any tissue other than the eye-antennal disk. In addition, *eyg* and *ey* are not required for each other's expression or ability to induce ectopic photoreceptor development, suggesting that these genes function in independent pathways. Nevertheless, animals doubly mutant for *ey* and *eyg* show synthetic lethality and co-expression of these genes causes synergistic induction of photoreceptor development (H. Sun, personal communication). Thus, while *eyg* plays a key role in normal eye development, it has not been possible to clearly position *eyg* relative to the other RD genes.

4.4 *homothorax* and *extradenticle*

Induction of mutant clones for either *extradenticle* (*exd*) or *homothorax* (*hth*) leads to the production of ectopic eyes on ventral head regions in *Drosophila* (Gonzalez-Crespo and Morata 1995; Pai et al. 1998). Both genes encode homeoproteins that are expressed anterior to the furrow in the eye disk and at the posterior and lateral margins of the eye disk which are destined to become head cuticle (Gonzalez-Crespo and Morata 1995; Pai et al. 1998). Hth is required for the nuclear localization of Exd, thus providing a means for the regulation of Exd function during development (Pai et al. 1998). Ectopic expression of *hth* in the eye disk suppresses eye development while ectopic expression of *exd* alone has no effect. These data suggest that *hth* and *exd* are required to suppress eye development, thereby delimiting retinal and cuticle fields (Pai et al. 1998). There is no reported evidence that either *hth* or *exd* play a positive role in specification or determination of retinal cell fate or that these genes interact in the eye with other members of the RD network. In the leg imaginal disk, *dac* and *hth* appear to mutually antagonize the expression of each other and this contributes to the elaboration of the proximal-distal axis of the limb (Abu-Shaar and Mann 1998). However, this relationship is not maintained in the eye disk since *hth* and *dac* expression extensively overlap anterior to the furrow.

5 Patterning Genes and Retinal Determination

A set of three secreted factors encoded by *hedgehog* (*hh*), *decapentaplegic* (*dpp*) and *wingless* (*wg*) are required for normal development of all imaginal disks, including the eye. In general, these factors are required to control patterning of each disk, acting as morphogens that regulate cell fate in a concentration-dependent manner. However, these genes do not specify cell fate directly. In contrast to the RD genes, ectopic expression of *hh*, *dpp* or *wg* does not change the fate of one disk type to another (e.g., a leg to an eye). Instead, their misexpression causes pattern duplications specific to each disk type. Thus, the patterning genes may cooperate with tissue-specific factors to determine cell fates during development. Although the molecular basis for such cooperation is largely unknown, recent studies have revealed a molecular mechanism linking patterning and retinal specification.

5.1 *hedgehog*

Mutations in *Drosophila hedgehog* (*hh*) were initially isolated in a genetic screen for defects in embryonic pattern formation (Nusslein-Volhard and Wieschaus 1980). *hh* encodes a secreted signaling molecule that controls patterning of most tissues during development and is highly conserved throughout phylogeny (reviewed in Chuang and Kornberg 2000; Mohler and Vani 1992). *hh* is expressed at the posterior margin of the eye imaginal disk prior to furrow initiation and in all photoreceptor cells posterior to the furrow during progression (Ma et al. 1993; Dominguez and Hafen 1997; Borod and Heberlein 1998). Loss of *hh* function at any time during eye development blocks morphogenetic furrow movement but is not required specifically for photoreceptor differentiation (Heberlein et al. 1993; Ma et al. 1993; Dominguez and Hafen 1997; Borod and Heberlein 1998). Since *hh* encodes a secreted factor, small clones of *hh* mutant tissue have little or no effect on eye development. However, large *hh* mutant clones that include the posterior margin of the eye disk block furrow initiation (Heberlein et al. 1993). As expected, posterior margin clones mutant for *smoothened* (*smo*), a cell autonomous transducer of the Hh signal, also block furrow initiation (Curtiss and Mlodzik 2000). Such clones also lack Eya and Dac expression, suggesting that Eya and Dac may be targets of *hh* signaling during initiation of the morphogenetic furrow (Curtiss and Mlodzik 2000). *hh* is required for *decapentaplegic* (*dpp*) expression at the posterior margin of the eye disk prior to furrow initiation, placing *hh* upstream of *dpp* in eye development (Borod and Heberlein 1998).

5.2 *decapentaplegic*

dpp is a *Drosophila* homologue of the TGFβ superfamily of secreted signaling molecules that patterns many adult and embryonic structures in a concentration-dependent fashion (reviewed in Podos and Ferguson 1999; Padgett et al. 1987). *dpp* is expressed at the posterior margin of the early eye disk and is required for furrow initiation (Masucci et al. 1990; Chanut and Heberlein 1997). Similarly, loss-of-function mutations in *Mothers against dpp* (*Mad*), *thick veins* (*tkv*) and *punt*, cell autonomous components of the *dpp* signaling pathway, block initiation of the morphogenetic furrow (Burke and Basler 1996; Wiersdorff et al. 1996). Forced expression of *dpp* at the anterior margin of the eye disk leads to the initiation of ectopic furrows (Chanut and Heberlein 1997; Pignoni and Zipursky 1997). These data demonstrate that *dpp* is both necessary and sufficient to initiate retinal morphogenesis in the eye disk.

Although ectopic expression of *dpp* alone is not sufficient to induce photoreceptor development in tissues other than the eye, recent studies have shown that *dpp* can cooperate with other members of the RD network to direct ectopic eye development (Chen et al. 1999). Targeted expression of *ey* in the wing disk is able to induce *eya*, *so*, and *dac* expression and photoreceptor development, but only in the vicinity of endogenous *dpp* (Halder et al. 1998; Chen et al. 1999). Coexpression of *dpp* with *ey* greatly expands the domain of RD gene induction and ectopic eye development, demonstrating that *ey* and *dpp* can cooperate to determine retinal cell fate (Chen et al. 1999). These results suggest that *dpp* may be required for RD gene expression during normal eye development. Indeed, *dpp* is required in the eye for *eya*, *so*, and *dac* expression, but not *ey* (Chen et al. 1999; Curtiss and Mlodzik 2000). Thus, *dpp* and *ey* either act together or in parallel to determine retinal fate in the eye disk.

5.3 *wingless*

wingless (*wg*) encodes a secreted signaling protein that belongs to the Wnt family of growth factors in vertebrates (reviewed in Nusse and Varmus 1992). In *Drosophila*, *wg* is required during embryogenesis for segmentation and development of imaginal disks (Baker 1988a). *wg* is expressed in the dorsal and ventral margins of the anterior eye disk, areas that will eventually give rise to head cuticle (Baker 1988b). Loss of *wg* in these marginal zones leads to ectopic *dpp* expression and premature furrow initiation (Ma and Moses 1995). Conversely, ectopic expression of *wg* at the posterior margin of the eye disk prevents furrow initiation. These results indicate that the function of *wg* in the eye may be to antagonize *dpp* signaling, thereby helping to shape the furrow as a linear wave and allowing the orderly assembly of ommatidia (Treisman and Rubin 1995). Although it is not clear if *wg* signaling has any regulatory input into the RD network, *wg* is ectopically expressed along the entire posterior margin of the eye disk in *dac* null mutants, possibly accounting for the failure of furrow initiation in *dac* mutants (Treisman and Rubin 1995). However, since clones lacking both *dac* and *wg* function still fail to initiate the furrow, *wg* misexpression cannot be the sole cause of this *dac* mutant phenotype (Chen and Mardon, unpubl. observ.).

6 Conclusions and Future Directions

While the work described above represents a huge leap in our understanding of *Drosophila* eye development, it also appears to be the tip of the iceberg. Downstream targets of the RD genes during specification and determination are not known and the link between the retinal determination network and differentiation is unclear. Moreover, since more than 3000 genes may be required to construct the visual system, it would be an understatement to say that many more genes functioning during early retinal specification and determination remain to be identified (Thaker and Kankel 1992). One clear and fascinating aspect of work in the last decade is the emerging paradigm of a phylogenetically conserved retinal specification machinery. The ease of genetic manipulation and the frequent development of new molecular and genetic tools have made *Drosophila* the model system of choice to decipher conserved mechanisms of eye development. If the past decade is any indication of the future, many surprises and much excitement await us.

Acknowledgements. We thank Benjamin Frankfort, Richard Davis, Beril Tavsanli, and Edwin Ostrin for critical reading of the manuscript and other members of the laboratory for stimulating discussions. Research in the laboratory is supported by the National Eye Institute, the Retina Research Foundation and the Moran Foundation.

References

Abu-Shaar M, Mann RS (1998) Generation of multiple antagonistic domains along the proximodistal axis during *Drosophila* leg development. Development 125:3821–3830

Baker NE (1988a) Embryonic and imaginal requirements for *wingless*, a segment-polarity gene in *Drosophila*. Dev Biol 125:96–108

Baker NE (1988b) Transcription of the segment-polarity gene *wingless* in the imaginal discs of *Drosophila*, and the phenotype of a pupal-lethal *wg* mutation. Development 102:489–497

Bonini NM, Leiserson WM, Benzer S (1993) The *eyes absent* gene: genetic control of cell survival and differentiation in the developing *Drosophila* eye. Cell 72:379–395

Bonini NM, Bui QT, Grayboard GL, Warrick JM (1997) The *Drosophila eyes absent* gene directs ectopic eye formation in a pathway conserved between flies and vertebrates. Development 124:4819–4826

Bonini NM, Leiserson WM, Benzer S (1998) Multiple roles of the *eyes absent* gene in *Drosophila*. Dev Biol 196:42–57

Borod ER, Heberlein U (1998) Mutual regulation of *decapentaplegic* and *hedgehog* during the initiation of differentiation in the *Drosophila* retina. Dev Biol 197:187–197

Brand AH, Perrimon N (1993) Targeted gene expression as a means of altering cell fates and generating dominant phenotypes. Development 118:401–415

Bui QT, Zimmerman JE, Liu H, Bonini NM (2000) Molecular analysis of *Drosophila eyes absent* mutants reveals features of the conserved eya domain. Genetics 155:709–720

Burke R, Basler K (1996) Hedgehog-dependent patterning in the *Drosophila* eye can occur in the absence of dpp signaling. Dev Biol 179:360–368

Caubit X, Thangarajah R, Theil T, Wirth J, Nothwang HG, Ruther U, Krauss S (1999) Mouse Dac, a novel nuclear factor with homology to *Drosophila* dachshund shows a dynamic expression in the neural crest, the eye, the neocortex, and the limb bud. Dev Dyn 214:66–80

Chanut F, Heberlein U (1997) Role of *decapentaplegic* in initiation and progression of the morphogenetic furrow in the developing *Drosophila* retina. Development 124:559–567

Chen R, Amoui M, Zhang ZH, Mardon G (1997) Dachshund and Eyes Absent proteins form a complex and function synergistically to induce ectopic eye development in *Drosophila*. Cell 91:893–903

Chen R, Halder G, Zhang Z, Mardon G (1999) Signaling by the TGF-β homolog *decapentaplegic* functions reiteratively within the network of genes controlling retinal cell fate determination in *Drosophila*. Development 126:935–943

Cheyette BN, Green PJ, Martin K, Garren H, Hartenstein V, Zipursky SL (1994) The *Drosophila sine oculis* locus encodes a homeodomain-containing protein required for the development of the entire visual system. Neuron 12:977–996

Chuang PT, Kornberg TB (2000) On the range of Hedgehog signaling. Curr Opin Genet Dev 10:515–522

Cohen SB, Nicol R, Stavnezer E (1998) A domain necessary for the transforming activity of SnoN is required for specific DNA binding, transcriptional repression and interaction with TAF(II)110. Oncogene 17:2505–2513

Curtiss J, Mlodzik M (2000) Morphogenetic furrow initiation and progression during eye development in *Drosophila*: the roles of *decapentaplegic*, *hedgehog* and *eyes absent*. Development 127:1325–1336

Czerny T, Halder G, Kloter U, Souabni A, Gehring WJ, Busslinger M (1999) *twin of eyeless*, a second *Pax-6* gene of *Drosophila*, acts upstream of *eyeless* in the control of eye development. Mol Cell 3:297–307

Dahl E, Koseki H, Balling R (1997) *Pax* genes and organogenesis. Bioessays 19:755–765

Daniel A, Dumstrei K, Lengyel JA, Hartenstein V (1999) The control of cell fate in the embryonic visual system by *atonal*, *tailless* and EGFR signaling. Development 126:2945–2954

Davis RJ, Shen W, Heanue TA, Mardon G (1999) Mouse *Dach*, a homologue of *Drosophila dachshund*, is expressed in the developing retina, brain and limbs. Dev Genes Evol 209:526–536

Desplan C (1997) Eye development – governed by a dictator or a junta? Cell 91:861–864

Dominguez M, Hafen E (1997) *hedgehog* directly controls initiation and propagation of retinal differentiation in the *Drosophila* eye. Genes Dev 11:3254–3264

Fasano L, Roder L, Core N, Alexandre E, Vola C, Jacq B, Kerridge S (1991) The gene *teashirt* is required for the development of *Drosophila* embryonic trunk segments and encodes a protein with widely spaced zinc finger motifs. Cell 64:63–79

Gallet A, Angelats C, Kerridge S, Therond PP (2000) Cubitus interruptus-independent transduction of the Hedgehog signal in *Drosophila*. Development 127:5509–5522

Garcia-Bellido A, Merriam JR (1969) Cell lineage of the imaginal discs in *Drosophila* gynandromorphs. J Exp Zool 170:61–75

Gonzalez-Crespo S, Morata G (1995) Control of *Drosophila* adult pattern by *extradenticle*. Development 121:2117–2125

Halder G, Callaerts P, Flister S, Walldorf U, Kloter U, Gehring WJ (1998) *eyeless* initiates the expression of both *sine oculis* and *eyes absent* during *Drosophila* compound eye development. Development 125:2181–2191

Hammond KL, Hanson IM, Brown AG, Lettice LA, Hill RE (1998) Mammalian and *Drosophila dachshund* genes are related to the *Ski* proto- oncogene and are expressed in eye and limb. Mech Dev 74:121–131

Hazelett DJ, Bourouis M, Walldorf U, Treisman JE (1998) *decapentaplegic* and *wingless* are regulated by *eyes absent* and *eyegone* and interact to direct the pattern of retinal differentiation in the eye disc. Development 125:3741–3751

Heberlein U, Wolff T, Rubin GM (1993) The *TGF beta* homolog *dpp* and the segment polarity gene *hedgehog* are required for propagation of a morphogenetic wave in the *Drosophila* retina. Cell 75:913–926

Jarman AP, Grell EH, Ackerman L, Jan LY, Jan YN (1994) *atonal* is the proneural gene for *Drosophila* photoreceptors. Nature 369:398–400

Jarman AP, Sun Y, Jan LY, Jan YN (1995) Role of the proneural gene, *atonal*, in formation of *Drosophila* chordotonal organs and photoreceptors. Development 121:2019–2030

Kozmik Z, Pfeffer P, Kralova J, Paces J, Paces V, Kalousova A, Cvekl A (1999) Molecular cloning and expression of the human and mouse homologues of the *Drosophila* dachshund gene. Dev Genes Evol 209:537–545

Land MF, Fernald RD (1992) The evolution of eyes. Annu Rev Neurosci 15:1–29

Lebovitz RM, Ready DF (1986) Ommatidial development in *Drosophila* eye disc fragments. Dev Biol 117:663–671

Ma C, Moses K (1995) *wingless* and *patched* are negative regulators of the morphogenetic furrow and can affect tissue polarity in the developing *Drosophila* compound eye. Development 121:2279–2289

Ma C, Zhou Y, Beachy PA, Moses K (1993) The segment polarity gene *hedgehog* is required for progression of the morphogenetic furrow in the developing *Drosophila* eye. Cell 75:927–938

Mardon G, Solomon NM, Rubin GM (1994) *dachshund* encodes a nuclear protein required for normal eye and leg development in *Drosophila*. Development 120:3473–3486

Masucci JD, Miltenberger RJ, Hoffmann FM (1990) Pattern-specific expression of the *Drosophila decapentaplegic* gene in imaginal disks is regulated by 3′ *cis*-regulatory elements. Genes Dev 4:2011–2023

Mohler J, Vani K (1992) Molecular organization and embryonic expression of the *hedgehog* gene involved in cell-cell communication in segmental patterning of *Drosophila*. Development 115:957–971

Niimi T, Seimiya M, Kloter U, Flister S, Gehring WJ (1999) Direct regulatory interaction of the Eyeless protein with an eye-specific enhancer in the *sine oculis* gene during eye induction in *Drosophila*. Development 126:2253–2260

Nusse R, Varmus HE (1992) *Wnt* genes. Cell 69:1073–87

Nusslein-Volhard C, Wieschaus E (1980) Mutations affecting segment number and polarity in *Drosophila*. Nature 287:795–801

Padgett RW, St Johnston RD, Gelbart WM (1987) A transcript from a *Drosophila* pattern gene predicts a protein homologous to the transforming growth factor-beta family. Nature 325:81–84

Pai CY, Kuo TS, Jaw TJ, Kurant E, Chen CT, Bessarab DA, Salzberg A, Sun YH (1998) The Homothorax homeoprotein activates the nuclear localization of another homeoprotein, Extradenticle, and suppresses eye development in *Drosophila*. Genes Dev 12:435–446

Pan D, Rubin GM (1998) Targeted expression of *teashirt* induces ectopic eyes in *Drosophila*. Proc Natl Acad Sci USA 95:15508–15512

Pignoni F, Zipursky SL (1997) Induction of *Drosophila* eye development by *decapentaplegic*. Development 124:271–278

Pignoni F, Hu B, Zavitz KH, Xiao J, Garrity PA, Zipursky SL (1997) The eye-specification proteins So and Eya form a complex and regulate multiple steps in *Drosophila* eye development. Cell 91:881–891

Podos SD, Ferguson EL (1999) Morphogen gradients: new insights from DPP. Trends Genet 15:396–402

Quiring R, Walldorf U, Kloter U, Gehring WJ (1994) Homology of the *eyeless* gene of *Drosophila* to the *Small eye* gene in mice and *Aniridia* in humans [see comments]. Science 265:785–789

Ready DF, Hanson TE, Benzer S (1976) Development of the *Drosophila* retina, a neurocrystalline lattice. Dev Biol 53:217–240

Seimiya M, Gehring WJ (2000) The *Drosophila* homeobox gene *optix* is capable of inducing ectopic eyes by an *eyeless* independent mechanism. Development 127:1879–1886

Shen W, Mardon G (1997) Ectopic eye development in *Drosophila* induced by directed *dachshund* expression. Development 124:45–52

Suzuki T, Saigo K (2000) Transcriptional regulation of *atonal* required for *Drosophila* larval eye development by concerted action of *eyes absent*, *sine oculis* and Hedgehog signaling independent of Fused kinase and Cubitus interruptus. Development 127:1531–1540

Thaker HM, Kankel DR (1992) Mosaic analysis gives an estimate of the extent of genomic involvement in the development of the visual system in *Drosophila melanogaster*. Genetics 131: 883–894

Toy J, Yang JM, Leppert GS, Sundin OH (1998) The *Optx2* homeobox gene is expressed in early precursors of the eye and activates retina-specific genes. Proc Natl Acad Sci USA 95: 10643–10648

Treisman JE, Rubin GM (1995) *wingless* inhibits morphogenetic furrow movement in the *Drosophila* eye disc. Development 121:3519–3527

Wiersdorff V, Lecuit T, Cohen SM, Mlodzik M (1996) Mad acts downstream of dpp receptors, revealing a differential requirement for dpp signaling in initiation and propagation of morphogenesis in the *Drosophila* eye. Development 122:2153–2162

Wolff T, Ready DF (1993) Pattern formation in the *Drosophila* retina. In: Bate M, Arias AM (ed) Cold Spring Harbor Laboratory Press, Plainview, NY, pp 1277–1326

Zheng G, Blumenthal KM, Ji Y, Shardy DL, Cohen SB, Stavnezer E (1997a) High affinity dimerization by Ski involves parallel pairing of a novel bipartite alpha-helical domain. J Biol Chem 272:31855–31864

Zheng G, Teumer J, Colmenares C, Richmond C, Stavnezer E (1997b) Identification of a core functional and structural domain of the v-Ski oncoprotein responsible for both transformation and myogenesis. Oncogene 15:459–471

Zimmerman JE, Bui QT, Liu H, Bonini NM (2000) Molecular genetic analysis of *Drosophila eyes absent* mutants reveals an eye enhancer element. Genetics 154:237–246

Regulators of the Morphogenetic Furrow

Jeffrey D. Lee and Jessica E. Treisman[1]

1 Introduction

Unlike other imaginal discs, the *Drosophila* eye disc has a progressive pattern of differentiation. Photoreceptor clusters begin to form at the posterior margin of the eye disc in the third larval instar, and more anterior rows of clusters then differentiate in succession (Ready et al. 1976). Just prior to their differentiation, cells undergo an apical constriction and apical-basal contraction that produces an indentation in the disc known as the morphogenetic furrow (MF; Ready et al. 1976). Cells anterior to the MF divide actively and appear unpatterned. Just posterior to the MF, cells assemble into evenly spaced rosettes; slightly more posteriorly these transform into arcs, and the arcs then close to produce five-cell preclusters (Wolff and Ready 1991). Concurrently, these cells initiate a program of gene expression resulting in the appearance of neural-specific markers in a defined sequence in the cells of each cluster (Tomlinson and Ready 1987). Cells in the MF are arrested in the G1 phase of the cell cycle (Thomas et al. 1994); posterior to the MF, cells excluded from the preclusters undergo one more round of division, the second mitotic wave, to generate the remaining cells of each ommatidium (Ready et al. 1976; Wolff and Ready 1991). This orderly and sequential pattern of differentiation, proliferation, and morphogenesis is organized by a set of signaling molecules that also direct many other developmental processes. The expression patterns and interactions of these molecules in the developing eye disc are shown in Fig. 1.

2 Notch Activation Defines the Initiation Point

Photoreceptor differentiation initiates at the intersection of the dorsal-ventral (D-V) midline of the disc with the posterior margin. Determination of the D-V midline, or equator, is thus critical to delimit the initiation site. The dorsal side of the eye disc is defined during embryogenesis by its expression of the GATA transcription factor encoded by *pannier* (*pnr*; Ramain et al. 1993; Heitzler et al. 1996; Maurel-Zaffran and Treisman 2000). *pnr* is required for

[1] Skirball Institute of Biomolecular Medicine and Department of Cell Biology, NYU School of Medicine, 540 First Avenue, New York, New York 10016, USA

Results and Problems in Cell Differentiation, Vol. 37
K. Moses (Ed.): Drosophila Eye Development

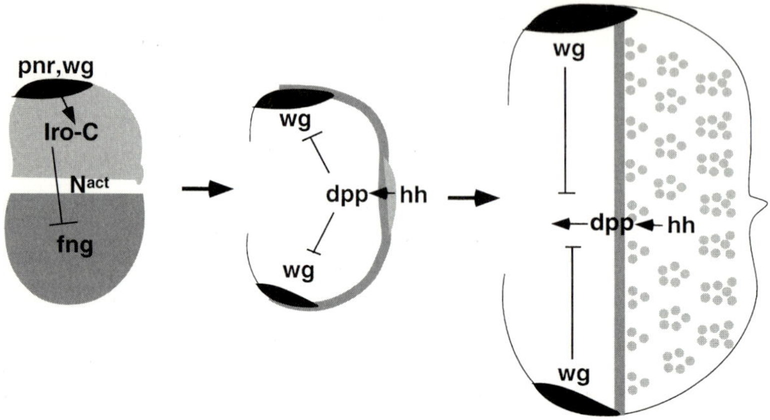

Fig. 1. Expression and interactions of some of the molecules used to pattern the eye disc. Before MF initiation, *pnr* activates *wg* expression on the dorsal side of the eye disc, leading to the activation of the *Iro-C* genes in the dorsal compartment. The *Iro-C* genes repress *fng*, and the boundary of *fng* expression leads to N activation at the dorso-ventral midline. The initiation point is specified by the combination of activated N with *hh* and *dpp*, which are expressed at the posterior margin. *wg* prevents ectopic initiation from the lateral margins. Later in development, Hh present in the differentiating photoreceptors activates *dpp* expression and differentiation in more anterior cells

the expression of *wingless* (*wg*), a member of the *Wnt* gene family (Cadigan and Nusse 1997), at the dorsal margin of the eye disc (Maurel-Zaffran and Treisman 2000). Together with the secreted protein Hedgehog (Hh), Wg then activates the expression of the three homeobox genes of the *Iroquois* complex (*Iro-C*), *mirror* (*mirr*), *araucan* (*ara*), and *caupolican* (*caup*; Gomez-Skarmeta et al. 1996; McNeill et al. 1997; Heberlein et al. 1998; Cavodeassi et al. 1999; Maurel-Zaffran and Treisman 2000). Expression of the *Iro-C* genes fills the dorsal compartment of the disc but ends sharply at the D-V midline; the mechanism by which this sharp expression boundary is established is not understood. The JAK/STAT pathway ligand Unpaired, which is present at the center of the posterior margin, appears to contribute to the ventral repression of these genes (Zeidler et al. 1999). This repression is maintained by chromatin-mediated mechanisms requiring the *Polycomb* group of genes (Netter et al. 1998).

The dorsally expressed Iro-C proteins repress the expression of *fringe* (*fng*), limiting it to the ventral compartment of the eye disc (Cho and Choi 1998; Dominguez and de Celis 1998; Cavodeassi et al. 1999; Yang et al. 1999). Fng is a glycosyltransferase that modifies the transmembrane receptor Notch (N) by adding N-acetylglucosamine to O-linked fucose residues (Bruckner et al. 2000; Moloney et al. 2000). This modification increases the affinity of N for its ligand Delta (Dl; Bruckner et al. 2000); in vivo, *fng*-expressing cells also appear less sensitive to an alternative ligand, Serrate (Ser; Panin et al. 1997). *Dl* expression is restricted to the dorsal side of the early eye disc and *Ser* to the ventral side

(Cho and Choi 1998; Dominguez and de Celis 1998; Papayannopoulos et al. 1998). The presence of Fng in the ventral compartment thus limits N activation to the midline, where modified ventral N is exposed to dorsal Dl and unmodified dorsal N is exposed to ventral Ser. The initial domains of *Dl* and *Ser* expression may be controlled by Hh and Wg signaling from peripodial membrane cells (Cho et al. 2000).

A boundary of Fng expression is essential to trigger N activation and define a central initiation point; either loss of *fng* function or ubiquitous *fng* expression causes a loss of the eye that can be rescued by expressing an activated form of N (Cho and Choi 1998; Dominguez and de Celis 1998; Papayannopoulos et al. 1998). Similarly, ubiquitous expression of *caup* or *mirr* can abolish the eye (Cho and Choi 1998; Dominguez and de Celis 1998), while removal of all three *Iro-C* genes from clones of dorsal cells induces ectopic eyes composed of both mutant and wild-type cells (Cavodeassi et al. 1999; Pichaud and Casares 2000). Because *fng* is misexpressed in the mutant cells, N becomes activated at the clonal boundary and may promote the initiation of an ectopic MF. Removal of the upstream gene *pnr* from clones of cells also leads to ectopic dorsal eye development, and ubiquitous expression of an activated form of *pnr* prevents MF initiation (Maurel-Zaffran and Treisman 2000). These results show that activation of N at a line source is both necessary and sufficient, in combination with signals present at the disc margin, to initiate differentiation.

3 Hedgehog is Essential for Morphogenetic Furrow Movement

The *hh* gene is critical for initiation and progression of the MF. *hh* encodes a secreted protein related to vertebrate proteins of the Sonic hedgehog family (Fietz et al. 1994). Although *hh* has a very dynamic pattern of expression in the eye disc, its upstream regulators are unknown. During the second instar *hh* is expressed predominantly in the peripodial membrane, where it shifts from a ventral to a dorsal domain (Cho et al. 2000) that has been implicated in the regulation of D-V polarity (see above). In early third instar eye discs, *hh* is expressed at the center of the posterior margin, where it is required for the onset of photoreceptor differentiation (Dominguez and Hafen 1997; Royet and Finkelstein 1997; Borod and Heberlein 1998). Using a temperature-sensitive allele, Borod and Heberlein (1998) determined that *hh* function was required for differentiation at the time of MF initiation, but not earlier.

During MF progression, *hh* is strongly expressed in the newly differentiating R2 and R5 photoreceptors, and more weakly in the other cells of the forming precluster (Heberlein et al. 1993; Ma et al. 1993). Hh secreted by these cells is essential to promote the differentiation of more anterior cells. Removal of *hh* function from the entire disc using eye-specific or temperature-sensitive alleles results in an arrest of MF progression that can be visualized by the presence of a full complement of photoreceptors in the most anterior row of ommatidia (Heberlein et al. 1993; Ma et al. 1993). When misexpressed in the anterior domain of the eye disc, Hh is sufficient to produce ectopic radially oriented morphogenetic furrows (Heberlein et al. 1995; Dominguez 1999), and clones of cells lacking the negative regulators of Hh signaling *ptc* or *PKA* have the same effect (Chanut and Heberlein 1995; Ma and Moses 1995; Pan and Rubin 1995; Strutt et al. 1995; Wehrli and Tomlinson 1995). The range of Hh signaling in the eye is limited in part by the apical constriction of cells as they enter the MF; cells that lack the actin binding protein Act up do not undergo apical constriction, allowing Hh to move further anteriorly and trigger precocious differentiation (Benlali et al. 2000). Despite a global requirement for Hh signaling during MF progression, clones of cells mutant for *smoothened* (*smo*), which encodes a cell autonomous receptor component, show only a delay in progression; this implies that Hh functions at least in part by activating a secondary signal (Strutt and Mlodzik 1997; Dominguez 1999; Greenwood and Struhl 1999). This signal is probably the bone morphogenetic protein (BMP) family member encoded by *decapentaplegic* (*dpp*; see below).

Hh promotes differentiation by activating the expression of the proneural gene *atonal (ato*; Borod and Heberlein 1998; Greenwood and Struhl 1999). *ato* encodes a basic helix-loop-helix (bHLH) protein that is required for the formation of the "founder" R8 photoreceptor (Jarman et al. 1994). A broad stripe of *ato* expression just anterior to the MF is later refined first to a proneural cluster of cells and then to individual R8 cells (Jarman et al. 1994, 1995; Dokucu et al. 1996). *ato* appears to be a direct target of Hh signaling, since it is autonomously lost in *smo* clones; however, *smo* is also required for the subsequent down-regulation of *ato* between the proneural clusters (Dominguez 1999; Greenwood and Struhl 1999). The repressive effect of Hh on *ato* may be mediated by the homeodomain protein Rough (Ro; Kimmel et al. 1990). *ro* and *ato* are expressed in complementary patterns, and Ro has been shown to repress the transcription of *ato* (Dokucu et al. 1996). Hh signaling is required for *ro* expression (Dominguez 1999), and the gain of function mutation ro^{DOM} interferes with upregulation of *ato* by Hh (Heberlein et al. 1993; Chanut et al. 2000). Thus Hh regulates both the formation and the spacing of the R8 photoreceptors.

In addition to regulating genes that promote MF progression, *hh* lies upstream of another bHLH protein encoded by *hairy* (*h*; Ma et al. 1993; Heberlein et al. 1995; Pan and Rubin 1995). H is present in a stripe anterior to the stripe of Ato, where it acts to inhibit differentiation in cooperation with the HLH protein Extramacrochaete (Brown et al. 1995). Hh signaling is required for *h* expression and is sufficient to ectopically activate *h* (Ma et al. 1993; Heberlein et al. 1995; Pan and Rubin 1995); however, its effect on *h* is probably mediated by Dpp (see below).

Recent work has suggested that *hh* homologues may play a similar role in patterning the differentiation of neurons in the vertebrate retina (Jensen and Wallace 1997; Levine et al. 1997; Neumann and Nüsslein-Volhard 2000; Stenkamp et al. 2000). Retinal differentiation begins near the optic stalk and proceeds outward in a centrifugal pattern (Burrill and Easter 1995; Raymond et al. 1995; Schmitt and Dowling 1996; McCabe et al. 1999). The *hh* homologues *Sonic hedgehog (Shh)* and *Tiggywinkle hedgehog (twhh)* are expressed in the zebrafish ganglion cell layer at the time of ganglion cell differentiation and in the adjacent retinal pigmented epithelium at the time of photoreceptor differentiation (Neumann and Nüsslein-Volhard 2000; Stenkamp et al. 2000). Reduction of Shh and Twhh production by antisense oligonucleotide injection, or using null mutations in the *shh* gene *sonic you (syu)*, reduces retinal ganglion cell and rod photoreceptor differentiation, while blocking all Hh family signaling with cyclopamine can prevent ganglion cell neurogenesis (Neumann and Nüsslein-Volhard 2000; Stenkamp et al. 2000). In addition, the expression of a *shh* reporter is lost in *syu* mutants (Neumann and Nüsslein-Volhard 2000), suggesting that, as in the fly retina, *hh* expression requires reception of the Hh signal.

4 Decapentaplegic Promotes Morphogenetic Furrow Movement

dpp, which encodes a homologue of the secreted BMPs 2 and 4 (Padgett et al. 1987), acts downstream of Hh to perform a subset of its functions in MF movement. *dpp* is first expressed at the ventral margin of the first instar eye disc, and its expression subsequently expands to include the dorsal and posterior margins (Cho et al. 2000). Following the initiation of differentiation, *dpp* expression becomes restricted to a stripe of cells within the MF (Masucci et al. 1990). At both early and late stages, *dpp* expression is dependent on *hh* (Heberlein et al. 1993; Ma et al. 1993; Royet and Finkelstein 1997; Strutt and Mlodzik 1997; Borod and Heberlein 1998).

Dpp signaling is critical for the initiation of differentiation. Large clones of cells mutant for *dpp* can block the formation of posterior eye regions (Heberlein et al. 1993). An eye-specific enhancer mutation, *dpp*$^{d\text{-}blk}$, prevents initiation from occurring in the ventral region of the posterior margin (St. Johnston et al. 1990; Blackman et al. 1991; Wiersdorff et al. 1996; Chanut and Heberlein 1997a), and a similar effect can be seen using temperature-sensitive *dpp* alleles (Chanut and Heberlein 1997b). In addition, clones of cells mutant for genes encoding components of the Dpp signaling pathway, such as the type I receptor Thickveins (Tkv), the type II receptor Punt, or the SMADs Mothers against Dpp (Mad) and Medea, fail to differentiate when they contact the posterior margin of the disc (Burke and Basler 1996; Wiersdorff et al. 1996; Das et al. 1998). Consistent with a function specific to initiation, misexpression of Dpp can induce ectopic MF initiation from the anterior margin even at a distance, although it does not induce photoreceptor differentiation in internal regions of the disc (Chanut and Heberlein 1997b; Pignoni and Zipursky 1997).

In the absence of *dpp* or of the downstream component encoded by *Mad*, expression of the eye specification genes *eyes absent* (*eya*), *sine oculis* (*so*) and *dachshund* (*dac*; see G. Mardon, Chap. 1, this Vol.) is not induced despite the presence of the upstream Pax-6 homologues encoded by *eyeless* (*ey*) and *twin of eyeless* (Quiring et al. 1994; Chen et al. 1999; Czerny et al. 1999; Curtiss and Mlodzik 2000). This may explain the requirement for *dpp* in initiation, as loss of *Mad* function at the posterior margin can be rescued by supplying *eya* (Curtiss and Mlodzik 2000). Ectopic *dpp* induces expression of *eya*, *so* and *dac* at the anterior eye disc margin, and ectopic *ey* can induce expression of these genes in other imaginal discs only when *dpp* is also present, confirming a critical role for *dpp* at this stage (Pignoni and Zipursky 1997; Halder et al. 1998; Chen et al. 1999). However, *eya* and *so* are necessary to maintain the expression of *dpp*, suggesting that these genes act in an autoregulatory loop (Pignoni et al. 1997; Hazelett et al. 1998). Another important function for Dpp signaling may be to repress the expression of the homeoprotein Homothorax (Hth), which is present in all cells of the early eye disc and then becomes restricted to the anterior margin, where it blocks MF initiation (Pai et al. 1998; Pichaud and Casares 2000).

Progression of the MF does not require Dpp signaling as strongly as initiation, as clones of cells mutant for *Mad*, *punt* or *tkv* in internal regions of the eye disc are able to differentiate almost normally (Burke and Basler 1996; Wiersdorff et al. 1996) and to express *eya*, *so* and *dac* (Curtiss and Mlodzik 2000). Because Dpp signaling is required for cell growth, large clones completely lacking gene activity could not be analyzed; however, the alleles tested were able to completely block initiation while only delaying progression. Low levels of Dpp signaling may be required for MF progression, as complete removal of *dpp* function from the eye disc using a temperature-sensitive allele can arrest the MF (Chanut and Heberlein 1997b). At this stage the functions of *dpp* and *hh* are partially redundant, as some photoreceptor differentiation is observed in clones mutant for the Hh receptor encoded by *smo* (Strutt and Mlodzik 1997) but not in clones doubly mutant for *smo* and *Mad* or *tkv* (Greenwood and Struhl 1999; Curtiss and Mlodzik 2000).

Despite this redundancy, loss of Dpp signaling alone causes more subtle phenotypes than loss of Hh signaling. Dpp signaling is required for cells in the MF to arrest in the G1 phase of the cell cycle (Penton et al. 1997; Horsfield et al. 1998). In addition, upregulation of the negative regulator H does not occur in *tkv* mutant clones anterior to the MF (Greenwood and Struhl 1999). However, *smo* mutant clones show high levels of H, presumably due to Dpp diffusing in from neighboring wild-type cells (Greenwood and Struhl 1999). Hh may thus control progression by inducing some target genes, such as *ato*, directly and others, such as *h*, indirectly through *dpp*. However, redundancy between the two pathways requires *ato* and other Hh target genes to be activated by Dpp in the absence of Hh. In *smo* mutant clones, the first broad stripe of *ato* expression is missing, but expression in single R8 cells is present, while *tkv* mutant clones also show a reduction in the initial broad *ato* expression (Dominguez 1999; Greenwood and Struhl 1999). The most likely explanation of these results is that separate enhancers of *ato* can respond to Hh and Dpp signaling.

5 Wingless Inhibits Morphogenetic Furrow Movement

wg is expressed in a dorsal domain of the margin and peripodial membrane of the early eye disc (Cavodeassi et al. 1999; Cho et al. 2000), although its secreted protein product is more broadly distributed (Royet and Finkelstein 1997; Cho et al. 2000). Prior to MF initiation, *wg* expression becomes restricted to the anterior dorsal and ventral margins of the eye disc (Baker 1988). The dorsal expression domain of *wg* is established in the embryo by *pnr* (Maurel-Zaffran and Treisman 2000), while ventral *wg* expression is dependent on *hth* (Pichaud and Casares 2000). *dpp* is required to restrict *wg* expression to the anterior and to maintain its repression at the posterior margin (Wiersdorff et al. 1996; Royet and Finkelstein 1997).

An important role of *wg* is to prevent ectopic photoreceptor differentiation from initiating at the lateral margins. When *wg* activity is removed during the larval stages using a temperature-sensitive allele, a MF initiates at the dorsal margin and progresses toward the center of the disc; the ventral margin is more weakly affected (Ma and Moses 1995; Treisman and Rubin 1995). Clones of cells mutant for *dishevelled* (*dsh*), a downstream component of the *wg* pathway, likewise form ectopic photoreceptors in the region that normally gives rise to dorsal head cuticle (Heslip et al. 1997). Conversely, ectopic expression of *wg* or activation of the Wg pathway can block both initiation and progression of the MF (Treisman and Rubin 1995; Heslip et al. 1997).

It was originally proposed that *wg* repression is the only role of *dpp* in MF initiation, as cells in which the Hh pathway is ectopically activated can become photoreceptors if they also lack both *dpp* and *wg* (Wiersdorff et al. 1996; Dominguez and Hafen 1997). However, *Mad* mutant clones at the posterior margin fail to differentiate even if they also lack *wg*, indicating that *dpp* has additional functions in initiation (Hazelett et al. 1998). Conversely, although high levels of Wg signaling repress *dpp* (Heslip et al. 1997) a level of ectopic Wg that does not affect *dpp* expression can still block differentiation (Treisman and Rubin 1995). Activating the Dpp pathway at the level of the receptor Tkv is not sufficient to restore photoreceptor differentiation in the presence of Wg (Hazelett et al. 1998). Interestingly, Wg can upregulate the expression of *hth*, and ectopic Hth similarly blocks differentiation downstream of *dpp* expression (Pai et al. 1998; Pichaud and Casares 2000). Anterior/posterior patterning of the eye disc appears to depend on the balance between Wg and Dpp, as anteriorly expressed genes such as *ey* are activated by Wg signaling and repressed by Dpp signaling (Curtiss and Mlodzik 2000; JD Lee and JE Treisman, 2001).

Some of the effects of *wg* on MF initiation may be due to its influence on D-V patterning. The early restriction of *wg* expression to the dorsal side of the eye disc allows it to contribute to the activation of the *Iro-C* genes, which act as dorsal determinants (see above). However, the *Iro-C* genes probably do not mediate all the effects of *wg*. Although activation of the Wg pathway in the ventral domain is not sufficient to induce *Iro-C* gene expression (Cavodeassi et al. 1999), it does prevent MF initiation (Treisman and Rubin 1995). Ectopic eyes produced by lack of *Iro-C* include wild-type tissue and an ectopic equator is induced at the clonal boundary (Cavodeassi et al. 1999; Pichaud and Casares 2000), while ectopic eyes produced by *dsh* clones are entirely composed of mutant tissue (Heslip et al. 1997), and ectopic furrows in discs lacking *wg* function have no equator (Ma and Moses 1995).

A final signal that appears to be important for MF progression is the steroid hormone ecdysone. Inactivation of the *ecdysoneless* gene, which is required for ecdysone production, leads to a block in MF progression characterized by the loss of *hh* and *ato* expression (Brennan et al. 1998). However, the mechanism by which this signal is transduced is unclear. The known ecdysone receptor (EcR) and another hormone receptor encoded by DHR78 are not required for normal photoreceptor differentiation (Brennan et al. 2001), while loss of the coreceptor Ultraspiracle leads to an acceleration of MF movement (Zelhof et al. 1997). A delay in MF progression does occur in clones of cells mutant for some components of the downstream ecdysone-regulated *Broad-complex* (Brennan et al. 1998, 2001).

6 Conclusions

The molecules that are used to establish spatial pattern in the eye disc are also used to provide positional information to other imaginal discs; however, the details of their interactions are tailored to fit the progressive nature of eye development. N activation defines the D-V boundary of the wing disc (Diaz-Benjumea and Cohen 1995; de Celis et al. 1996; Doherty et al. 1996), as it does in the eye, but the selector gene that determines the dorsal compartment of the wing is *apterous* rather than the *Iro-C* genes (Diaz-Benjumea and Cohen 1993). Although *hh* in the posterior of the wing disc also activates *dpp* in more anterior cells (Basler and Struhl 1994; Tabata et al. 1995; Zecca et al. 1995), the domain of *hh* expression is stable rather than progressive and depends on the selector gene *engrailed* (Tabata et al. 1992; Zecca et al. 1995), which is not required in the eye disc (Strutt and Mlodzik 1996). The direct effects of *hh* on wing patterning are restricted to the region near the compartment boundary (Mullor et al. 1997; Strigini and Cohen 1997), while *dpp* organizes the long-range pattern (Lecuit et al. 1996; Nellen et al. 1996). In the eye, the progressive expansion of *hh* expression allows it to play a more significant direct role, taking over most of the functions of *dpp*. Eye development thus exhibits some interesting variations on the mechanisms known to specify spatial pattern.

Acknowledgements. We thank Franck Pichaud, Russ Collins and Florence Janody for their insightful comments on the manuscript. Research in our laboratory is supported by grants from the National Institutes of Health (GM56131) and the National Science Foundation (IBN-9728140 and IBN-9982093).

References

Baker NE (1988) Transcription of the segment-polarity gene *wingless* in the imaginal discs of *Drosophila*, and the phenotype of a pupal-lethal *wg* mutation. Development 102:489–497

Basler K, Struhl G (1994) Compartment boundaries and the control of *Drosophila* limb pattern by hedgehog protein. Nature 368:208–214

Benlali A, Draskovic I, Hazelett DJ, Treisman JE (2000) *act up* controls actin polymerization to alter cell shape and restrict Hedgehog signaling in the *Drosophila* eye disc. Cell 101:271–81

Blackman RK, Sanicola M, Raftery LA, Gillevet T, Gelbart WM (1991) An extensive 3′ *cis*-regulatory region directs the imaginal disc expression of *decapentaplegic*, a member of the TGF-β family in *Drosophila*. Development 111:657–665

Borod ER, Heberlein U (1998) Mutual regulation of *decapentaplegic* and *hedgehog* during the initiation of differentiation in the *Drosophila* retina. Dev Biol 197:187–197

Brennan CA, Ashburner M, Moses K (1998) Ecdysone pathway is required for furrow progression in the developing *Drosophila* eye. Development 125:2653–64

Brennan CA, Li TR, Bender M, Hsiung F, Moses K (2001) *Broad-complex*, but not *Ecdysone receptor*, is required for progression of the morphogenetic furrow in the *Drosophila* eye. Development 128:1–11

Brown NL, Sattler CA, Paddock SW, Carroll SB (1995) *Hairy* and *emc* negatively regulate morphogenetic furrow progression in the *Drosophila* eye. Cell 80:879–887

Bruckner K, Perez L, Clausen H, Cohen S (2000) Glycosyltransferase activity of Fringe modulates Notch-Delta interactions. Nature 406:411–415

Burke R, Basler K (1996) Hedgehog-dependent patterning in the *Drosophila* eye can occur in the absence of dpp signaling. Dev Biol 179:360–368

Burrill JD, Easter SS Jr (1995) The first retinal axons and their microenvironment in zebrafish: cryptic pioneers and the pretract. J Neurosci 15:2935–2947

Cadigan KM, Nusse R (1997) Wnt signaling: a common theme in animal development. Genes Dev 11:3286–3305

Cavodeassi F, Diez Del Corral R, Campuzano S, Dominguez M (1999) Compartments and organising boundaries in the *Drosophila* eye: the role of the homeodomain Iroquois proteins. Development 126:4933–4942

Chanut F, Heberlein U (1995) Role of the morphogenetic furrow in establishing polarity in the *Drosophila* eye. Development 121:4085–4094

Chanut F, Heberlein U (1997a) Retinal morphogenesis in *Drosophila*: hints from an eye-specific *decapentaplegic* allele. Dev Genet 20:197–207

Chanut F, Heberlein U (1997b) Role of decapentaplegic in initiation and progression of the morphogenetic furrow in the developing *Drosophila* retina. Development 124:559–567

Chanut F, Luk A, Heberlein U (2000) A screen for dominant modifiers of *ro*(*Dom*), a mutation that disrupts morphogenetic furrow progression in *Drosophila*, identifies *groucho* and *Hairless* as regulators of *atonal* expression. Genetics 156:1203–1217

Chen R, Halder G, Zhang Z, Mardon G (1999) Signaling by the TGF-β homolog *decapentaplegic* functions reiteratively within the network of genes controlling retinal cell fate determination in *Drosophila*. Development 126:935–943

Cho K-O, Choi K-W (1998) Fringe is essential for mirror symmetry and morphogenesis in the *Drosophila* eye. Nature 396:272–276

Cho K-O, Chern J, Izaddoost S, Choi K-W (2000) Novel signaling from the peripodial membrane is essential for eye disc patterning in *Drosophila*. Cell 103:331–342

Curtiss J, Mlodzik M (2000) Morphogenetic furrow initiation and progression during eye development in *Drosophila*: the roles of *decapentaplegic*, *hedgehog* and *eyes absent*. Development 127:1325–1336

Czerny T, Halder G, Kloter U, Souabni A, Gehring WJ, Busslinger M (1999) *twin of eyeless*, a second *Pax*-6 gene of *Drosophila*, acts upstream of *eyeless* in the control of eye development. Mol Cell 3:297–307

Das P, Maduzia LL, Wang H, Finelli AL, Cho SH, Smith MM, Padgett RW (1998) The *Drosophila* gene *Medea* demonstrates the requirement for different classes of Smads in dpp signaling. Development 125:1519–1528

De Celis JF, Garcia-Bellido A, Bray SJ (1996) Activation and function of *Notch* at the dorsal-ventral boundary of the wing imaginal disc. Development 122:359–369

Diaz-Benjumea FJ, Cohen SM (1993) Interaction between dorsal and ventral cells in the imaginal disc directs wing development in *Drosophila.* Cell 75:741–52

Diaz-Benjumea FJ, Cohen SM (1995) Serrate signals through Notch to establish a *wingless*-dependent organizer at the dorsal-ventral compartment boundary of the *Drosophila* wing. Development 121:4215–4225

Doherty D, Feger G, Younger-Shepherd S, Jan LY, Jan YN (1996) Delta is a ventral to dorsal signal complementary to Serrate, another Notch ligand, in *Drosophila* wing formation. Genes Dev 10:421–434

Dokucu ME, Zipursky SL, Cagan RL (1996) Atonal, Rough and the resolution of proneural clusters in the developing *Drosophila* retina. Development 122:4139–4147

Dominguez M (1999) Dual role for *hedgehog* in the regulation of the proneural gene *atonal* during ommatidia development. Development 126:2345–2353

Dominguez M, de Celis JF (1998) A dorsal/ventral boundary established by Notch controls growth and polarity in the *Drosophila* eye. Nature 396:276–278

Dominguez M, Hafen E (1997) Hedgehog directly controls initiation and propagation of retinal differentiation in the *Drosophila* eye. Genes Dev 11:3254–3264

Fietz MJ, Concordet JP, Barbosa R, Johnson R, Krauss S, McMahon AP, Tabin C, Ingham PW (1994) The *hedgehog* gene family in *Drosophila* and vertebrate development. Development (Suppl) 36:43–51

Gomez-Skarmeta JL, del Corral RD, de la Calle-Mustienes E, Ferre-Marco D, Modolell J (1996) *araucan* and *caupolican*, two members of the novel *iroquois* complex, encode homeoproteins that control proneural and vein-forming genes. Cell 85:95–105

Greenwood S, Struhl G (1999) Progression of the morphogenetic furrow in the *Drosophila* eye: the roles of Hedgehog, Decapentaplegic and the Raf pathway. Development 126:5795–5808

Halder G, Callaerts P, Flister S, Walldorf U, Kloter U, Gehring WJ (1998) Eyeless initiates the expression of both *sine oculis* and *eyes absent* during *Drosophila* compound eye development. Development 125:2181–2191

Hazelett DJ, Bourouis M, Walldorf U, Treisman JE (1998) *decapentaplegic* and *wingless* are regulated by *eyes absent* and *eyegone* and interact to direct the pattern of retinal differentiation in the eye disc. Development 125:3741–3751

Heberlein U, Wolff T, Rubin GM (1993) The TGFβ homolog *dpp* and the segment polarity gene *hedgehog* are required for propagation of a morphogenetic wave in the *Drosophila* retina. Cell 75:913–926

Heberlein U, Singh CM, Luk AY, Donohoe TJ (1995) Growth and differentiation in the *Drosophila* eye coordinated by *hedgehog.* Nature 373:709–711

Heberlein U, Borod ER, Chanut FA (1998) Dorsoventral patterning in the *Drosophila* retina by *wingless.* Development 125:567–577

Heitzler P, Haenlin M, Ramain P, Calleja M, Simpson P (1996) A genetic analysis of *pannier*, a gene necessary for viability of dorsal tissues and bristle positioning in *Drosophila.* Genetics 143: 1271–1286

Heslip TR, Theisen H, Walker H, Marsh JL (1997) SHAGGY and DISHEVELLED exert opposite effects on *wingless* and *decapentaplegic* expression and on positional identity in imaginal discs. Development 124:1069–1078

Horsfield J, Penton A, Secombe J, Hoffman FM, Richardson H (1998) *decapentaplegic* is required for arrest in G1 phase during *Drosophila* eye development. Development 125:5069–5078

Jarman AP, Grell EH, Ackerman L, Jan LY, Jan YN (1994) *atonal* is the proneural gene for *Drosophila* photoreceptors. Nature 369:398–400

Jarman AP, Sun Y, Jan LY, Jan YN (1995) Role of the proneural gene, *atonal*, in formation of *Drosophila* chordotonal organs and photoreceptors. Development 121:2019–2030

Jensen AM, Wallace VA (1997) Expression of Sonic hedgehog and its putative role as a precursor cell mitogen in the developing mouse retina. Development 124:363–371

Kimmel BE, Heberlein U, Rubin GM (1990) The homeodomain protein rough is expressed in a subset of cells in the developing *Drosophila* eye where it can specify photoreceptor cell subtype. Genes Dev 4:712–727

Lecuit T, Brook WJ, Ng M, Sun H, Cohen SM (1996) Two distinct mechanisms for long-range patterning by decapentaplegic in the *Drosophila* wing. Nature 381:387–393

Lee JD, Treisman JE (2001) The role of Wingless signaling in establishing the anterior-posterior and dorsal-ventral axes of the eye disc. Development 128:1519–1529.

Levine EM, Roelink H, Turner J, Reh TA (1997) Sonic hedgehog promotes rod photoreceptor differentiation in mammalian retinal cells in vitro. J Neurosci 17:6277–6288

Ma C, Moses K (1995) *wingless* and *patched* are negative regulators of the morphogenetic furrow and can affect tissue polarity in the developing *Drosophila* compound eye. Development 121:2279–2289

Ma C, Zhou Y, Beachy PA, Moses K (1993) The segment polarity gene *hedgehog* is required for progression of the morphogenetic furrow in the developing *Drosophila* eye. Cell 75:927–938

Masucci JD, Miltenberger RJ, Hoffmann FM (1990) Pattern-specific expression of the *Drosophila decapentaplegic* gene in imaginal discs is regulated by 3′ cis-regulatory elements. Genes Dev 4:2011–2023

Maurel-Zaffran C, Treisman JE (2000) *pannier* acts upstream of *wingless* to direct dorsal eye disc development in *Drosophila*. Development 127:1007–1016

McCabe KL, Gunther EC, Reh TA (1999) The development of the pattern of retinal ganglion cells in the chick retina: mechanisms that control differentiation. Development 126:5713–5724

McNeill H, Yang CH, Brodsky M, Ungos J, Simon MA (1997) *mirror* encodes a novel PBX-class homeoprotein that functions in the definition of the dorsal-ventral border in the *Drosophila* eye. Genes Dev 11:1073–1082

Moloney DJ, Panin VM, Johnston SH, Chen J, Shao L, Wilson R, Wang Y, Stanley P, Irvine KD, Haltiwanger RS, Vogt TF (2000) Fringe is a glycosyltransferase that modifies Notch. Nature 406:369–735

Mullor JL, Calleja M, Capdevila J, Guerrero I (1997) Hedgehog activity, independent of *decapentaplegic*, participates in wing disc patterning. Development 124:1227–1237

Nellen D, Burke R, Struhl G, Basler K (1996) Direct and long-range action of a dpp morphogen gradient. Cell 85:357–368

Netter S, Fauvarque MO, Diez del Corral R, Dura JM, Coen D (1998) *white*$^{+}$ transgene insertions presenting a dorsal/ventral pattern define a single cluster of homeobox genes that is silenced by the polycomb-group proteins in *Drosophila melanogaster*. Genetics 149:257–275

Neumann CJ, Nüsslein-Volhard C (2000) Patterning of the zebrafish retina by a wave of sonic hedgehog activity. Science 289:2137–2139

Padgett RW, St Johnston RD, Gelbart WM (1987) A transcript from a *Drosophila* pattern gene predicts a protein homologous to the transforming growth factor-beta family. Nature 325:81–84

Pai C-Y, Kuo T-S, Jaw TJ, Kurant E, Chen C-T, Bessarab DA, Salzberg A, Sun YH (1998) The Homothorax homeoprotein activates the nuclear localization of another homeoprotein, Extradenticle, and suppresses eye development in *Drosophila*. Genes Dev 12:435–446

Pan D, Rubin GM (1995) cAMP-dependent protein kinase and *hedgehog* act antagonistically in regulating *decapentaplegic* transcription in *Drosophila* imaginal discs. Cell 80:543–552

Panin VM, Papayannopoulos V, Wilson R, Irvine KD (1997) Fringe modulates Notch-ligand interactions. Nature 387:908–912

Papayannopoulos V, Tomlinson A, Panin VM, Rauskolb C, Irvine KD (1998) Dorsal-ventral signaling in the *Drosophila* eye. Science 281:2031–2034

Penton A, Selleck SB, Hoffmann FM (1997) Regulation of cell cycle synchronization by *decapentaplegic* during *Drosophila* eye development. Science 275:203–206

Pichaud F, Casares F (2000) *homothorax* and *iroquois-C* genes are required for the establishment of territories within the developing eye disc. Mech Dev 96:15–25

Pignoni F, Zipursky SL (1997) Induction of *Drosophila* eye development by Decapentaplegic. Development 124:271–278

Pignoni F, Hu B, Zavitz KH, Xiao J, Garrity PA, Zipursky SL (1997) The eye specification proteins so and eya form a complex and multiple steps in *Drosophila* eye development. Cell 91:881–892

Quiring R, Walldorf U, Kloter U, Gehring WJ (1994) Homology of the *eyeless* gene of *Drosophila* to the *Small eye* gene in mice and *Aniridia* in humans. Science 265:785–789

Ramain P, Heitzler P, Haenlin M, Simpson P (1993) *pannier*, a negative regulator of *achaete* and *scute* in *Drosophila*, encodes a zinc finger protein with homology to the vertebrate transcription factor GATA-1. Development 119:1277–1291

Raymond PA, Barthel LK, Curran GA (1995) Developmental patterning of rod and cone photoreceptors in embryonic zebrafish. J Comp Neurol 359:537–50

Ready DF, Hanson TE, Benzer S (1976) Development of the *Drosophila* retina, a neurocrystalline lattice. Dev Biol 53:217–240

Royet J, Finkelstein R (1997) Establishing primordia in the *Drosophila* eye-antennal imaginal disc: the roles of *decapentaplegic*, *wingless* and *hedgehog*. Development 124:4793–4800

Schmitt EA, Dowling JE (1996) Comparison of topographical patterns of ganglion and photoreceptor cell differentiation in the retina of the zebrafish, Danio rerio. J Comp Neurol 371: 222–234

Stenkamp DL, Frey RA, Prabhudesai SN, Raymond PA (2000) Function for Hedgehog genes in zebrafish retinal development. Dev Biol 220:238–252

St Johnston RD, Hoffmann FM, Blackman RK, Segal D, Grimaila R, Padgett RW, Irick HA, Gelbart WM (1990) Molecular organization of the *decapentaplegic* gene in *Drosophila melanogaster*. Genes Dev 4:1114–1127

Strigini M, Cohen SM (1997) A Hedgehog activity gradient contributes to AP axial patterning of the *Drosophila* wing. Development 124:4697–4705

Strutt DI, Mlodzik M (1996) The regulation of *hedgehog* and *decapentaplegic* during *Drosophila* eye imaginal disc development. MOD 58:39–50

Strutt DI, Mlodzik M (1997) Hedgehog is an indirect regulator of morphogenetic furrow progression in the *Drosophila* eye disc. Development 124:3233–3240

Strutt DI, Wiersdorff V, Mlodzik M (1995) Regulation of furrow progression in the *Drosophila* eye by cAMP-dependent protein kinase A. Nature 373:705–709

Tabata T, Eaton S, Kornberg TB (1992) The *Drosophila hedgehog* gene is expressed specifically in posterior compartment cells and is a target of *engrailed* regulation. Genes Dev 6:2635–2645

Tabata T, Schwartz C, Gustavson E, Ali Z, Kornberg TB (1995) Creating a *Drosophila* wing de novo, the role of *engrailed*, and the compartment border hypothesis. Development 121:3359–3369

Thomas BJ, Gunning DA, Cho J, Zipursky SL (1994) Cell cycle progression in the developing *Drosophila* eye: *roughex* encodes a novel protein required for the establishment of G1. Cell 77: 1003–1014

Tomlinson A, Ready DF (1987) Neuronal differentiation in the *Drosophila* ommatidium. Dev Biol 120:366–376

Treisman JE, Rubin GM (1995) *wingless* inhibits morphogenetic furrow movement in the *Drosophila* eye disc. Development 121:3519–3527

Wehrli M, Tomlinson A (1995) Epithelial planar polarity in the developing *Drosophila* eye. Development 121:2451–2459

Wiersdorff V, Lecuit T, Cohen SM, Mlodzik M (1996) *Mad* acts downstream of dpp receptors, revealing a differential requirement for *dpp* signaling in initiation and propagation of morphogenesis in the *Drosophila* eye. Development 122:2153–2162

Wolff T, Ready DF (1991) The beginning of pattern formation in the *Drosophila* compound eye: the morphogenetic furrow and the second mitotic wave. Development 113:841–850

Yang CH, Simon MA, McNeill H (1999) *mirror* controls planar polarity and equator formation through repression of *fringe* expression and through control of cell affinities. Development 126:5857–5866

Zecca M, Basler K, Struhl G (1995) Sequential organizing activities of *engrailed*, *hedgehog* and *decapentaplegic* in the *Drosophila* wing. Development 121:2265–2278

Zeidler MP, Perrimon N, Strutt DI (1999) Polarity determination in the *Drosophila* eye: a novel role for unpaired and JAK/STAT signaling. Genes Dev 13:1342–1353

Zelhof AC, Ghbeish N, Tsai C, Evans RM, McKeown M (1997) A role for *ultraspiracle*, the *Drosophila* RXR, in morphogenetic furrow movement and photoreceptor cluster formation. Development 124:2499–2506

NOTCH and the Patterning of Ommatidial Founder Cells in the Developing *Drosophila* Eye

Nicholas E. Baker[1]

1 Introduction

Each *Drosophila* compound eye is composed of hundreds of similar ommatidia, or unit eyes. Ommatidial differentiation begins at the posterior margin of the eye imaginal disc and extends more anteriorly as a "morphogenetic furrow" sweeps across the eye disc. Anterior to the position reached by the morphogenetic furrow, cells are proliferative and undifferentiated. Posterior to the morphogenetic furrow, ommatidia differentiate in dorso-ventral columns (Fig. 1; Wolff and Ready 1993).

The present review outlines the initiation and patterning of ommatidia, focusing on the specification of the founder R8 cells. R8 cells are the first cells to differentiate in each ommatidium, and are required for recruitment of other cells. Because of this founder role, the specification, positioning, and enumeration of R8 cells determine the number and position of all the ommatidial cells that will subsequently be formed, and so play a central role in retinal development.

Most ommatidial cells are recruited by short-range inductive signals between adjacent cells, but R8 specification is different. A pattern of regularly spaced single cells adopt R8 fate before their neighbors are specified (Fig. 1). Thus, R8-inductive signals are likely to act at longer range. In addition, the spacing pattern of R8s implies that one or more signals operate to prevent multiple nearby cells from adopting R8 fate. Both the mechanisms by which R8 specification is stimulated and by which "lateral inhibition" separates R8 precursors are important.

R8 fate specification depends on a basic Helix-Loop-Helix (bHLH) transcription factor encoded by the *atonal* gene. In normal development *atonal* is absolutely required for R8 specification. Null mutants for *ato* lack nearly all the eye because R8 cells are required for recruitment of other cells to each ommatidium (Jarman et al. 1994).

The *ato* expression pattern reveals the main features of R8 specification and patterning (Jarman et al. 1995). *ato* transcription begins in all cells just ahead of the morphogenetic furrow. Within the morphogenetic furrow the majority

[1] Department of Molecular Genetics, Albert Einstein College of Medicine, 1300 Morris Park Avenue, Bronx, New York 10461, USA

Results and Problems in Cell Differentiation, Vol. 37
K. Moses (Ed.): Drosophila Eye Development

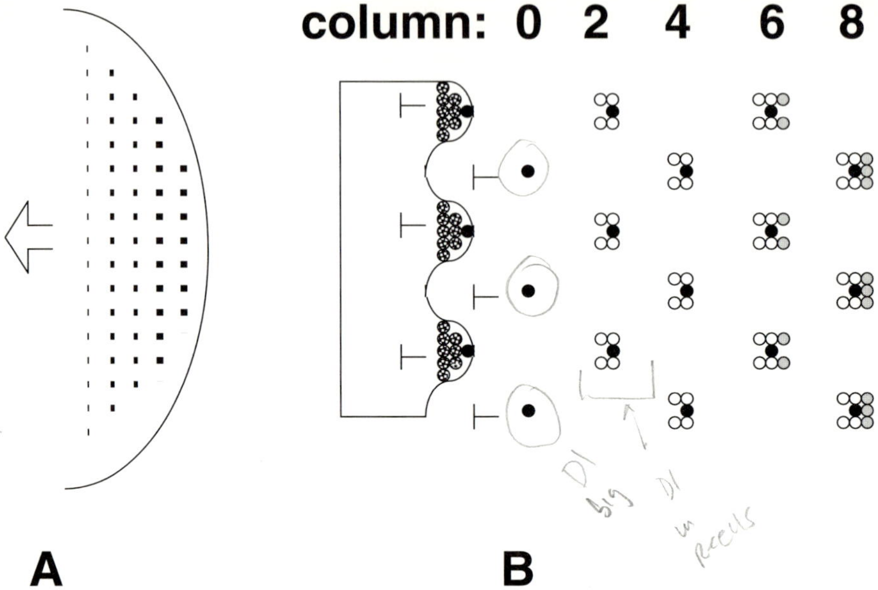

Fig. 1. Aspects of eye disc development (Wolff and Ready 1993). In this and the following figures, late third instar eye discs are shown with anterior to the left. **A** Differentiation of the eye imaginal disc begins at the posterior margin (*right*) and progressively spreads anteriorly (*left*), one new column of ommatidia initiating every 90–120 min. **B** The proneural gene *atonal* is expressed by all cells in a stripe anterior to any differentiation (*box on left*). Expression is lost from some cells but maintained by autoregulation in regularly spaced "intermediate groups" comprising ~10 cells each (*shaded nuclei*). Within these groups one cell will be specified as R8 (*solid shading*), maintain *ato* expression longest, differentiate and recruit other photoreceptor cells by expressing ligands for the EGFR and Bride of sevenless (BOSS) receptor tyrosine kinases. In wild type development the first column where R8 is the only cell retaining *ato* expression corresponds to column 0 in the nomenclature of Wolff and Ready (1993; Baker and Zitron 1995; Jarman et al. 1995). The alternating phasing of the ommatidial columns implies that each column is an inhibitory template for the next. The spacing factor has variously been proposed to be *scabrous* (Baker and Zitron 1995), *argos* (Spencer et al. 1998), *hedgehog* (Dominguez 1999), or an unidentified factor expressed in response to MAP Kinase activity (Chen and Chien 1999). (Adapted from Yang and Baker 2001)

of cells stop expressing Ato, which is retained only by single cells that are so defined as R8 precursor cells (Fig. 1). The evolution of *ato* expression occurs within one to two ommatidial columns, corresponding to a period of less than 3 h. During the transition, *ato* is briefly retained by groups of about ten cells, each of which is soon restricted to a single Ato-expressing R8 cell. Such "intermediate groups" confer a wavy appearance on the posterior margin of uniform Ato expression, and result from interactions between the posterior differentiating retina and the unpatterned cells ahead of the furrow (Fig. 1).

The Ato pattern reveals that patterning of R8 fate depends on selective *loss* of the critical regulatory gene expression, and also indicates at least two phases to this loss, corresponding first to the loss of *ato* expression from cells between each intermediate group, and subsequently to the loss of *ato* from all cells but one within each intermediate group. As will become clear below, many aspects of Ato expression and R8 specification are patterned by posterior to anterior signals originating from the differentiating posterior retina. The receptor protein Notch plays several roles in patterning *ato* expression. Pathways that regulate Ato expression are summarized in Fig. 6. Posterior to anterior signals that initiate *ato* transcription as the morphogenetic furrow progresses are discussed elsewhere (see Lee and Treisman, this Vol.).

2 The Discovery of Ommatidial Founder Cells

2.1 Founder Cells for Each Ommatidium

Historically, the existence and significance of ommatidial founder cells were not always appreciated. After lineage analysis had shown that cell fates must be assigned by postmitotic cell interactions, it was suggested that every unspecified cell would encounter a particular niche next to differentiating retinal cells; this lattice position would determine cell fate (Ready et al. 1976). In this model individual ommatidia would not depend on founder cells. The model was found wanting in a surgical test, however, because ordered retinal differentiation could occur in amputated disc fragments from which differentiating retina had been removed (Lebovitz and Ready 1986). This ruled out the crystal-growth model of fate specification. Still plausible was the possibility that cues radiated from differentiating cells in a diffusible manner, instead of being limited to the surfaces of differentiating cells.

The idea of diffusible signals gained support from electron microscope studies of ommatidial assembly. A stereotyped sequence of cell differentiation within each ommatidium began with R8 cell differentiation (Tomlinson and Ready 1987). Since R8 precursors were several cell diameters anterior to other differentiating cells, signals specifying R8 fate appeared to act diffusibly.

Further evidence for an ommatidial founder cell came from studies of the *Ellipse* mutations, hypermorphic alleles of the epidemial growth factor (EGF) receptor in which few ommatidia differentiate (Baker and Rubin 1989). The few isolated ommatidia can differentiate normally, indicating that all the essential information for determination and differentiation of cells is intrinsic to the ommatidium itself (Baker and Rubin 1992). More recently, identification of *ato* as a proneural gene specifically required for R8 has shown definitively that the R8 cell was essential for recruiting other cell types to each ommatidium, because specification of other retinal cells depends on the *ato* genotype of R8 cells only (Jarman et al. 1994). After founding by R8 cells, however, ommatidia seem to be almost self-organizing.

2.2 Spacing Patterns, Lateral Inhibition and Notch

The specification of single R8 cells in a regular array is an example of a spacing pattern. Simple ideas for how biological spacing patterns might arise were suggested half a century ago (Wigglesworth 1940; Turing 1952). In addition to positive inducers of R8 cell fate, diffusible inhibitors are expected that preclude R8 specification near one another (Meinhardt 1982). The term "lateral inhibition" has been adapted from visual physiology to describe the putative inhibitory signals and applied to spacing neurogenesis of *Drosophila* (Wilcox et al. 1973; Moscoso del Prado and Garcia-Bellido 1984).

In the embryonic nervous system, Notch (N) and the neurogenic genes had been found to be required for lateral inhibition (Campos-Ortega and Jan 1991). Excess neural differentiation occurs in the eye when N function is reduced, suggesting a role comparable to embryonic neurogenesis (Cagan and Ready 1989). Accordingly, additional R8 cells were specified when N function was reduced (Baker et al. 1990).

In most neural tissues the targets of lateral inhibition by Notch are bHLH proneural transcription factors encoded by the Achaete-Scute gene Complex (AS-C) (Heitzler and Simpson 1991; Jan and Jan 1993). AS-C is not required for ommatidial development (Jimenez and Campos-Ortega 1987). However, the *scabrous* gene which is transcribed in response to AS-C proneural gene function is also expressed during R8 specification (Mlodzik et al. 1990). Both the role of N in R8 specification and the expression of *scabrous* suggested that there should be another bHLH proneural gene for R8 cells, and this was later found to be *ato* (Baker et al. 1990; Mlodzik et al. 1990; Jarman et al. 1994). The *Drosophila* genome has the capacity to encode still further bHLH proteins of as yet unknown function (Moore et al. 2000). Perhaps some of these novel proteins are involved in specifying other eye cells.

Since it is clear that R8 specification uses some mechanisms that also apply elsewhere, studies of R8 specification are influenced by work on other lateral inhibitions mediated by Notch proteins, such as the nematode gonad or *Drosophila* bristle specification, and these are models for comparison to neurogenesis in vertebrates in turn (Greenwald and Rubin 1992). Conversely, R8 specification may serve as a model for other tissues. Although our understanding of R8 patterning is still incomplete, it appears that the actual mechanisms used in development are more complicated than anticipated theoretically. Patterning occurs in several successive stages, a number of distinct signals seem to be important, and lateral inhibition mediated by N is responsible only for a subset of R8 patterning.

3 R8 Cell Specification and Patterning

The source of patterning signals for R8 specification is usually developmentally more mature cells that are nearby to the posterior. It helps to discuss R8 specification in reverse temporal order, beginning with older cells that are the source of signals affecting more anterior regions.

3.1 Specification and Differentiation of R8 Precursors

R8 cells are differentiating and morphologically distinct by column 1 at the posterior of the morphogenetic furrow (Fig. 1; Tomlinson and Ready 1987). Certain neural antigens are detectable at this stage. Axon outgrowth begins. R8 specification must occur earlier, within column 0 when R8 cells alone retain Ato expression. R8 precursors can be identified within column 0 on the basis of Ato or Sca expression patterns (Baker and Zitron 1995; Jarman et al. 1995).

R2 and R5 photoreceptors differentiate alongside R8 precursors 2–3 h afterwards. R8 specification is required for recruitment of R2, R5 and further ommatidial cells, so that *atonal* null mutants lack nearly all the eye. Recruitment depends on the *ato* genotype only of R8 cells, confirming the organizing role of R8 (Jarman et al. 1994). It is thought that Ato activates Spitz secretion from R8 to begin EGFR-mediated recruitment of neighboring cells (Tio and Moses 1997; Baonza et al. 2001). Consequently, the requirement for atonal can be bypassed by experimental activation of EGFR (Dominguez et al. 1998). In this case many other ommatidial cells differentiate even though R8 cells are missing in the absence of the *ato* gene. Exceptionally, ommatidia can be formed in the apparent absence of differentiated R8 cells when *ato* mutants are partially rescued by expression of the related bHLH protein Scute (Sun et al. 2000). It may be that Sc mimics Ato sufficiently to trigger EGFR-mediated recruitment without causing R8-specific differentiation. Another possibility is that Sc mimics other, unknown bHLH genes that might be targets for EGFR signaling in R1–R7 differentiation.

Interestingly, Ato not only triggers recruitment by specifying R8 cells but also seems to act more directly in EGFR-mediated recruitment once R8 cells are already differentiating. Normally Ato expression is maintained in R8 precursor cells until column 3 or 4. Reduction in Ato expression levels in R8 leads to reduced recruitment of other R cells even for ommatidia where R8 is already differentiating. Conversely, overexpressing Ato in R8 leads to excess recruitment of additional photoreceptor cells (White and Jarman 2000). Apparently, Ato levels must control levels of EGFR signalling during recruitment to the cluster. The continued role of Ato within R8 precursors early in their differentiation may explain why R8 cell differentiation can be reversed by activated N expression, which abolishes Ato expression in the differentiating R8 precursor cells (Baker et al. 1996).

3.2 Mechanism of N Signaling During Lateral Inhibition of R8 Precursors

When Notch function is reduced, clusters of up to ten R8-like cells differentiate in the place of single R8 cells (Baker and Zitron 1995). Such results identify groups of cells that have the potential to differentiate as R8, in which N mediates lateral inhibition to restrict R8 fate specification. Without N, extra R8-like cells differentiate because Ato fails to be repressed from the intermediate groups. Once normal Ato repression has occurred at column 0, N is no longer required to repress Ato expression (Baker et al. 1996). Cells outside the intermediate groups lose Ato even when N activity is reduced (Lee et al. 1996).

Despite initial expression of Ato in all cells ahead of the furrow, only intermediate group cells show proneural behavior. This is partly due to inhibitory factors including Hairy and Emc that oppose Ato function ahead of the furrow (Brown et al. 1995), and partly due to activatory signals that elevate Ato function within intermediate groups (see below).

Intermediate groups and R8 cells show multiple signs of elevated Ato activity compared to other cells. They express *sca* (Lee et al. 1996), and elevate levels of Daughterless protein (Da), the ubiquitous bHLH partner required for Ato function (Brown et al. 1996). Ato expression in intermediate groups is autoregulatory and dependent on *ato* function, whereas earlier initiation of Ato expression in all cells is independent of *ato* function (Jarman et al. 1995). Ato-sensitive reporter constructs derived from *ato* gene regulatory sequences are active only in intermediate groups and R8 cells (Sun et al. 1998). A Zn-finger protein encoded by *senseless* that is required for Ato function is also expressed only in intermediate groups and R8 cells (Nolo et al. 2000). These observations support the view that Ato protein becomes able to activate transcription of itself and of other genes, and to direct R8 specification, at and after the intermediate group stage, when N represses *ato* autoregulation (Baker et al. 1996).

Studies with a N^{ts} allele indicate that implementing N signaling takes most of the ~100 min lifetime of each intermediate group (Baker and Yu 1998). Ato protein is first lost from the anterior cells of intermediate groups, and lastly from cells adjacent to the future R8 cell, which usually arises close to the posterior apex of the intermediate group (Baker et al. 1996; Dokucu et al. 1996).

The molecular pathway of Notch function has been elucidated from studies of lateral inhibition in many tissues and organisms (Fig. 2). Most of the same genes are required to repress *ato* in intermediate groups indicating that the same general pathway of lateral inhibition is at work (Baker and Zitron 1995; Parks et al. 1995; Treisman et al. 1997; Ligoxygakis et al. 1998; Li and Baker 2001). One exception is the transcriptional co-activator *mastermind* (*mam*), which appears to be completely dispensable during eye development (Li and Baker 2001). Mam is expressed in the eye, however, and *mam* mutant alleles have been recovered from genetic screens as enhancers or suppressors of other eye defects (Karim et al. 1996; Treisman and Rubin 1996; Verheyen et al. 1996).

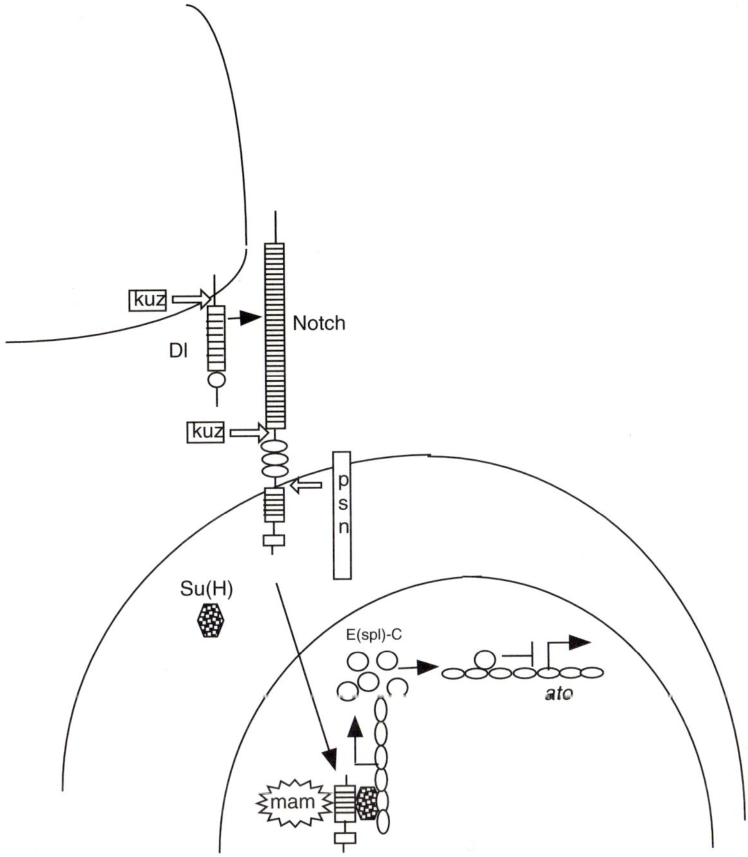

Fig. 2. Notch signal transduction in lateral inhibition. Ligand (Delta) binding to the extracellular domain EGF repeats 11–12 leads to receptor (Notch) activation (Rebay et al. 1991). *Delta* has been found to be cleaved by the metalloprotease *kuzbanian*. Activation of Notch by ligand leads to proteolytic cleavages, releasing the intracellular domain of the Notch protein (Kidd et al. 1998; Lecourtois and Schweisguth 1998; Schroeter et al. 1998; Struhl and Adachi 1998). The metalloprotease Kuzbanian may be involved in processing both Dl and N (Qi et al. 1999; Brou et al. 2000; Mumm et al. 2000). N processing inside the membrane requires the *presenilin* gene, which is therefore required for Notch signalling activity (Struhl and Greenwald 1999; Ye et al. 1999). Released intracellular domain can enter the nucleus, where it acts as a coactivator for the site-specific DNA binding protein encoded by *Suppressor of Hairless*, promoting the transcription of specific target genes (Jarriault et al. 1995; Kidd et al. 1998; Lecourtois and Schweisguth 1998; Schroeter et al. 1998; Struhl and Adachi 1998). *mastermind* might also be an essential component of the activation complex (Wu et al. 2000). Seven linked genes of the Enhancer of split gene Complex are important target genes during neural development (Jennings et al. 1994). These bHLH proteins inhibit the transcription and function of proneural genes (such as ato) that promote neural fate specification (Lieber et al. 1993; Nakao and Campos-Ortega 1996; Ligoxygakis et al. 1998). The mechanisms by which other neurogenic genes including *big brain* and *neuralized* remain to be determined. (Adapted from Li and Baker 2001)

One explanation may be that the role of Mam in eye development is redundant with another, unidentified protein. In addition, the role of *big brain* (*bib*) in R8 specification seems to be quite minor. The transmembrane protein Bib plays a general but unidentified role in lateral N function. Only a small number of ectopic R8 cells differentiate in clones of *bib* mutant cells (Li and Baker 2001). The most important E(spl) protein during lateral inhibition of R8s may be E(spl)-mδ, judging from its dramatic phenotype on overexpression (Ligoxygakis et al. 1998, 1999). Since the *mδ* transcription unit can be deleted without affecting eye development, other E(spl) proteins must also be important (The et al. 1997). One of them is likely to be E(spl) m8, which is altered in the original *E(spl)*^{D} mutation whose eye phenotype led to the discovery of the locus (Campos-Ortega and Knust 1990).

3.3 Selection of Particular Cells Within Each Intermediate Group as R8 Precursors

It is not certain how particular proneural cells avoid lateral inhibition and become neural precursors, in the case of either R8 cells or neural precursors elsewhere in the nervous system. There are two, possibly related, aspects to the question. One is, how is it that Notch does not become activated to inhibit *ato* in every cell of the intermediate group. The second is what patterning mechanism specifies the particular cell that escapes inhibition (Baker 2000).

The first question is prompted by the observation that both N and Dl are expressed relatively uniformly during R8 specification, so that different protein levels at the cell surface do not seem to explain why one cell is not inhibited (Baker and Yu 1998). The second question follows from observations suggesting that particular intermediate group cells are not equally likely to become R8. First is the apparently asymmetric location of future R8 cells close to the posterior apex of intermediate groups (Baker et al. 1996; Dokucu et al. 1996). Secondly, if intermediate group cells were equally proneural, one would expect that a spatial bias in N signaling capacity introduced through mosaicism for N gene dose would influence the choice of the neural precursor cell. The R8 precursor would be expected always to be a cell with lower N gene dose (Heitzler and Simpson 1991). In contrast to this prediction, R8 selection within an intermediate group does not seem to be influenced by mosaicism for Notch gene dose (Baker and Yu 1998). Therefore intermediate groups must contain an intrinsic bias to select certain cells as R8 precursors.

How is the intermediate group biased into cells with distinct potentials? It has been suggested that intermediate groups first shrink to about three cells (the "R8 equivalence group"), and that additional genes are necessary to eliminate Ato from these three cells (Fig. 3A; Cagan 1993). Evidence for the three cell equivalence group came from the *rough* gene (*ro*), where null mutations transform nearby cells into R8, but not cells elsewhere in the intermediate groups. It is not known how Ro acts on *ato* transcription or function.

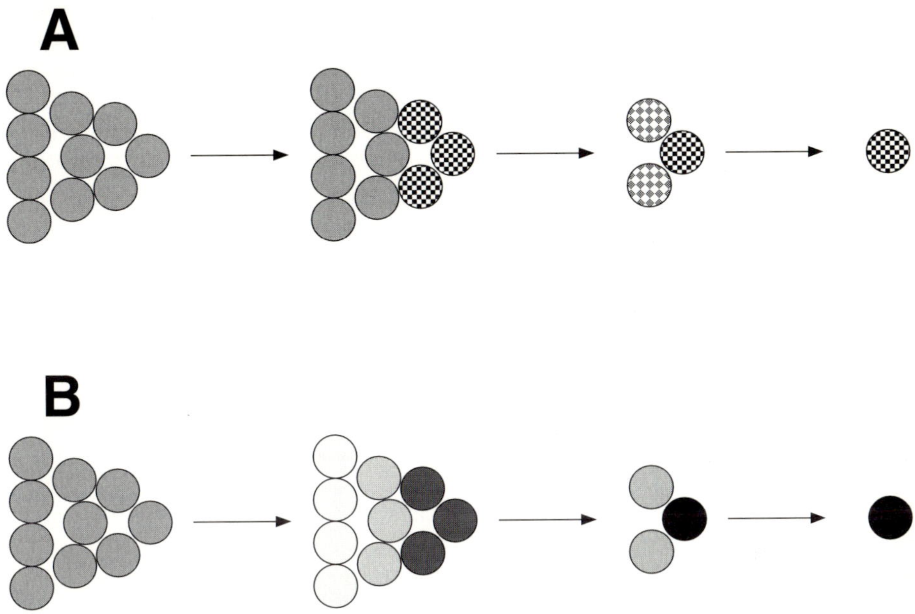

Fig. 3. Two models of R8 specification. **A** Single R8 cells may derive from a qualitatively distinct subset of intermediate groups cells (the "R8 equivalence group" – *checkered cells* in **A**). N signaling first eliminates Ato from the other intermediate group cells, then acts within the small R8 equivalence group, along with other molecules, to restrict R8 fate to single cells. **B** Intermediate group cells may differ from one another only qualitatively, so that different levels of Ato expression or N signaling are responsible for progressive resolution to single R8 cells. In this view, cells near the R8 precursor are qualitatively similar to other intermediate group cells, are inhibited later and so more easily transformed to R8 fate

Ro is expressed in a pattern identical to several proteins from the E(spl) complex that appear important mediators of N signaling in intermediate groups (Dokucu et al. 1996). This is interesting because E(spl) expression depends on N signaling, whereas Rough depends on EGFR signaling (Jennings et al. 1994; Baker and Yu 1997; Dominguez et al. 1998). The correspondence between Ro and E(spl) gene expression therefore suggests that activity of N and the EGFR is linked somehow, and perhaps functionally reinforcing.

An alternative interpretation of the *ro* phenotype is that all the intermediate group cells are part of a single equivalence group but that the two or three last cells to lose Ato expression might be most sensitive to reduced lateral inhibition, and so most easily transformed into R8 by mutations such as *ro* (Fig. 3B). These models make different predictions for the types of mutations that should be found. If all the intermediate group cells are part of a single equivalence group, one would predict that mutations affecting Ato inhibition in the whole intermediate group would most readily transform cells adjacent to the R8 cell into R8. By contrast, if there is a qualitatively distinct late equivalence group that can replace R8 through mechanisms not shared by the intermediate group as a whole, one would predict that there would be genes required *only* in this equivalence group to prevent specification of more than one R8 cell.

Subsequent work shows that *ro* mutations retard loss of Ato from all intermediate group cells, with the consequence that those cells near to the R8 fail to lose Ato in time to avoid R8 fate (Dokucu et al. 1996). Several other mutations have been now been described that lead to specification of multiple adjacent R8 cells from intermediate groups. These include *scabrous*, null alleles of the EGFR (or of Ras or Raf, components of EGFR signal transduction), and *big brain* (Baker and Zitron 1995; Dominguez et al. 1998; Li and Baker 2001; Powell et al. 2001; Yang and Baker 2001). All these genes seem to function throughout intermediate groups. For example EGFR activates mitogen-activated protein kinase (MAPK) in all intermediate group cells, and *bib* is involved generally in N signaling (Dominguez et al. 1998; Kumar et al. 1998; Lesokhin et al. 1999). The discovery of multiple mutations that affect the whole intermediate group but lead to a small number of adjacent R8 cells, and the failure to identify mutations specifically affecting subsets of intermediate group cells, are consistent with the model of a single R8 equivalence group including all the intermediate group cells. It is an intriguing possibility that intermediate groups might be first subdivided prior to R8 specification within one particular subset (Fig. 3A). Any future discovery of genes functioning differently in subsets of intermediate group cells would require this conclusion. So far, the data remain consistent with the simpler model also (Fig. 3B).

4 Intermediate Group Specification and Patterning

4.1 Making an Intermediate Group – Proneural Enhancement

One column before N inhibits Ato expression from the intermediate groups, intermediate groups themselves emerge from uniform Ato expression preceding the morphogenetic furrow by the loss of Ato from other cells. It is important to appreciate that *ato* expression is a composite of two independent transcription programs, each associated with distinct regulatory enhancers, and that intermediate groups appear at the transition between these transcriptional programs (Fig. 4; Baker et al. 1996; Sun et al. 1998). Uniform initiation of *ato* transcription ahead of the furrow is a response to the morphogens Hh and Dpp, diffusing forwards from the posterior differentiating portion of the eye and acting redundantly through an enhancer 3′ to the *ato* transcription unit (Lee and Treisman, this Vol.; Curtiss and Mlodzik 2000; Greenwood and Struhl 1999; Sun et al. 1998). Cells maintain *ato* transcription through the 3′ enhancer until the morphogenetic furrow, when they become insensitive to morphogens. An autoregulatory circuit now takes over maintenance of *ato* transcription, acting through an enhancer 5′ to the *ato* transcription unit (Fig.

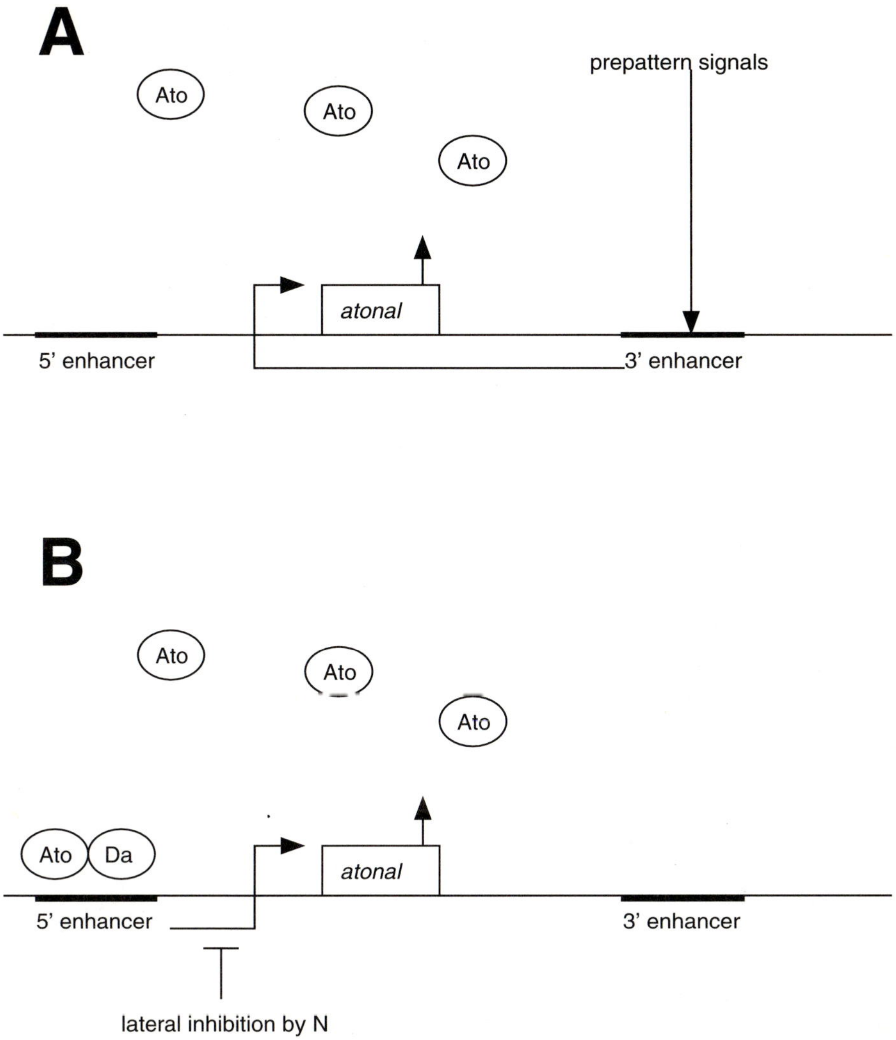

Fig. 4. Different modes of *ato* transcription. **A** Prepattern expression. Ato transcription begins ahead of the morphogenetic furrow in response to Hh and Dpp signaling. Prepattern signals act through a 3′ enhancer and are independent of functional Ato protein. **B** Autoregulatory expression. Beginning in the intermediate groups and continuing in R8 cells from column 0–3, *ato* transcription is dependent on autoregulation acting through a 5′ enhancer, and is not regulated by prepattern signals. Autoregulation is the target of lateral inhibition by Notch (see Sun et al. 1998)

4; Jarman et al. 1995; Sun et al. 1998). After this point there is no known mechanism for *ato* transcription to reinitiate once lost from a cell.

One possible mechanism underlying Ato maintenance in intermediate groups is that Ato or Da proteins become posttranslationally modified to promote *ato* autoregulation (or subject to inhibitory modification in other cells). At present, there is no evidence for, or against the possibility of proneural protein modification. It is likely, however, that function of Ato as a transcription factor also depends in part on protein expression levels.

The first evidence that Ato levels are important came from studies of mutations in *h* and *emc*, two genes encoding HLH proteins that repress neurogenesis (Brown et al. 1995). Hairy is a bHLH DNA-binding protein that represses transcription from proneural genes. Emc is an HLH protein that inhibits proneural protein function through formation of inactive heterodimers. In *h emc* double mutant clones, premature morphogenetic furrow progression and retinal differentiation occur, presumably due to precocious Ato activity (Brown et al. 1995). Since Hairy and Emc proteins are thought to act by different mechanisms, the simplest explanation of their redundancy is that cumulative derepression of Ato reaches levels sufficient for transcriptional activation in the *h emc* double mutant clones.

Although the *h emc* mutant phenotype suggests that Ato levels are important, there is little evidence that Hairy or Emc pattern intermediate groups in vivo. Emc protein levels do not change as *ato* autoregulates (Brown et al. 1995). Hairy expression terminates abruptly exactly as Ato levels elevate (Li and Baker 2001), but neither mutation of *h* nor of the Dpp pathway that induces *h* expression have much effect on intermediate groups (Brown et al. 1991; Greenwood and Struhl 1999). Another pathway that is both necessary and sufficient for Ato activation in intermediate groups is proneural enhancement.

The proneural enhancement aspect of Notch signaling was revealed by the unexpected phenotypes of N and Dl null mutations (Baker and Yu 1997). If the first role of N was lateral inhibition within intermediate groups, then the effect of complete loss of *N* function should be R8-like differentiation of all intermediate group cells. By contrast, in the absence of *N* or *Dl* function, both Ato expression and R8 specification were much reduced compared to wild type. These findings implied that in addition to lateral inhibition, Notch signaling was required for Ato to reach high levels and activate the proneural program in intermediate groups, before lateral inhibition could begin. Conversely it was found that ectopic expression of the N intracellular (N-intra) domain would elevate Ato expression and transcriptional activator function prematurely (Baker and Yu 1997). Targeted Dl expression anterior to the furrow leads to premature morphogenetic furrow progression and retinal differentiation (Li and Baker 2001).

Proneural Notch signaling acts through the DNA binding protein Su(H), already well known for targeting N-intra transactivation to E(spl)-C genes during lateral inhibition (Fig. 5). Su(H) can also mediate transcriptional repression through interactions with corepressor proteins (Morel and Schweisguth 2000). Proneural N signaling disrupts such Su(H)-mediated gene repression. Ahead of the furrow, Su(H) is acting to repress Ato function; N activation by Dl is necessary and sufficient to lift repression and elevate Ato function. Deleting the *Su(H)* gene has the same effect, so that Ato expression and retinal differentiation occur precociously in clones null for *Su(H)*, owing

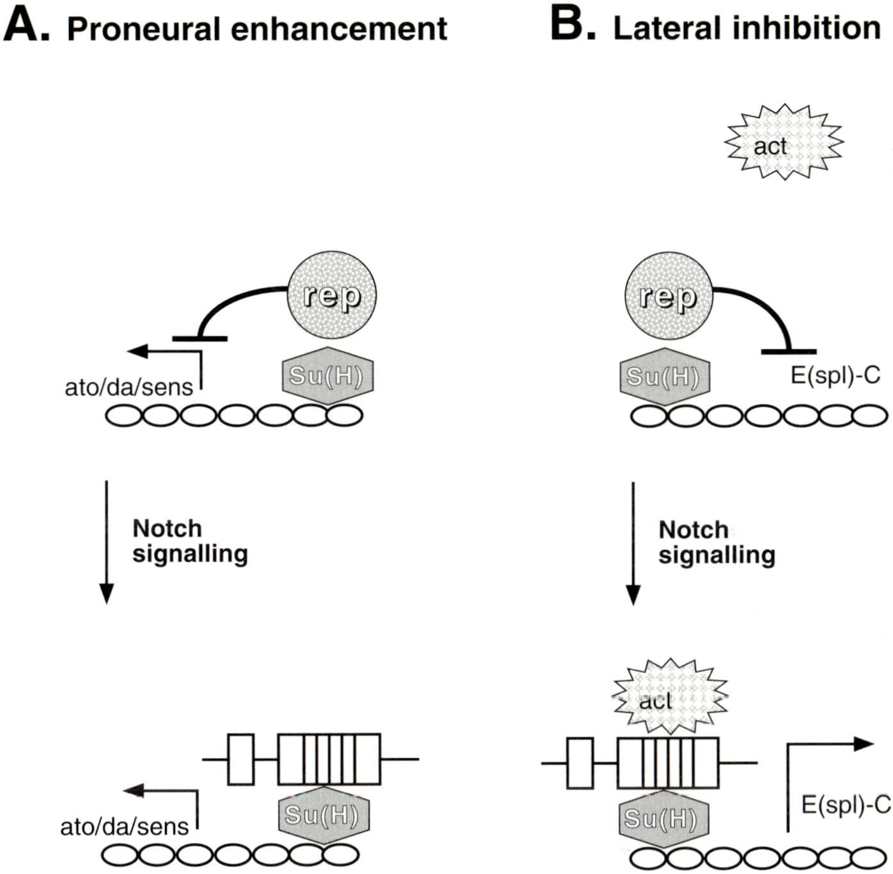

Fig. 5. Su(H) function in lateral inhibition and proneural enhancement. **A** Proneural enhancement. In unstimulated cells *ato*, *sens*, and perhaps *da* are repressed by Su(H) and repressor molecules. *H* encodes a candidate repressor. Ligand binding releases the N intracellular domain, which binds Su(H) to derepress transcription. The fate of the repressor is not known and it is not shown here. Su(H) may remain associated with DNA as shown, or might dissociate. **B** Lateral inhibition. In unstimulated cells, E(spl)-C transcription is repressed by Su(H) and unidentified repressor molecules. Ligand binding releases the N intracellular domain, which binds Su(H) and transcribes E(spl)-C genes, perhaps in association with an activator such as *mastermind*. The fate of the repressor is not known and it is not shown here. E(spl) proteins subsequently repress *ato* expression and function. (Adapted from Li and Baker 2001)

to the absence of the Su(H)-mediated repression that is normally overcome by proneural N signaling (Fig. 5; Li and Baker 2001).

Although the identity of the corepressor that acts along with Su(H) is not certain, Hairless protein is known to bind Su(H) and genetic evidence makes H a candidate to be involved in Su(H)-mediated repression (Brou et al. 1994; Furriols and Bray 2000). Consistent with this model, cells mutant for some *H* alleles show precocious Ato expression and retinal differentiation ahead of the furrow like that seen in *Su(H)* clones (Chanut et al. 2000).

The simplest model is that Ato itself is repressed by Su(H) until N signaling becomes active (Fig. 5). Regulatory sequences from the *ato* gene have been studied to identify activating regions, but regions mediating transcriptional repression may not have been identified (Sun et al. 1998). Other candidate target genes for repression include Sens and Da, however both of which are thought to be required for transcriptional activation by Ato (Brown et al. 1996; Nolo et al. 2000). Some evidence suggests there may be yet further genes required for Ato function and autoregulation, which might be targets for repression by Su(H). This is the finding of multiple enhancers mediating *ato* autoregulation, each specific for distinct tissues (Sun et al. 1998). Such specificity can only be explained by further, tissue-specific co-factors for Ato function.

4.2 Spacing the Intermediate Groups

The spaced appearance of intermediate groups, alternating with intervening groups of cells from which Ato is lost without reaching autoregulatory levels, suggests that proneural enhancement might be localized to pattern intermediate group formation. The data are inconclusive on this point. The expression of Dl, the ligand for proneural enhancement, is not obviously patterned. Dl protein appears on cell surfaces several cell diameters anterior to column 0, consistent with mosaic analysis showing that proneural enhancement depends on posterior-to-anterior signals (Baker and Yu 1997, 1998). Such Dl expression is uniform across the eye disc, expressed equally in intermediate groups and in the intervening cells, and provides no evidence that proneural enhancement is patterned by differential ligand expression. If Dl activity is patterned then the mechanism must act posttranslationally. There is transient E(spl) expression that might indicate patterned N activation just anterior to intermediate groups (Baker et al. 1996).

An alternative possibility that has stronger support is that intermediate groups could be patterned by inhibitory signals (Fig. 1). In this view Ato maintenance is not permitted in the cells between intermediate groups, leaving the intermediate groups to continue autoregulatory Ato expression. Signals could act in one or both of two ways (Fig. 4). Signals could inhibit Ato autoregulation through it's 5′ enhancer, so preventing Ato expression being maintained in intervening cells. Signals could prematurely terminate transcription from the 3′ prepattern enhancer; lack of Ato protein would then be the reason that autoregulation could not proceed.

There is direct evidence for regulation of 3′ enhancer activity. First, this enhancer drives reporter gene expression in a modulated pattern that foreshadows the positions of intermediate group formation (Sun et al. 1998). Secondly, 3′-enhancer activity is subject to short range inhibition by a signal from posterior, differentiating wild-type cells. It has been suggested that this signal is dependent on MAPK activation by the EGFR, because ectopic activation of the EGFR/Ras/Raf/MAPK pathway decreases Ato expression and R8 specification (Chen and Chien 1999). Some other overexpression studies have reported different results, however, underlining the caution that must be applied to overexpression experiments (Spencer et al. 1998; Greenwood and Struhl 1999).

The spatial location of the intervening cells where Ato is lost, near to the resolving intermediate groups of the previous column, suggests the intermediate groups as a potential source of negative signals (Fig. 1). A number of potential signals have been suggested to originate from intermediate group cells. One is Scabrous, which is secreted from intermediate group cells in response to Ato function and can be detected in vivo diffusing several cell diameters anterior to them (Baker and Zitron 1995; Lee et al. 1996). Others are Argos, also secreted from intermediate groups (Spencer et al. 1998), and Hedgehog, secreted by differentiating photoreceptor cells (Dominguez 1999). No doubt many other secreted and transmembrane proteins become expressed in similar patterns as ommatidial assembly and differentiation begin and could play signaling roles, such as the unidentified factor proposed to be secreted in response to EGFR activation (Chen and Chien 1999).

Spacing of intermediate groups is affected in *sca* loss-of-function mutants (Baker et al. 1990). When lateral inhibition is reduced as well, through reduction of *N* function in addition, a continuous stripe of R8 cells differentiates, suggesting that Sca inhibits Ato expression between intermediate groups to complement the role of lateral inhibition within them (Baker and Zitron 1995). Although it is not certain whether Sca binds N directly, Sca appears to be present in a complex with N in vivo (Lee and Baker 1996; Powell et al. 2001). In misexpression experiments Sca protein inhibits Dl and N function, so one plausible hypothesis is that secreted Sca inhibits proneural enhancement just anterior to intermediate groups so that subsequent intermediate groups appear maximally distant from them (Lee et al. 2000). Further experiments are required to confirm that this is how Sca acts in normal development, however.

An analogous model has been proposed in which Argos plays the role of intermediate group spacing (Spencer et al. 1998). Argos is a secreted antagonist of the EGFR (Schweitzer et al. 1995; Jin et al. 2000). It was originally suggested that Argos inhibited putative essential roles for EGFR in R8 specification and intermediate group maintenance (Spencer et al. 1998). Studies with null mutations show that EGFR is not essential for R8 specification or intermediate group maintenance, and that Argos is dispensable for intermediate group spacing, refuting the model in its original form (Dominguez et al. 1998; Kumar et al. 1998; Lesokhin et al. 1999; Baonza et al. 2001; Yang and Baker 2001). However, *argos* does have a demonstrated effect on spacing in a particular mutant background (*Elp*; Lesokhin et al. 1999; see next section). It remains possible that Argos might have a redundant role in some process as yet poorly and erstood.

4.3 Role of the EGF Receptor

The *Drosophila* EGF receptor is important for proper R8 patterning. In fact, *Elp* alleles of the EGFR were the first mutations described to affect ommatidial founder cells. The mechanisms by which EGFR regulates Ato expression are not yet known in detail.

In *Elp*/*Elp* homozygotes, R8 specification and ommatidium formation are almost completely blocked (Baker and Rubin 1989). Ato expression ahead of the furrow is reduced and intermediate groups never form, so that only a few sporadic cells maintain autoregulatory Ato expression to become R8 cells (Baker and Rubin 1992; Jarman et al. 1995). *Elp* mutations are hypermorphic, so an initial model was that EGFR signaling must be elevated between ommatidia in wild type, so that higher signaling in *Elp* transformed more cells away from ommatidial fates (Baker and Rubin 1989). In contrast to this prediction, following EGFR activity with antibodies to activate MAPK (dpERK) instead shows that EGFR is most active in intermediate group cells (Kumar et al. 1998; Spencer et al. 1998; Lesokhin et al. 1999). This led to the model that EGFR might be involved in lateral inhibition within intermediate groups (Lesokhin et al. 1999). An earlier finding that R8 cells were absent from *egfr* null clones is probably explicable through *egfr* function in R8 survival (Xu and Rubin 1993; Dominguez et al. 1998; Kumar et al. 1998; Baker and Yu 2001; Baonza et al. 2001).

Null mutant clones for EGFR show delayed loss of Ato and extra R8 cells (Dominguez et al. 1998; Greenwood and Struhl 1999; Lesokhin et al. 1999). This would be consistent with a role for EGFR in lateral inhibition within intermediate groups. The *egfr* mutant phenotype cannot be entirely explained this way, however, because aspects of the EGFR null and *Elp* mutant phenotypes are nonautonomous and appear to act via Ato 3′ regulatory sequences, implying that EGFR activity acts in part through further secreted signals that affect the uniform Ato expression that precedes intermediate groups (Chen and Chien 1999; Lesokhin et al. 1999).

In the case of *Elp* it appears that little ommatidial loss is caused directly by elevated EGFR signaling; most is caused indirectly by the Aos protein that is secreted by cells in response to EGFR signaling (Lesokhin et al. 1999). It may be temporal deregulation of Aos that is critical, with premature low-level Aos exposure of cells ahead of the furrow responsible for the phenotype. What the ectopic Aos does is unknown. Aos is not required for spacing intermediate groups in normal development, and complete loss of EGFR does not mimic the *Elp* phenotype, which would be expected if ectopic Aos acted to inhibit EGFR function (Baonza et al. 2001; Yang and Baker 2001).

One possibility is that Argos has a role in spacing redundant with that of an unknown protein. If this were the case, one would predict that ectopic Argos expression should block ommatidium formation. This is the case in the *Elp* mutant background (Lesokhin et al. 1999). In wild type, however, both decreased or increased R8 specification have been reported, alternatively supporting or contradicting the model (Spencer et al. 1998; Chen and Chien 1999).

If Aos does act redundantly to repress R8 specification, then it is uncertain whether Aos and it's putative partner act by reducing EGFR signaling. That is because *egfr* loss of function does not repress R8 specification. It is unlikely that EGFR acts redundantly with another tyrosine kinase receptor, because mutations in Ras and Raf resemble *egfr* mutations phenotypically, although they should prevent most RTK signaling (Yang and Baker 2001). It has been suggested that the *egfr*, *ras* and *raf* null mutants may not mimic any wild type situation, where all cells may show some EGFR, Ras and Raf activity, and that elevating EGFR activity above such minimal levels may be necessary for R8 specification (Lesokhin et al. 1999). At present, there has been no further experimental test of this hypothesis.

In the case of *egfr* null mutants, part of the spacing defect has been attributed to a nonautonomous signal downstream of MAPK activation that acts via 3′ regulatory sequences to inhibit Ato (Chen and Chien 1999). This may be involved in spacing intermediate groups by reducing Ato expression before autoregulation can be established in intermediate groups (Fig. 6). It has been shown that this hypothesized signal is not Sca, and could only be Argos if Argos acts redundantly (Baonza et al. 2001; Yang and Baker 2001).

The apparent role of EGFR in patterning R8 specification remains a subject for ongoing research. It will be informative to determine precisely which aspects of Ato expression are affected and in which cells, to distinguish direct, autonomous responses to EGFR activity from responses mediated nonautonomously by secondary signals, and to identify the specific ligands responsible for these aspects of EGFR function, their sources and ranges of action. Such data will be required before the position of EGFR can be finalized in the model shown in Fig. 6.

5 Conclusions

5.1 Summary of R8 Specification and Patterning

Features of R8 specification are summarized in Fig. 6. R8 cells are the founder cells that organize each ommatidium. R8 specification depends on the proneural gene *ato*, and is patterned in at least two steps. First, ubiquitous prepattern *ato* expression is replaced by intermediate groups in which *ato* autoregulates. Intermediate groups arise through proneural enhancement and other mechanisms that positively and negatively regulate how prepattern *ato* expression is lost and autoregulation becomes sufficient for maintenance of *ato* expression. Subsequently, lateral inhibition by *N* selects single R8 precursor cells. Although lateral inhibition might theoretically be able to select single cells from an entire ommatidia column, this is not what happens in normal development. Instead lateral inhibition by N occurs only in intermediate

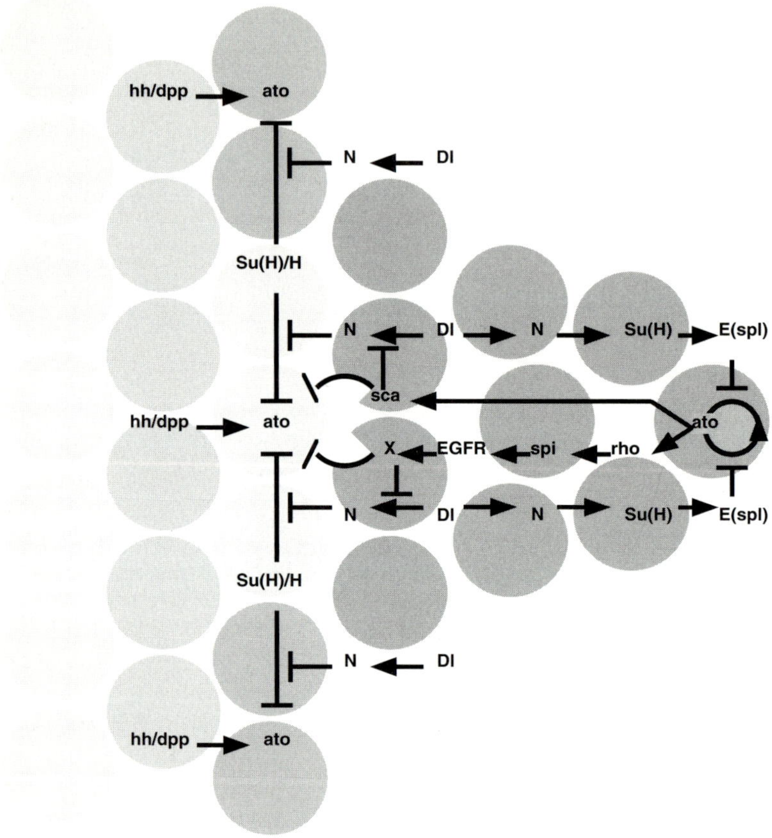

Fig. 6. Features of R8 specification. The diagram shows known signaling pathways superimposed on the outline of an intermediate group in the morphogenetic furrow. *Discs* represent individual Ato-expressing nuclei. *Darker shading* represents a higher Ato level. Ahead of the morphogenetic furrow (*left*), uniform Ato expression in response to Hh and Dpp intensifies until the autoregulatory intermediate groups appear. Within intermediate groups, Dl triggers N signaling to activate Su(H)dependent E(spl) transcription (lateral inhibition). E(spl) proteins inhibit autoregulatory *ato* transcription and Ato function, ensuring that only one R8 cell develops from each intermediate group. Unknown mechanisms that select one particular cell as the R8 precursor, and protect this cell from lateral inhibition. Ahead of the furrow, Ato levels are held in check by the Emc and H proteins (not shown). In addition, Ato is repressed by Su(H) and H, either directly through repression of *ato* transcription, or indirectly through repression of cofactors such as Da or Sens. Such repression is overcome by proneural enhancement, in which Dl activates N to disrupt Su(H)-mediated repression. Dl expression appears spatially uniform, so it is possible that proneural enhancement becomes localized to future intermediate groups by negative signals that preclude proneural enhancement just anterior to existing intermediate groups but permit proneural enhancement more distantly. The secreted protein Sca appears to be an inhibitory signal for intermediate group formation. Sca is secreted in response to Ato activity. Misexpression studies suggest Sca might act by inhibiting proneural enhancement. Alternatively, Sca might inhibit *ato* expression by an N-independent mechanism. Another unidentified signal (X) is secreted in response to EGFR activity. Signal X is distinct from Sca, and is probably not Argos (unless Argos can act redundantly and on a novel target receptor). It is not known whether Signal X acts on proneural enhancement or independently of N, and whether Signal X is independent of Dl

groups that each produce a single R8 cell. Within intermediate groups, unknown mechanisms pattern N signaling so as to select particular cells as R8 precursors, and to protect this cell from inhibition itself. It is possible that this outline of R8 specification will serve as a model for selection of neural cells in other tissues, where some of the processes have yet to be described.

5.2 Comparisons with Other Proneural Groups

It is useful to compare our limited understanding of R8 patterning in the retina with other examples of lateral inhibition in neurogenesis. One feature revealed by studies of R8 specification is that N signaling is particularly effective at inhibiting autoregulatory Ato expression (Baker et al. 1996; Sun et al. 1998). Proneural gene autoregulation now seems also to be the target of lateral inhibition during thoracic bristle specification (Culi and Modolell 1998). Proneural autoregulation only occurs in the selected macrochaete progenitor cell. The contrast with autoregulation in the whole intermediate group in the eye might only be superficial. In the eye, N signaling may begin after autoregulation, but in bristle specification, N signaling may already be active before autoregulation starts.

In some proneural regions, lateral inhibition appears to select a particular neural cell at random (Heitzler and Simpson 1991). It has been speculated that this might not be true for all proneural regions, e.g., macrochaete (bristle) selection might not be random (Simpson 1997). For the eye, N gene dosage mosaics show directly that R8 specification is not random but subject to intrinsic biases within intermediate groups (Baker and Yu 1998). The nature of the bias is presently unknown (but might involve EGFR signaling).

Proneural enhancement by N was first described during eye development, where it was surprising to find N and Dl playing positive roles in R8 specification (Baker and Yu 1997). Proneural enhancement has not yet been reported for any other neural tissues. N and Dl are not generally required for neurogenesis, and in most tissues *N* and *Dl* mutations are neurogenic. Several recent observations suggest that proneural enhancement may have a role in neural tissues other than the eye, however. First, proneural enhancement reflects a role of N signaling in overcoming Su(H)-mediated gene repression (Fig. 5; Li and Baker 2001). Recent work shows that N also overcomes Su(H)-mediated gene repression in development of the embryonic midline, and of the dorsal-ventral margin of the *Drosophila* wing (Furriols and Bray 2000; Klein et al. 2000; Morel and Schweisguth 2000). This seems increasingly likely to be a general feature of N signaling that may also be important in other neural tissues. If this is the case, why are *N* and *Dl* not essential for neural determination in all tissues? Studies of eye development in hypomorphic alleles of *N* suggest an explanation. Alleles that reduce proneural enhancement without preventing it give rise to a neurogenic phenotype (Li and Baker 2001). This raises the possibility that in neural tissues where *N* or *Dl* null mutants appear neurogenic, proneural enhancement might contribute to neurogenesis without being as essential as in the eye. Such a model predicts that where this occurs, *N* or *Dl* null phenotypes should be found to be quantitatively less neurogenic than E(spl) mutants.

In principle, the mechanism of lateral inhibition should be capable of spacing an array of neural precursor cells simultaneously (Greenwald and Rubin 1992). In the eye, however, each R8 precursor is from one intermediate group separated earlier by mechanisms different from those acting within each intermediate group (Figs. 1, 6). Successive patterning by different signaling pathways is possible in part because of the evolving transcriptional regulation of Ato from the prepattern to the autoregulatory stage (Fig. 4; Baker et al. 1996; Sun et al. 1998), and relies on the presence nearby of more mature neural cells from earlier ommatidial columns. The patterning role of extant neural cells may not be unique to the *Drosophila* eye. In grasshopper embryos ablation experiments show that neuroblast specification is inhibited by much older neural cells nearby, that have already begun stem cell divisions (Doe and Goodman 1985). The classic studies of Wigglesworth likewise addressed spacing of new (non-neural) structures by older ones during moults of the bug *Rhodnius* (Wigglesworth 1940).

In being required both for R8 specification and for aspects of R8 differentiation, Ato plays roles that are devolved onto multiple proteins in vertebrate neurogenesis, where a succession of bHLH determination and differentiation factors are expressed (Lee 1997). In *Drosophila*, too, many neural precursor cells replace proneural gene expression with *asense* once lateral inhibition has occurred (Jarman et al. 1993). Through its evolving expression pattern, Ato may be seen as playing the role of an R8 determination gene, a proneural gene, and, earlier still, as a prepattern gene. The importance of autoregulatory expression is typical of fate-determining genes. Autoregulation confers independence from extracellular signals, by definition cell determination. Conversely, initial regulatory gene expression that is *not* autoregulatory may be a general molecular feature of prepatterns.

Acknowledgements. I thank Yanxia Li, David Tyler, and Lihui Yang for comments. Work on Notch and R8 cells in the author's laboratory has been funded by the National Institutes of Health, the March of Dimes Birth Defects Foundation, the American Heart Association (New York City Affiliate) and the American Cancer Society.

References

Baker NE (2000) Notch signalling in the nervous system. Pieces still missing from the puzzle. BioEssays 22:264–273

Baker NE, Rubin GM (1989) Effect on eye development of dominant mutations in *Drosophila* homologue of EGF receptor. Nature 340:150–153

Baker NE, Rubin GM (1992) *Ellipse* mutations in the *Drosophila* homologue of the EGF receptor affect pattern formation, cell division, and cell death in eye imaginal discs. Dev Biol 150: 381–396

Baker NE, Yu S (1997). Proneural function of neurogenic genes in the developing *Drosophila* eye. Curr Biol 7:122–132

Baker NE, Yu SY (1998) The R8-photoreceptor equivalence group in *Drosophila*: fate choice precedes regulated *Delta* transcription and is independent of *Notch* gene dose. Mech Dev 74:3–14

Baker NE, Yu SY (2001) The EGF receptor defines domains of cell cycle progression and survival to regulate cell number in the developing *Drosophila* eye. Cell 104:699–708

Baker NE, Zitron AE (1995) *Drosophila* eye development: *Notch* and *Delta* amplify a neurogenic pattern conferred on the morphogenetic furrow by *scabrous*. Mech Dev 49:173–189

Baker NE, Mlodzik M, Rubin GM (1990) Spacing differentiation in the developing *Drosophila* eye: a fibrinogen-related lateral inhibitor encoded by *scabrous*. Science 250:1370–1377

Baker NE, Yu S, Han D (1996) Evolution of proneural atonal expression during distinct regulatory phases in the developing *Drosophila* eye. Curr Biol 6:1290–1301

Baonza A, Casci T, Freeman M (2001) A primary role for the EGF receptor in ommatidial spacing in the *Drosophila* eye. Curr Biol 396–404

Brou C, Logeat F, Lecourtois M, Vandekerckhove J, Kourilsky P, Schweisguth F, Israel A (1994) Inhibition of the DNA-binding activity of *Drosophila* Suppressor of Hairless and of its human homolog, KBF2/RBP-J kappa, by direct protein-protein interaction with *Drosophila* Hairless. Genes Dev 8:2491–2503

Brou C, Logeat F, Gupta N, Bessia C, LeBail O, Doedens JR, Cumano A, Roux P, Black RA, Israel A (2000) A novel proteolytic cleavage involved in Notch signaling: the role of the disintegrin-metalloprotease TACE. Mol Cell 5:207–216

Brown NL, Paddock SW, Markey DR, Carroll SB (1991) *hairy* gene function in the *Drosophila* eye: normal expression is dispensable but ectopic expression alters cell fates. Development 113: 1245–1256

Brown NL, Sattler SA, Paddock SW, Carroll SB (1995) *hairy* and *emc* negatively regulate morphogenetic furrow progression in the developing *Drosophila* eye. Cell 80:879–887

Brown NL, Paddock SW, Sattler CA, Cronmiller C, Thomas BJ, Carroll SB (1996) *daughterless* is required for *Drosophila* photoreceptor cell determination, eye morphogenesis, and cell cycle progression. Dev Biol 179:65–78

Cagan R (1993) Cell fate specification in the developing *Drosophila* retina. Dev Suppl:19–28

Cagan R, Ready D (1989) *Notch* is required for successive cell decisions in the developing *Drosophila* eye. Genes Dev 3:1099–1112

Campos-Ortega JA, Jan YN (1991) Genetic and molecular bases of neurogenesis in *Drosophila melanogaster*. Annu Rev Neurosci 14:399–420

Campos-Ortega JA, Knust E (1990) Defective ommatidial cell assembly leads to defective morphogenesis: a phenotypic analysis of the $E(spl)^D$ mutation of *Drosophila melanogaster*. Roux's Arch Dev Biol 198:286–294

Chanut F, Luk A, Heberlein U (2000) A screen for dominant modifiers of ro^{Dom}, a mutation that disrupts morphogenetic furrow progression in *Drosophila*, identifies *groucho* and *Hairless* as regulators of *atonal* expression. Genetics 156:1203–1217

Chen C-K, Chien C-T (1999) Negative regulation of *atonal* in proneural cluster formation of *Drosophila* R8 photoreceptors. Proc Natl Acad Sci USA 96:5055–5060

Culi J, Modolell J (1998) Proneural gene self-stimulation in neural precursors: an essential mechanism for sense organ development that is regulated by Notch signaling. Genes Dev 12:2036–2047

Curtiss J, Mlodzik M (2000) Morphogenetic furrow initiation and progression during eye development in *Drosophila*: the roles of decapentaplegic, hedgehog and eyes absent. Development 127:1325–1336

Doe CQ, Goodman CS (1985) Early events in insect neurogenesis. II. The role of cell interactions and cell lineage in the development of neuronal precursor cells. Dev Biol 111:206–219

Dokucu ME, Zipursky SL, Cagan RL (1996) Atonal, Rough and the resolution of proneural clusters in the developing *Drosophila* retina. Development 122:4139–4147

Dominguez M (1999) Dual role for Hedgehog in the regulation of the proneural gene *atonal* during ommatidia development. Development 126:2345–2353

Dominguez M, Wassarman JD, Freeman M (1998) Multiple functions of the EGF receptor in *Drosophila* eye development. Curr Biol 8:1039–1048

Furriols M, Bray S (2000) Dissecting the mechanisms of Suppressor of Hairless function. Dev Biol 227:520–532

Greenwald I, Rubin GM (1992) Making a difference: the role of cell-cell interactions in establishing separate identities for equivalent cells. Cell 68:271–281

Greenwood S, Struhl G (1999) Progression of the morphogenetic furrow in the *Drosophila* eye: the roles of Hedgehog, Decapentaplegic and the Raf pathway. Development 126:5795–5808

Heitzler P, Simpson P (1991) The choice of cell fate in the epidermis of *Drosophila*. Cell 64:1083–1092

Jan YN, Jan LY (1993) HLH proteins, fly neurogenesis, and vertebrate myogenesis. Cell 75:827–830

Jarman AP, Brand M, Jan LY, Jan YN (1993) The regulation and function of the helix-loop-helix gene, *asense*, in *Drosophila* neural precursors. Development 119:19–29

Jarman AP, Grell EH, Ackerman L, Jan LY, Jan YN (1994) *atonal* is the proneural gene for *Drosophila* photoreceptors. Nature 369:398–400

Jarman AP, Sun Y, Jan LY, Jan YN (1995) Role of the proneural gene, *atonal*, in formation of *Drosophila* chordotonal organs and photoreceptors. Development 121:2019–2030

Jarriault S, Brou C, Logeat F, Schroeter EH, Kopan R, Israel A (1995) Signalling downstream of activated mammalian Notch. Nature 377:355–358

Jennings B, Preiss A, Delidakis C, Bray S (1994) The Notch signalling pathway is required for *Enhancer of split* bHLH protein expression during neurogenesis in the *Drosophila* embryo. Development 120:3537–3548

Jimenez F, Campos-Ortega JA (1987) Genes in subdivision 1B of the *Drosophila melanogaster* X-chromosome and their influence on neural development. J Neurogenet 4:179–200

Jin M-H, Sawamoto K, Ito M, Okano H (2000) The interaction between the *Drosophila* secreted protein Argos and the epidermal growth factor receptor inhibits dimerization of the receptor and binding of secreted Spitz to the receptor. Mol Cell Biol 20:2098–2107

Karim FD, Chang HC, Therrien M, Wassarman DA, Laverty T, Rubin GM (1996) A screen for genes that function downstream of Ras1 during *Drosophila* eye development. Genetics 143:315–329

Kidd S, Lieber T, Young MW (1998) Ligand-induced cleavage and regulation of nuclear entry of Notch in *Drosophila melanogaster* embryos. Genes Dev 12:3728–3740

Klein T, Seugnet L, Haenlin M, Martinez-Arias A (2000) Two different activities of Suppressor of Hairless during wing development in *Drosophila*. Development 127:3553–3566

Kumar JP, Tio M, Hsiung F, Akopyan S, Seger R, Shilo B-Z, Moses K (1998) Dissecting the roles of the *Drosophila* EGF receptor in eye development and MAP kinase activation. Development 125:3875–3885

Lebovitz RM, Ready DF (1986) Ommatidial development in *Drosophila* eye disc fragments. Dev Biol 117:663–671

Lecourtois F, Schweisguth F (1998) Indirect evidence for Delta-dependent intracellular processing of Notch in *Drosophila* embryos. Curr Biol 8:771–774

Lee E, Baker NE (1996) GP300sca is not a high affinity ligand for Notch. Biochem Biophy Res Comm 225:720–725

Lee E-C, Hu X, Yu SY, Baker NE (1996) The *scabrous* gene encodes a secreted glycoprotein dimer and regulates proneural development in *Drosophila* eyes. Mol Cell Biol 16:1179–1188

Lee E-C, Yu S-Y, Baker NE (2000) The SCABROUS protein can act as an extracellular antagonist of Notch signaling in the *Drosophila* wing. Curr Biol 10:931–934

Lee JE (1997) Basic helix-loop-helix genes in neural development. Curr Opin Neurobiol 7:13–20

Lesokhin A, Yu S-Y, Katz J, Baker NE (1999) Several levels of EGF Receptor signalling during photoreceptor specification in *Ellipse*, wild type, and null mutant *Drosophila*. Dev Biol 205:129–144

Li Y, Baker NE (2001) Proneural enhancement by Notch overcomes Suppressor-of-Hairless-repressor function in the developing *Drosophila* eye. Curr Biol 11:330–338

Lieber T, Kidd S, Alcamo E, Corbin V, Young MW (1993) Antineurogenic phenotypes induced by truncated Notch proteins indicate a role in signal transduction and may point to a novel function of *Notch* in nuclei. Genes Dev 7:1949–1965

Ligoxygakis P, Yu S-Y, Delidakis C, Baker NE (1998) A subset of N functions during *Drosophila* eye development require Su(H) and the E(spl) gene complex. Development 125:2893–2900

Ligoxygakis P, Bray SJ, Apidianakis Y, Delidakis C (1999) Ectopic expression of individual *E(spl)* genes has differential effects in different cell fate decisions and underscores the biphasic requirement for Notch activity in wing margin establishment in *Drosophila*. Development 126:2205–2214

Meinhardt H (1982) Models of biological pattern formation. Academic Press, London

Mlodzik M, Baker NE, Rubin GM (1990) Isolation and expression of *scabrous*, a gene regulating neurogenesis in *Drosophila*. Genes Dev 4:1848–1861

Moore AW, Barbel S, Jan LY, Jan YN (2000) A genome-wide survey of basic helix-loop-helix factors in *Drosophila*. Proc Natl Acad Sci USA 97:10436–10441

Morel V, Schweisguth F (2000) Repression by Suppressor of Hairless and activation by Notch are required to define a single row of *single-minded* expressing cells in the *Drosophila* embryo. Genes Dev 14:377–388

Moscoso del Prado J, Garcia-Bellido A (1984) Cell interactions in the generation of chaetae pattern in *Drosophila*. Roux's Arch Dev Biol 193:246–251

Mumm JS, Schroeter EH, Saxena MT, Griesemer A, Tian X, Pan DJ, Ray WJ, Kopan R (2000) A ligand-induced extracellular cleavage regulates gamma-secretase-like proteolytic activation of Notch1. Mol Cell 207:197–206

Nakao K, Campos-Ortega JA (1996) Persistent expression of genes of the *Enhancer of split* complex suppresses neural development in *Drosophila*. Neuron 16:275–286

Nolo R, Abbot LA, Bellen HJ (2000) Senseless, a Zn finger transcription factor, is necessary and sufficient for sensory organ development in *Drosophila*. Cell 102:349–362

Parks AL, Turner RF, Muskavitch MAT (1995) Relationships between complex Delta expression and the specification of retinal fates during *Drosophila* eye development. Mech Dev 50: 201–216

Powell PA, Wesley CS, Spencer S, Cagan RL (2001) Scabrous complexes with Notch to mediate boundary formation. Nature 409:626–630

Qi H, Rand MD, Wu X, Sestan N, Wang W, Rakic P, Xu T, Artavanis-Tsakonas S (1999) Processing of the Notch ligand Delta by the metalloprotease Kuzbanian. Science 283:91–94

Ready DF, Hanson TE, Benzer S (1976) Development of the *Drosophila* retina, a neurocrystalline lattice. Dev Biol 53:217–240

Rebay I, Fleming RJ, Fehon RG, Cherbas L, Cherbas P, Artavanis-Tsakonas S (1991) Specific EGF repeats of *Notch* mediate interactions with *Delta* and *Serrate*: implications for *Notch* as a multifunctional receptor. Cell 67:687–699

Schroeter EH, Kisslinger JA, Kopan R (1998) Notch-1 signalling requires ligand-induced proteolytic release of intracellular domain. Nature 373:304–305

Schweitzer R, Howes R, Smith R, Shilo BZ, Freeman M (1995) Inhibition of *Drosophila* EGF receptor activation by the secreted protein Argos. Nature 376:699–702

Simpson P (1997) Notch signalling in development: on equivalence groups and asymmetric developmental potential. Curr Opin Genet Dev 7:537–542

Spencer SA, Powell PA, Miller DT, Cagan RL (1998) Regulation of EGF receptor signaling establishes pattern across the developing *Drosophila* retina. Development 125:4777–4790

Struhl G, Adachi A (1998) Nuclear access and action of Notch in vivo. Cell 93 649–660

Struhl G, Greenwald I (1999) Presenilin is required for activity and nuclear access of Notch in *Drosophila*. Nat Genet 398:522–525

Sun Y, Jan LY, Jan YN (1998) Transcriptional regulation of *atonal* during development of the *Drosophila* peripheral nervous system. Development 125:3731–3740

Sun Y, Jan LY, Jan YN (2000) Ectopic Scute induces *Drosophila* ommatidia development without R8 founder photoreceptors. Proc Natl Acad Sci USA 97:6815–6819

The I, Hannigan GE, Cowley GS, Reginald S, Zhong Y, Gusella JF, Hariharan IK, Bernards A (1997) Rescue of a *Drosophila* NF1 mutant phenotype by protein kinase A. Science 276:791–794

Tio M, Moses K (1997) The *Drosophila* TGFalpha homolog Spitz acts in photoreceptor recruitment in the developing retina. Development 124:343–351

Tomlinson A, Ready DF (1987) Neuronal differentiation in the *Drosophila* ommatidium. Dev Biol 120:366–376

Treisman JE, Rubin GM (1996) Targets of *glass* regulation in the *Drosophila* eye disc. Mech Dev 56:17–24

Treisman JE, Luk A, Rubin GM, Heberlein U (1997) *eyelid* antagonizes *wingless* signaling during *Drosophila* development and has homology to the Bright family of DNA-binding proteins. Genes Dev 11:1949–1962

Turing AM (1952) The chemical basis of morphogenesis. Phil Trans R Soc Lond B 237:37–72

Verheyen EM, Purcell KJ, Fortini ME, Artavanis-Tsakonas S (1996) Analysis of dominant enhancers and suppressors of activated Notch in *Drosophila*. Genetics 144:1127–1141

White NM, Jarman AP (2000) *Drosophila* Atonal controls photoreceptor R8-specific properties and modulates both receptor tyrosine kinase and Hedgehog signalling. Development 127: 1681–1689

Wigglesworth VB (1940) Local and general factors in the development of "pattern" in *Rhodnius prolixus* (*Hemiptera*). J Exp Zool 17:180–220

Wilcox M, Mitchison GJ, Smith RJ (1973) Pattern formation in the bluegreen alga, *Anabaena*. J Cell Sci 12:707–723

Wolff T, Ready DF (1993) Pattern formation in the *Drosophila* retina. In: Bate M, Martinez Arias A (eds) The Development of *Drosophila melanogaster*, vol 2. Cold Spring Harbor Laboratory Press, Cold Spring Harbor

Wu L, Aster JC, Blacklow SC, Lake R, Artavanis-Tsakonas S, Griffin D (2000) MAML1, a human homologue of *Drosophila* mastermind, is a transcriptional co-activator for NOTCH receptors. Nat Genet 26:484–489

Xu T, Rubin GM (1993) Analysis of genetic mosaics in the developing and adult *Drosophila* tissues. Development 117:1223–1236

Yang L, Baker NE (2001) Role of the EGFR/Ras/Raf pathway in specification of photoreceptor cells in the *Drosophila* retina. Development 128:1183–1191

Ye Y, Lukinova N, Fortini ME (1999) Neurogenic phenotypes and altered Notch processing in *Drosophila presenilin* mutants. Nat Genet 398:525–529

The Epidermal Growth Factor Receptor in *Drosophila* Eye Development

Justin P. Kumar[1]

1 Introduction

The Epidermal Growth Factor Receptor (Egfr) has been implicated in an extraordinarily vast number of developmental processes during the development of both invertebrates such as the fruit fly and vertebrates such as mouse and man. Hardly a month goes by without a new report surfacing that describes hitherto unknown roles for the Egfr pathway in new developmental contexts. The Egfr pathway has been shown to function in (among other things) cell fate specification, cell proliferation, cell cycle control, cell maintenance, programmed cell death, pattern formation and organ development. Furthermore, mutations that alter the activity and/or expression of this pathway have been shown to be the underlying cause for a wide ranging field of cancers within humans. This myriad of roles for the Egfr pathway is both a blessing and a curse in that, while it seemingly provides an unending set of opportunities to study the receptor, it is often difficult to attribute which function of the receptor is responsible for the observed phenotype. The simple structure and stereotyped development of the fly compound eye makes it an attractive model system for studying the Egfr pathway and untangling each of its many functions from one another. The goal of this review is to (1) summarize what is known about the established roles of Egfr signaling in cell fate specification and cell death and (2) to describe recent exciting results on the role of the Egfr in early eye development.

2 EGFR Gene Organization, Protein Structure and Mutant Classes

Prior to the isolation of the *Drosophila Egfr* gene it had been shown that the human EGF receptor was quite similar in sequence to the avian erythroblastosis virus oncogene, v-*erbB*, that is a member of the *src* oncogene family. A screen of a *Drosophila* genomic library using sequences that encoded for the v-*erbB* kinase domain yielded a single fly homologue of the receptor, which

[1] Department of Cell Biology, Emory University School of Medicine, 1648 Pierce Drive, Atlanta, Georgia 30322-3030, USA

Results and Problems in Cell Differentiation, Vol. 37
K. Moses (Ed.): Drosophila Eye Development

mapped to band 57F on the right arm of the second chromosome (Livneh et al. 1985). The *Egfr* locus encodes two polypeptides (type I and type II), which differ only at the extreme N-terminal by using alternate 5′ exons (Schejter et al. 1986). The *Drosophila Egfr* is a member of the receptor tyrosine kinase (RTK) family of proteins (Carpenter and Cohen 1979). Each member of this family contains an extracellular ligand binding domain, a single transmembrane domain and a cytoplasmic region containing the kinase domain (Livneh et al. 1985; Weiss et al. 1997; Hubbard 1999; Tan and Kim 1999).

Egfr mutants fall into three phenotypic classes, which were in fact isolated independently and given different names. The *torpedo (top)* alleles were isolated in a genetic screen for recessive female sterile mutations (Schupbach 1987). The *faint little ball (flb)* alleles were recovered in a large scale screen for recessive zygotic mutations that altered the embryonic cuticle pattern (Nusslein-Volhard et al. 1984). The *Ellipse (Elp)* mutants are dominant "eye specific" alleles isolated by their deleterious effect on compound eye and wing structure (Grell 1960; Baker and Rubin 1989). All three classes were later shown to be allelic to each other and were mutants of the *Egfr* locus (Baker and Rubin 1989; Price et al. 1989; Schejter and Shilo 1989). Several of the loss of function alleles have been identified as true null alleles and both *Elp* alleles are known to be biochemically hypermorphic for activity (Clifford and Schupbach 1989, 1994; Raz and Shilo 1993; Lesokhin et al. 1999). Recently, a tightly regulated temperature-sensitive allele of *Egfr* has been isolated which has allowed for the selective temporal removal of Egfr signaling (Kumar et al. 1998). Complementing these naturally occurring mutants are genetically engineered flies that harbor dominant negative and constitutively activated versions of the receptor that are under the control of inducible promoters. These constructs have allowed for the role of Egfr signaling in eye development to be dissected on a cell by cell basis (Freeman 1996; Queenan et al. 1997).

3 EGFR Pathway

The EGF receptor belongs to the superfamily of membrane receptors with tyrosine kinase activity. Studies in vertebrate models and tissue culture cells had indicated that the activation of an RTK via extracellular ligand binding leads to, among other things, the activation of the canonical RAS/MAPK (mitogen-activated protein kinase) signaling cascade. This pathway has been subsequently shown to be downstream of all known vertebrate and *Drosophila* RTKs including the fly Egfr (Diaz-Benjumea and Hafen 1994). While Egfr signaling is known to function in the development of nearly all tissues in the fly, the identification of the fly orthologs of the RAS/MAPK cascade were born from studies that focused not on *Egfr* but on another RTK, Sevenless (Sev), that affected the development of a single cell, the R7 photoreceptor. Inactivation of Sev results in the deletion of the R7 neuron from each ommatidium while constitutively active Sev receptors transform the developing cone cells into extra

R7 photoreceptors (Dickson and Hafen 1993). Genetic screens were carried out to isolate second site mutants that modified these phenotypes. The first seminal screens isolated both Ras itself and Sos, a guanine nucleotide exchange factor. Similar genetic screens isolated several additional components such as the adapter protein Drk, the ras GTPase-activating protein Gap1 and the three cytoplasmic protein kinases that lie in a linear path downstream of Ras signaling: Raf1, Dsor and MAPK (Wassarman et al. 1995). Each of these components has been shown to function downstream of the Egfr pathway (Diaz-Benjumea and Hafen 1994). Within the cytoplasm MAPK is non-phosphorylated and presumably inactive. Upon dual phosphorylation by MAP kinase kinase (MEK), MAPK is transported into the nucleus where it modulates transcription by phosphorylating nuclear factors. Such proteins with identifiable MAPK phosphorylation sites include Pointed (Pnt) and Anterior Open (Aop), (Brunner et al. 1994; Rebay and Rubin 1995). In this way, the EGFR/RAS pathway functions as a bridge for the transmission of instructions received at the cell surface in the form of diffusible ligands to the nucleus.

4 Origin and Structure of the Eye

The compound eye of *Drosophila melanogaster* consists of approximately 800 unit eyes or ommatidia (Ready et al. 1976). Within each ommatidium lie a set of eight photoreceptors and 12 accessory cone and pigment cells. The position of each cell within an ommatidium is precisely stereotyped so that each unit eye is an exact replica of its neighbors. In sections of adult compound eyes, this uniformity is most evident in the arrangement of the light capturing organelle of the photoreceptor, the rhabdomere. Each photoreceptor projects the rhabdomere into a precise position within the core of the ommatidium which gives rise to the now famous asymmetric trapezoidal pattern (Dietrich 1909). The cone cells lie above the photoreceptors and secrete the overlying corneal lens which can be seen in surface views of the adult eyes while the pigment cells ensheathe the photoreceptors and optically insulate each ommatidium from its adjacent neighbors (Waddington and Perry 1960).

Classical embryology and clonal analysis have traced the origins of the adult eye to the developing embryo. During embryogenesis, small groups of cells are set aside, apart from the developing embryo. These cells continually divide to produce monolayer epithelia called imaginal discs that are the predecessors of adult epidermal structures (Cohen 1993; Wolff and Ready 1993). During embryogenesis and the first two larval instars, the cells within the eye imaginal disc are unpatterned and undifferentiated. Midway through the third larval instar stage, pattern formation is initiated at the intersection of the midline and the posterior margin of the eye disc and sweeps towards the anterior edge of the epithelium (Ready et al. 1976; Wolff and Ready 1993). The midpoint of pupal eye development marks the completion of pattern formation and the beginning of the terminal differentiation of each cell type (Cagan and

Ready 1989a; Longley and Ready 1995). The role of Egfr signaling in eye specification, cell fate specification and cell death will be discussed below.

5 Eye Specification

The early development of the eye imaginal disc is marked by the expression of a set of nuclear factors: *twin of eyeless (toy)*, *eyeless (ey)*, *eyes absent (eya)*, *sine oculis (so)*, *dachshund (dac)*, *eye gone (eyg)* and *optix (opt)* (Treisman and Heberlein 1998; Heberlein and Treisman 2000; Seimiya and Gehring 2000). These factors are interwoven into a complex regulatory network and have been classified as "master control" genes for the eye based on two fundamental properties: (1) mutations within these genes result in flies that completely lack compound eyes and (2) ectopic expression of these genes (with the notable exception of *so*) is sufficient to generate ectopic eyes in normally non-eye tissue. But despite these impressive traits, it is becoming increasingly clear that the term "master control" as applied to these nuclear factors is somewhat of a misnomer. Mutations within the genes of the *Bithorax* and *Antennapedia* complexes result in homeotic transformations, an effect that prompted Lewis to coin and apply the phrase "master control gene" (Lewis 1992). This dramatic effect is not evidenced in any of the eye specification loss of function mutants. The eye primordium, in fact, develops reasonably well in all cases, albeit in a reduced form. It undergoes high levels of cell death just prior to the initiation of the morphogenetic furrow, preventing the assembly of the eye. In addition, the overexpression of the "eye specification" genes throughout the other imaginal discs is apparently capable of transforming only subregions of imaginal discs into eyes. The limiting factor(s) appear to be the expression of other signaling/patterning pathways. It has been recently shown that the ability of *ey* to induce ectopic eyes is heavily dependent upon *Dpp* signaling (Chen et al. 1999). Furthermore, since these genes are all nuclear factors it is unlikely that they reside at the very top of the regulatory hierarchy that controls eye specification. It is much more likely that they take their instructions from at least one, if not several, upstream signal transduction cascades.

In a set of very recent experiments, an effort was made to identify the upstream signals that direct the activity of the eye specification genes. Quite surprising is the identification of the Egfr and Notch signaling cascades as upstream regulators of eye formation. Removal of Egfr signaling in the eye and antennal primordia results in the deletion of these structures suggesting that Egfr activity is required for the formation of both structures. Even more astonishing is the finding that either hyperactivation of Egfr activity or downregulation of Notch signaling, leads to the complete homeotic transformation of the eye into an antenna (Fig. 1, Kumar and Moses 2001). The newly transformed antenna no longer express the "eye specification" genes and now express at least two of the "antennal specification" genes in appropriate patterns (*Distal-less*, *Dll* and *Spalt Major*, *SalM*). This suggests that during the

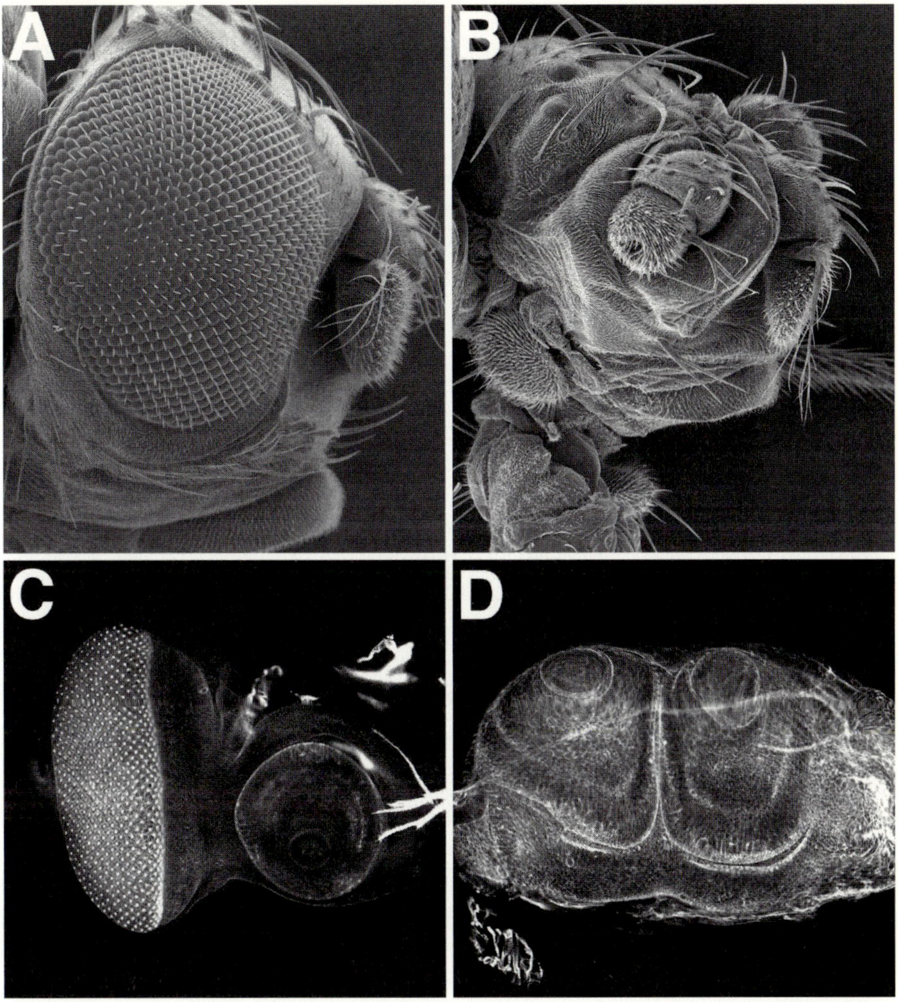

Fig. 1. Homeotic transformation of the eye into an antenna. **A, B** SEM images of adult heads viewed from the side. **C, D** 3rd instar imaginal discs. **A, C** Wild type. **B, D** A representative example of an eye to antenna transformation. In comparing **A, C** to **B, D**, note that the eye has been transformed into an antenna. This effect can be achieved by modulating the expression of either Egfr or Notch pathways within the eye primordium (see text for details). Anterior to the *right*

determination of eye and antennal disc fate, differential Egfr activity is required for the formation of both the eye and antenna while Notch function is essential for the specification of the eye. This effect by Egfr and Notch, in contrast to "eye specification" genes, seems to be more in line with the effects seen in homeotic mutants and may represent true eye "master control" genes as set forth by Lewis.

Where do the Egfr and Notch pathways intersect with the eye specification genes? It appears that they reside upstream of all known eye specification genes since the transcription of these elements, including the most upstream factor *toy* are all dramatically reduced to levels below detection in the newly transformed discs (Kumar and Moses 2001). The complex interactions among the eye specification genes do pose a potential caveat to this claim however. An alternative scenario to the direct regulation of *toy* is that Egfr and/or Notch signaling could "touch" an eye specification factor downstream of *toy*. This initial contact could then be propagated through the network of interactions among the eye specification genes leading to the determination of the eye. Distinguishing between these two alternatives will most likely require direct transcriptional and biochemical assays.

This observation that the eye can be transformed into an antenna by manipulating Egfr/Notch signaling is further surprising by its uniqueness. While sporadic reports of somatic mutants in which the eye field has been partially transformed into antennal tissue exist (Edmondson 1951), the effect described by Kumar and Moses is the first report of such a complete and reproducible transformation between the eye and the antenna. To date, this effect has not been observed with any of the existing homeotic mutants nor with any of the serial transplantation experiments conducted on developing eye-antennal imaginal discs (Hadorn 1978). The serial transplantation of imaginal discs led to a number of interesting developmental observations. First, among the imaginal discs there appeared to be a differential propensity to undergo transdetermination. Genital discs transdetermine into antennal discs at a significantly higher frequency than do leg discs. Second, imaginal discs also appeared restricted in their ability to be transformed. It was relatively easy to isolate antenna to wing transdeterminations but that same antennal disc was never observed to transdetermine into an eye (Hadorn 1978). Finally, the process of transdetermination appeared to be decidedly directional. While genital discs were never seen to transdetermine directly into wing discs they could however, first be transformed into leg discs which could then subsequently be respecified into wing tissue (Hadorn 1978). The extant homeotic mutants, by and large, mimic the phenotypes seen in the serial transplantation experiments. The lack of a more diverse set of homeotic mutants and transdetermining events has remained an enigma. Several theoretical models on how tissues are specified have been spawned in response to these apparent restrictions on the development of imaginal discs (Kauffman 1993). The eye to antenna transformations suggest that there may be fewer restrictions on imaginal disc development than once thought. The homeotic transformation of the eye to antenna was achieved by specifically manipulating Egfr and Notch signaling solely within the eye-antennal primordia via the upstream activator sequence (UAS)/GAL4 system. These and other signaling pathways are used throughout development. It is unlikely that one will isolate homeotic mutants in these pathways by traditional means since such mutants will have wide-ranging developmental consequences. In order to fully explore all aspects of imaginal disc fate in flies it is highly likely that the UAS/GAL4 system of specific spatial and temporal expression will become the technique of choice.

One additional finding by Kumar and Moses is that the eye disc does not seem to be determined during embryogenesis as previously thought, but rather this event takes place at the latter half of the second larval stage. While the seven "eye specification" genes are required to construct the eye, only *toy*, *ey* and *eyg* are expressed together in the eye primordium during embryogenesis. *Eya* joins during the first larval instar while the remaining factors are added later. It is only during the second larval stage that all seven factors have overlapping expression patterns in the eye imaginal disc. This suggests that the decision to adopt an "eye fate" is not finalized until this point in time. The authors build upon this idea by determining the phenocritical period for the eye to antenna transformation and interestingly it also takes place during the latter half of the second instar (Kumar and Moses 2001). A similar phenomenon has been demonstrated for the subdivision of the wing and notum compartments of the developing wing disc (Baonza et al. 2000). Loss of Egfr signaling within the developing notum leads to its transformation into wing tissue. Together these two reports suggest that Egfr signaling is critical for the establishment of individual disc fates and that all discs, like that of the eye, receive specification cues from the Egfr cascade during the second instar just prior to the onset of pattern formation.

6 Cell Fate Specification

The first suggestion that Egfr signaling played a role in eye development came from an analysis of *Elp* mutants (Baker and Rubin 1989, 1992). The eyes of *Elp* mutants are quite small with disorganized and widely spaced ommatidia. These ommatidia, although isolated, were shown to be properly constructed suggesting that Egfr signaling played a role not in cell fate specification per se, but rather in the spacing of ommatidia within the developing eye disc. At that time it was hypothesized that the *Egfr* could be the receptor for a spacing signal emanating from adjacent developing clusters. The phenotype of the *Elp* mutants fit quite nicely with this model, hyperactive Egfr signaling could result in more widely spaced ommatidia (Baker and Rubin 1989). This model made a simple prediction: if increased Egfr signaling led to increased spatial distance between ommatidia, then loss of Egfr activity should decrease the distance between clusters of photoreceptors. Since *Egfr* null mutants die as embryos, mosaic clones of *Egfr* null tissue were generated in the developing eye but the results were puzzling. Instead of observing closely packed ommatidia, *Egfr* null clones were shown to be completely void of any and all photoreceptors (Xu and Rubin 1993). Not only did this observation conflict with any possible role for Egfr signaling in ommatidial spacing, it clearly suggested that Egfr activity was essential to the formation of each cell type including photoreceptors within the assembling unit eye. This clear contradiction threw those interested in cell fate specification and signal transduction into a quandary that took several years to sort out.

A role for Egfr signaling in photoreceptor recruitment was subsequently supported by a series of papers that looked at the consequences of removing the ligand Spitz (Spi) during ommatidial assembly. *Spi* is the fly homologue of human TGFα and is known to activate Egfr signaling in a variety of tissues within the fly. An analysis of adult mosaic *spi* clones indicated that Spi function is required to varying degrees in each cell (R8, R2/R5 and R3/R4) of the precluster (Freeman 1994; Tio et al. 1994). An examination of developing eye imaginal discs shows that removal of *spi* does not affect the spacing between developing clusters or the specification of the founder cell R8. The first visible defect is an inhibition in the recruitment of the R2/R5 pair of photoreceptors. This has a domino effect on ommatidial assembly as all subsequent recruitment steps are blocked (Tio and Moses 1997). In *spi* clones, rows and rows of single R8 neurons are seen adjacent to wild-type tissue containing more mature clusters of three, five, seven and eight photoreceptor cells. Although this observation clearly indicates a requirement for Egfr signaling in the R2/R5 class of photoreceptors, it remained an open question whether Egfr activity is necessary for the specification of the remaining photoreceptors and accessory cells. The apparent *Egfr*-independent specification of the R8 neuron was also subject to some doubt. It is possible that another ligand (unknown at the time) activated Egfr signaling in the R8 precursor. Only by removal of Egfr signaling itself could these issues be conclusively addressed.

A number of independent efforts were made to remove Egfr function from the developing eye. Alternate methods to null mosaic clones were sought for removing Egfr signaling, since such clones are unhealthy, extremely small and rarely recovered (Baker and Rubin 1989; Xu and Rubin 1993). A novel method of mis-expressing dominant negative receptors in cell-specific patterns via the UAS/GAL4 system was used to inactivate Egfr signaling during ommatidial assembly. Removal of Egfr activity from all cells posterior to the furrow did not interfere with the initial specification of the R8 neuron, but did inhibit the recruitment of all other photoreceptors (Freeman 1996). The later recruitment of the accessory cone and pigment cells was shown to be also dependent upon Egfr activity. Like photoreceptors, the accessory cells are added to the developing clusters in a sequential pattern. The cone cells are added first followed, in order, by the 1°, 2° and 3° pigment cells (Cagan and Ready 1989b). Removal of Egfr signaling during successive developmental windows results in the sequential deletion of each individual cell type. Conversely, the successive addition of *Spi* results in the recruitment of extra accessory cells (Freeman 1996). This led to a model in which the sequential recruitment of every photoreceptor and accessory cell within the developing ommatidium was dependent upon successive rounds of Egfr activation (Freeman, 1996, 1997). This model was attractive because it provided an explanation for why only two RTKs that affected eye development had been identified but there are twenty different cell types. The previous model, the combinatorial code, had predicted that each cell type would be recruited by the expression of cell-specific signal transduction pathways and transcription factors. While the model seemed to explain several aspects of eye development there were several questions left unanswered.

One such question was whether or not Egfr signaling was necessary for the specification of the founder cell R8. For technical reasons the targeting strategies used in the misexpression studies were unable to address this question. Egfr signaling would have to be removed at earlier stages in ommatidial assembly. To do this a temperature-sensitive allele of the *Egfr* was isolated and was shown to mimic the phenotypic effects of null alleles in a variety of genetic tests at the restrictive temperature. Removal of Egfr signaling throughout the eye using this conditional allele did not affect the specification or the spacing of the R8 photoreceptor neuron, conclusively showing that no requirement for Egfr function exists during initial R8 recruitment. Interestingly, it was shown that R8 neurons require Egfr signaling only for its maintenance after the proneural gene *atonal* is downregulated, a function that is separable from roles in specification (Kumar et al. 1998). Technical innovations have made recent efforts to obtain null mutant clones more successful and the R8 photoreceptor is specified and patterned correctly in these clones confirming results obtained with the conditional allele (Dominguez et al. 1998; Lesokhin et al. 1999). Despite these reports there has been some suggestion that Egfr signaling is in fact required for R8 specification. Clones in which the modulating protein Rhomboid (Rho) and Vein are removed appear to lack R8 photoreceptors (Spencer et al. 1998). However, the analysis of these clones suffers from the same problems that plagued the earlier analysis of Egfr null clones. The clones are significantly smaller than the wild-type twin spots and rarely recovered suggesting that there are problems with cell proliferation and cell viability. This is similar to early attempts to generate Egfr null clones. Egfr clones have more recently been generated in a Minute background. This has allowed for the generation of large clones where a more reliable analysis can be done. In all likelihood, R8 specification will be normal in rho vn double clones generated in a Minute background. The simplest interpretation of all the data so far, is that with the exception of the recruitment of the R8 founder cell, Egfr signaling is necessary for the specification of every other cell type within the ommatidium.

It does still remain to be explained how each ommatidial precursor uses the same receptor but then adopts unique cell fates. As more and more mutants that affect cell fate have been examined, glimpses of potential regulation of cell fate can be seen. First, several positive and negative ligands for Egfr signaling have been identified and it has been shown that several developmental processes controlled by Egfr signaling are dependent upon specific ligands (Perrimon and Perkins 1997; Guichard et al. 1999; Kumar and Moses 2001). Second, a growing list of examples exists in which the Notch, Hedgehog (Hh), Decapentaplegic (Dpp) and Wingless (Wg) pathways are integrated with Egfr signaling to produce several different types of signals (Freeman 1998).

Finally, each cell within the developing eye expresses a combinatorial set of transcription factors that appears essential for the unique identity of each cell (Kumar and Moses 1997). By making use of these three types of regulation, cells within the developing ommatidium can turn a general signal to differentiate as an "ommatidial cell", into a specific instruction to adopt the identity of one of the 20 unique cell types that make up each ommatidium.

7 Programmed Cell Death

The final phase of pattern formation in the eye is characterized by the removal of surplus cells. This allows for the imposition of a near crystalline pattern upon a system that initially produced an excess and variable number of cells. In normal eye development a low level of cell death first occurs in a tight band just ahead of the advancing morphogenetic furrow followed by a scattering of death several columns behind (Wolff and Ready 1991). The vast majority of surplus cells are eliminated in a burst of death midway through pupal development (Cagan and Ready, 1989a, 1989b; Wolff and Ready 1991). Mutations that result in substantial loss of eye tissue have been shown to be accompanied by significant levels of elevated programmed cell death. A cluster of three genes, *grim (grim)*, *reaper (rpr)*, and *head involution defective (hid)* have been shown to control virtually all cell death in the developing embryo (White et al. 1994). While the loss of any of these three genes leads to the suppression of cell death within the embryo (Kurada and White 1998) ectopic expression of any of these genes within the developing eye leads to the ablation of the retina (Grether et al. 1995; Hay et al. 1995; Chen et al. 1996; White et al. 1996).

It has been long suspected that Egfr signaling functions in cell proliferation and cell viability. This suspicion stems from the difficulty in obtaining mosaic clones that are null for Egfr, Ras or MAPK activity. Such clones are rarely recovered and are very small compared to the wild-type twin spot. Two recent studies have demonstrated that Egfr/Ras signaling is required to suppress programmed cell death in the eye (Bergmann et al. 1998; Kurada and White 1998). The ablation of the eye resulting from the overexpression of either *hid* or *rpr* could be enhanced by reducing Egfr/Ras signaling. The Hid protein was shown to contain five MAPK phosphorylation sites. Alteration of those sites rendered the protein insensitive to Egfr/Ras signaling. From these studies a model has emerged in which cells within the developing imaginal disc are provided a "survival" signal from the Egfr/Ras cascade. This signal comes in the form of an inhibition upon the cell death machinery (Bergmann et al. 1998; Kurada and White 1998). It has been shown that Egfr signaling is also crucial to the observed epoch of cell death in the pupal retina (Miller and Cagan 1998). Expression of activated *Egfr* or *Ras* constructs, in cells normally destined to die, in the pupal retina survive and differentiate as extra pigment cells, a phenotype strikingly similar to that seen in *hid* mutants. As it appears quite clear that Egfr/Ras signaling blocks cell death, it will be very interesting to identify the corresponding signaling cascades that initiate cell death in the developing eye. It will also be interesting to correlate the expression of such components with the timings of cell death observed in the developing imaginal and pupal eye discs.

8 Concluding Remarks

It would be a mistake to think that the role for Egfr signaling is limited to what has been described here. The receptor is thought to function in a number of additional developmental processes such as the regulation of the cell cycle, cell size and cell number along with communication between the cytoskeleton and extracellular matrix. It has also been implicated as a potential regulator of the morphogenetic furrow. As an increasing number of roles for Egfr signaling are described, it becomes more and more urgent that we understand how a single receptor can regulate so many developmental decisions. These areas are under active research and promise to make exciting stories for the future.

References

Baker NE, Rubin GM (1989) Effect on eye development of dominant mutations in *Drosophila* homologue of the EGF receptor. Nature 340:150–153

Baker NE, Rubin GM (1992) *Ellipse* mutations in the *Drosophila* homologue of the EGF receptor affect pattern formation, cell division, and cell death in eye imaginal disc. Dev Biol 150: 381–396

Baonza A, Roch F, Martin-Blanco E (2000) DER signaling restricts the boundaries of the wing field during *Drosophila* development. Proc Natl Acad Sci USA 97:7331–7335

Bergmann A, Agapite J, McCall K, Steller H (1998) The *Drosophila* gene hid is a direct molecular target of ras-dependent survival signaling. Cell 95:331–341

Brunner D, Ducker K, Oellers N, Hafen E, Scholz H, Klambt C (1994) The ETS domain protein pointed-P2 is a target of MAP kinase in the Sevenless signal transduction pathway. Nature 370:386–389

Cagan RL, Ready DF (1989a) The emergence of order in the *Drosophila* pupal retina. Dev Biol 136:346–362

Cagan RL, Ready DF (1989b) Notch is required for successive cell decisions in the developing *Drosophila* retina. Genes Dev 3:1099–1112

Carpenter G, Cohen S (1979) Epidermal growth factor. Annu Rev Biochem 48:193–216

Chen P, Nordstrom W, Gish B, Abrams JM (1996) grim, a novel cell death gene in *Drosophila*. Genes Dev 10:1773–1782

Chen R, Halder G, Zhang Z, Mardon G (1999) Signaling by the TGF-b homolog decapentaplegic functions reiteratively within the network of genes controlling retinal cell fate determination in *Drosophila*. Development 126:935–943

Clifford RJ, Schupach T (1989) Coordinately and differentially mutable activities of torpedo, the *Drosophila melanogaster* homolog of the vertebrate EGF receptor gene. Genetics 123:771–787

Clifford R, Schupbach T (1994) Molecular analysis of the *Drosophila* EGF receptor homolog reveals that several genetically defined classes of alleles cluster in subdomains of the receptor protein. Genetics 137:531–550

Cohen SM (1993) Imaginal disc development. In: Bate M, Martinez Arias A (eds) The development of *Drosophila*. Cold Spring Harbor Laboratory Press, Cold Spring Harbor, pp 747–841

Diaz-Benjumea FJ, Hafen E (1994) The sevenless signalling cassette mediates *Drosophila* EGF receptor function during epidermal development. Development 120:569–578

Dickson B, Hafen E (1993) Genetic dissection of eye development in *Drosophila*. In: Bate M, Martinez Arias A (eds) The development of *Drosophila melanogaster*. Cold Spring Harbor Laboratory Press, Cold Spring Harbor, pp 1327–1362

Dietrich W (1909) Die Fazettenaugen der Dipteren. Z Wiss Zool 92:465–539

Dominguez M, Wasserman JD, Freeman M (1998) Multiple functions of the EGF receptor in *Drosophila* eye development. Curr Biol 8:1039–1048

Edmondson M (1951) Interchange of eye and antennal tissue during development. DIS 25:105

Freeman M (1994) The spitz gene is required for photoreceptor determination in the *Drosophila* eye where it interacts with the EGF receptor. Mech Dev 48:25–33

Freeman M (1996) Reiterative use of the EGF Receptor triggers differentiation of all cell types in the *Drosophila* eye. Cell 87:651–660

Freeman M (1997) Cell determination strategies in the *Drosophila* eye. Development 124:261–270

Freeman M (1998) Complexity of EGF receptor signaling revealed in *Drosophila*. Curr Opin Genet Dev 8:407–411

Grell EH (1960) New mutants report. DIS 34:50

Grether ME, Abrams JM, Agapite J, White K, Steller H (1995) The head involution defective gene of *Drosophila melanogaster* functions in programmed cell death. Genes Dev 9:1694–1708

Guichard A, Biehs B, Sturtevant MA, Wickline L, Chacko J, Howard K, Bier E (1999) rhomboid and Star interact synergistically to promote EGFR/MAPK signaling during *Drosophila* wing vein development. Development 126:2663–2676

Hadorn E (1978) Transdetermination. In: Ashburner M, Wright TE (eds) The genetics and biology of *Drosophila*, vol 2c. Academic Press, New York, pp 555–617

Hay BA, Wassarman DA, Rubin GM (1995) *Drosophila* homologs of baculovirus inhibitor of apoptosis proteins function to block cell death. Cell 83:1253–1262

Heberlein U, Treisman JE (2000) Early retinal development in *Drosophila*. In: Fini ME (ed) Vertebrate eye development. Springer, Berlin Heidelberg New York, pp 288

Hubbard SR (1999) Structural analysis of receptor tyrosine kinases. Prog Biophys Mol Biol 71:343–358

Kauffman SA (1993) The origins of order. Oxford University Press, New York

Kumar JP, Moses K (1997) Transcription factors in eye development: a georgeous mosaic? Genes Dev 11:2023–2028

Kumar JP, Moses K (2001) EGF Receptor and Notch signaling act upstream of Eyeless/Pax6 to control eye specification. Cell 104:687–697

Kumar JP, Tio M, Hsiung F, Akopyan S, Gabay L, Seger R, Shilo BZ, Moses K (1998) Dissecting the roles of the *Drosophila* EGF receptor in eye development and MAP kinase activation. Development 125:3875–3885

Kurada P, White K (1998) Ras promotes cell survival in *Drosophila* by downregulating hid expression. Cell 95:319–329

Lesokhin AM, Yu SY, Katz J, Baker NE (1999) Several levels of EGF receptor signaling during photoreceptor specification in wild-type Ellipse, and null mutant *Drosophila*. Dev Biol 205:129–144

Lewis EB (1992) The 1991 Albert Lasker Medical Awards. Clusters of master control genes regulate the development of higher organisms. Jama 267:1524–1531

Livneh E, Glazer L, Segal D, Schlessinger J, Shilo BZ (1985) The *Drosophila* EGF receptor gene homolog: conservation of both hormone binding and kinase domains. Cell 40:599–607

Longley RLJ, Ready DF (1995) Integrins and the development of three-dimensional structure in the *Drosophila* compound eye. Dev Biol 171:415–433

Miller DT, Cagan RL (1998) Local induction of patterning and programmed cell death in the developing *Drosophila* retina. Development 125:2327–2335

Nusslein-Volhard C, Wieschaus E, Kluding H (1984) Mutations affecting the pattern of the larval cuticle in *Drosophila melanogaster*. I. Zygotic loci on the second chromosome Roux's Arch Dev Biol 193:267–282

Perrimon N, Perkins LA (1997) There must be 50 ways to rule the signal: the case of the *Drosophila* EGF receptor. Cell 89:13–16

Price JV, Clifford RJ, Schupbach T (1989) The maternal ventralizing locus torpedo is allelic to faint little ball, an embryonic lethal, and encodes the *Drosophila* EGF receptor homolog. Cell 56:1085–1092

Queenan AM, Ghabrial A, Schupbach T (1997) Ectopic activation of torpedo/Egfr, a *Drosophila* receptor tyrosine kinase, dorsalizes both the eggshell and the embryo. Development 124: 3871–3880

Raz E, Shilo BZ (1993) Establishment of ventral cell fates in the *Drosophila* embryonic ectoderm requires DER, the EGF receptor homolog. Genes Dev 7:1937–1948

Ready DF, Hanson TE, Benzer S (1976) Development of the *Drosophila* retina, a neurocrystalline lattice. Dev Biol 53:217–240

Rebay I, Rubin GM (1995) Yan functions as a general inhibitor of differentiation and is negatively regulated by activation of Ras1/MAPK pathway. Cell 81:857–866

Schejter ED, Shilo BZ (1989) The *Drosophila* EGF receptor homolog (DER) gene is allelic to faint little ball, a locus essential for embryonic development. Cell 56:1093–1104

Schejter ED, Segal D, Glazer L, Shilo BZ (1986) Alternative 5′ exons and tissue specific expression of the *Drosophila* EGF receptor homolog transcripts. Cell 46:1091–1101

Schupbach T (1987) Germ line and soma cooperate during oogenesis to establish the dorsoventral pattern of egg shell and embryo in *Drosophila melanogaster*. Cell 49:699–707

Seimiya M, Gehring WJ (2000) The *Drosophila* homeobox gene optix is capable of inducing ectopic eyes by an eyeless-independent mechanism. Development 127:1879–1886

Spencer SA, Powell PA, Miller DT, Cagan RL (1998) Regulation of EGF receptor signaling establishes pattern across the developing retina. Development 125:4777–4790

Tan PB, Kim SK (1999) Signaling specificity: the RTK/RAS/MAP kinase pathway in metazoans. Trends Genet 15:145–149

Tio M, Moses K (1997) The *Drosophila* TGFa homolog Spitz acts in photoreceptor recruitment in the developing retina. Development 124:343–351

Tio M, Ma C, Moses K (1994) spitz, a *Drosophila* homolog of transforming growth factor-a, is required in the founding photoreceptor cells of the compound eye facets. Mech Dev 48:13–23

Treisman JE, Heberlein U (1998) Eye development in *Drosophila*: formation of the eye field and control of differentiation. Curr Topics Dev Biol 39:119–157

Waddington CH, Perry MM (1960) The ultrastructure of the developing eye of *Drosophila*. Proc Roy Soc Biol Sci 153:155–178

Wassarman DA, Therrien M, Rubin GM (1995) The Ras signaling pathway in *Drosophila*. Curr Opin Genet Dev 5:44–50

Weiss FU, Daub H, Ullrich A (1997) Novel mechanisms of RTK signal generation. Curr Opin Genet Dev 7:80–86

White K, Grether ME, Abrams JM, Young L, Farrell K, Steller H (1994) Genetic control of programmed cell death in *Drosophila* [see comments]. Science 264:677–683

White K, Tahaoglu E, Steller H (1996) Cell killing by the *Drosophila* gene reaper. Science 271:805–807

Wolff T, Ready DF (1991) Cell death in normal and rough eye mutants of *Drosophila*. Development 113:825–839

Wolff T, Ready DF (1993) Pattern formation in the *Drosophila* retina. In: Bate M, Martinez Arias A (eds) Cold Spring Harbor Laboratory Press, Cold Spring Harbor, pp 1277–1326

Xu T, Rubin GM (1993) Analysis of genetic mosaics in developing and adult *Drosophila* tissues. Development 117:1223–1237

Cell Fate Specification in the *Drosophila* Eye

Raghavendra Nagaraj, Jude Canon, and Utpal Banerjee[1]

1 Introduction

With the passing of the furrow over a uniformly equivalent group of cells, a spectacular array of cell types, each different in structure and function arises in the developing eye disc of *Drosophila*. A small number of ubiquitously expressed transcription factors combine with an even smaller number of signals emanating from the furrow and the developing clusters and generate cell-specific expression of a panoply of transcription factors (reviewed in Kumar and Moses 1997). The challenge is to understand how the cell-specific transcription factors are placed in their respective cells and how they then function in assigning different identities to each cell type. Here, we have presented our current understanding of this process. The aim is to be eclectic rather than comprehensive, and we apologize in advance to those investigators whose work we have not fully cited.

2 Early Events in Cell Fate Specification

The specification of the eye disc primordium occurs during stages 14–15 of embryonic development (Younossi-Hartenstein et al. 1996). Early events of cell fate specification continue through the third larval instar when the morphogenetic furrow forms. These processes have been described in detail in other chapters and will therefore be only briefly mentioned here.

The specification of the eye disc primoridium requires a set of early acting transcription factors including, Twin of eyeless, Eyeless, Eyegone, Optix, Sine oculis, Dachsund and Eyes absent. Loss of function mutations in genes encoding these proteins result in loss of the eye primordium, and ectopic expression of these genes is sufficient to induce ectopic eyes (Bonini et al. 1993; Cheyette et al. 1994; Halder et al. 1995; Shen and Mardon 1997; Jun et al. 1998; Czerny et al. 1999; Heberlein and Treisman 2000; Seimiya and Gehring 2000). The complex genetic epistases between these genes suggest that they might function as part of a regulatory loop in the specification of the eye primordium and

[1] Departments of Molecular, Cell and Developmental Biology, Biological Chemistry and Human Genetics. Molecular Biology Institute, University of California, Los Angeles, California 90095, USA

Results and Problems in Cell Differentiation, Vol. 37
K. Moses (Ed.): Drosophila Eye Development

in its subsequent differentiation. In vitro studies suggest that these proteins form large complexes that serve to define the identity of the eye tissue (reviewed in Desplan 1997; Gehring and Ikeo 1999; Pichaud et al. 2001).

During the second larval instar, Notch and Epidermal Growth Factor Receptor (EGFR) signaling pathways function antagonistically to each other in sorting out the eye tissue from that which will give rise to the antenna (Kumar and Moses 2001). Also at this stage, a dorsal/ventral (D/V) axis is specified in the eye disc (reviewed in Irvine 1999). Cells within and across the D/V midline are morphologically indistinguishable from one another, but genetic analysis involving molecular markers suggests that midline specification requires the function of the Notch signaling pathway. The D/V midline in the eye disc primordium will play a critical role in the proliferation of the precursor cells and will determine polarity of the future ommatidia (Strutt and Strutt 1999). The specification of the D/V axis also requires a signal from the peripodial membrane, an epithelial layer of cells that covers the eye disc proper. The cells in the peripodial membrane express such signaling molecules as Wingless (Wg), Decapentaplegic (Dpp), and Hedgehog (Hh). Hh expression in these cells is required for the expression of Serrate (Ser) at the D/V boundary of the eye disc. This ligand then activates the Notch receptor at the boundary (Cho and Choi 1998; Papayannopoulos et al. 1998). Additionally, Ser is found in the cells of the peripodial membrane, and loss of Ser expression in these cells results in defects in cell growth and pattern formation in the eye disc proper (Gibson and Schubiger 2000). These studies suggest the requirement of functional interactions between opposing epithelial layers in the eye disc for proper cell fate specification.

Initiation of the morphogenetic furrow at the posterior end of the eye disc occurs at the early third larval instar and requires the interplay between Hh, Dpp and Wg signaling. In the late second instar, Hh is expressed in the posterior end of the eye disc. Loss of either Hh or Dpp prevents initiation of the furrow (Dominguez and Hafen 1997; Royet and Finkelstein 1997). In contrast, Wg is expressed along the lateral edges of the eye disc and prevents furrow initiation from the lateral margins, (Ma and Moses 1995; Treisman and Rubin 1995).

The expression of Hh and Dpp at the furrow allows patterning and cell cycle synchronization of cells anterior to the furrow. These signals function over long ranges to convert naive undifferentiated cells well in advance of the furrow to a discrete "pre-proneural" fate (Greenwood and Struhl 1999). This preparatory fate has been considered essential for any cell within the retinal primordium to differentiate as a photoreceptor in response to extracellular signals. Dpp and Hh also cause cells ahead of the furrow to be synchronized for their G_1 phase arrest (Greenwood and Struhl 1999). In mutant conditions where this arrest is lost, these cells undergo two or more extra divisions before responding to differentiation signals. This results in a disorganized retinal array (Thomas et al. 1994; Penton et al. 1997).

Cells arrest at the G_1 stage of the cell cycle within the furrow (Thomas et al. 1994). Subsequently, some of these cells exit the cell cycle and differentiate as

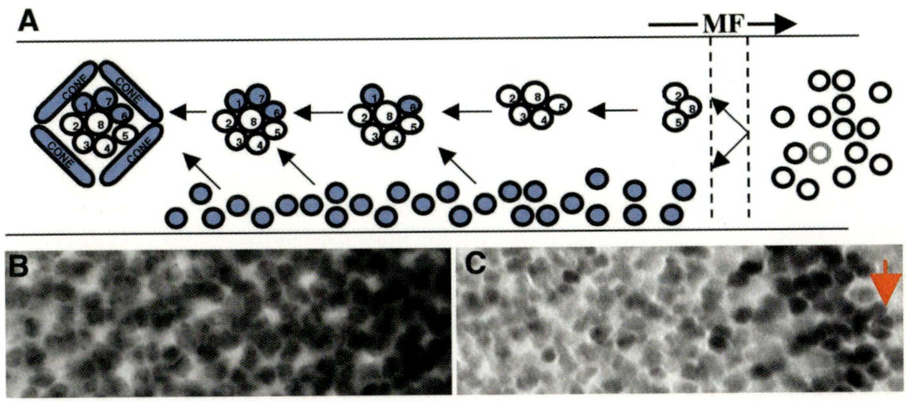

Fig. 1. A Two waves of morphogenesis. The morphogenetic furrow (*MF*) moves across the eye imaginal disc from posterior to anterior marking the onset of cell fate specification. First, the precluster cells (R8, R2/R5, R3/R4) form from the pool of undifferentiated cells anterior to the furrow (*open circles*). A round of mitosis generates a new pool of undifferentiated precursors posterior to the furrow (*blue circles*). These will give rise to the remaining ommatidial cells [R1/R6, R7, cone cells, pigment cells (not shown)]. **B** Lozenge and **C** Yan are expressed in this second pool of undifferentiated precursors. Both are expressed at their highest level near the furrow (*red arrow* in **C**)

one of the first five photoreceptors (R8, R2/R5, R3/R4) collectively called the "precluster". The rest of the cells undergo a synchronous round of mitosis and form a new group of undifferentiated cells that will be the source of all other cell types in a complete ommatidium (R1/R6, R7, cone cells and pigment cells). Recently, it has been proposed that EGFR signaling arising from the preclusters is necessary for this second round of mitosis in the rest of the cells. This signal upregulates the cell-cycle regulator String in the undifferentiated cells, which is required for their G_2 to M transition (Baker and Yu 2001). Although this second round of cell division seems obligatory in all cells that do not join the precluster, its role in patterning of the resulting cells is unclear. If mitosis is blocked at the furrow using a human-derived cell cycle inhibitor, cell fate in the developing ommatidium is not affected (de Nooij and Hariharan 1995). It remains possible that the sole purpose of this cell division is to generate enough cells to complete the clusters. In this review, we will largely concentrate on the cells arising from the second wave of mitosis. The process of determination of these cells is referred to as the second wave of morphogenesis in the eye disc (Fig. 1A).

3 Undifferentiated Cells

As mentioned before, a number of transcription factors are expressed in cells anterior to the furrow. These eye-specifying proteins are largely eliminated from the undifferentiated cells posterior to the furrow. Instead, cells generated from the second wave initiate the expression of new transcription factors that act as prepatterning molecules and prepare these cells for subsequent specification events. Yan, a downstream effector of the EGFR pathway, Lozenge (Lz), a Runt Domain protein, and Tramtrack (Ttk88) are all transcriptionally activated in the undifferentiated cells posterior to the furrow (Fig. 1B, C), but are absent from cells anterior to the furrow (Lai and Rubin 1992; Lai et al. 1996; Flores et al. 1998). Thus, although the cells ahead of and behind the furrow can both be undifferentiated and pluripotent, they are significantly different cell types. Yan is a global repressor of transcription in undifferentiated cells behind the furrow and is essential for maintaining their pluripotent nature (Lai and Rubin 1992). Similarly, Ttk88 is a negative regulator of neuronal differentiation that prevents these cells from prematurely assuming a neuronal fate (Xiong and Montell 1993). Lz promotes differentiation of all cell types in the second wave of morphogenesis. It can do so by either activating or repressing transcription of different genes in the precursors (Daga et al. 1996). The mechanisms of transcriptional regulation of genes such as *yan*, *ttk*, and *lz* are not well understood. Unraveling such processes will be very important in understanding the differences between the cells ahead of and behind the morphogenetic furrow.

4 The Precluster

4.1 R8 Specification

R8 is the first photoreceptor to be specified during eye development. The process of R8 specification begins well within the morphogenetic furrow. This specification event requires the function of Atonal (Ato), a transcription factor of the helix-loop-helix class (Jarman et al. 1994). As a heterodimeric complex with Daughterless, Ato was first shown to be required for chordotonal organ specification (Jarman et al. 1993). In the eye disc, Ato expression initiates in a stripe of cells at the anterior edge of the morphogenetic furrow in response to the Hh signal from the differentiated photoreceptors in clusters posterior to the furrow (reviewed in Treisman and Heberlein 1998). As the furrow progresses, Ato expression resolves into evenly spaced clusters of about 12 cells termed intermediate clusters (Jarman et al. 1994; Baker et al. 1996; Dokucu et al. 1996). This is followed by further resolution to three cells which have been called the "R8 equivalence group" from which R8, R2 and R5 are specified (Dokucu et al. 1996). Still later, *ato* expression is lost in two of the three cells and the single Ato positive cell assumes the R8 fate (Fig. 2A). The two remaining cells differentiate into R2 and R5. In Ato mutants, R8 specification is abnormal, and as a consequence, further cell fate specification within the ommatidium is stalled.

The EGFR pathway might be involved in some way in the specification and maintenance of R8 fate, although the literature is equivocal on its exact role

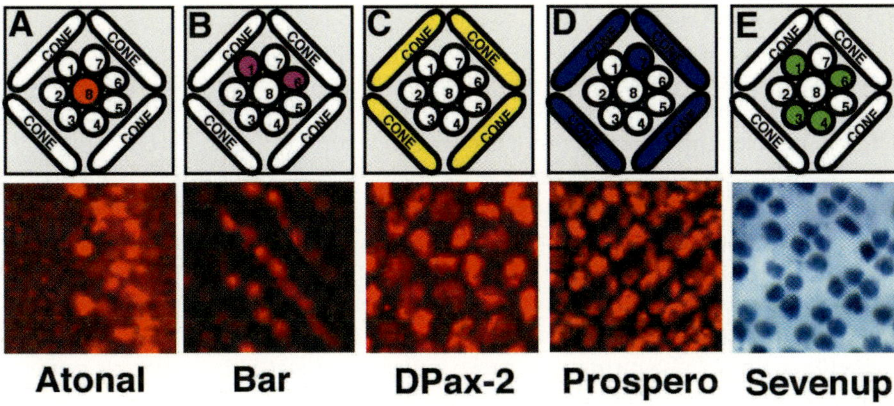

Fig. 2. Expression patterns of cell specific transcription factors. *Top panels* show the expression pattern of a representative set of transcription factors in schematic form. *Bottom panels* are the corresponding microscope images of protein expression in the eye disc. **A** Atonal is expressed at the furrow and then ultimately resolves to the R8 cell. **B** Bar is expressed in R1/R6. **C** DPax-2 is expressed in the four cone cells. **D** Prospero is expressed in R7 and the four cone cells. **E** A Sevenup reporter is expressed in R3/R4, R1/R6

(Dominguez et al. 1998; Kumar et al. 1998; Spencer et al. 1998). The Notch signaling pathway clearly plays a critical role in the specification of R8 and in the final resolution of Ato expression to this cell during eye development. In Notch mutants, an abnormally large number of cells assume an R8 fate at the furrow. This neural hypertrophy phenotype is due to the loss of Notch-mediated lateral inhibition (Baker and Zitron 1995; Cagan and Ready 1989) similar to that seen in the embryo.

4.2 R2/R5 Specification

R2/R5 cell fate specification requires the function of the *rough* gene. Rough is a homeobox domain transcription factor which functions exclusively in the eye imaginal disc. The expression of the Rough protein is first observed in a broad band of cells within the morphogenetic furrow and later in R2 and R5, and at lower levels in R3 and R4. Genetic mosaic studies suggest that Rough function is needed for the proper development of R2 and R5 (Saint et al. 1988; Tomlinson et al. 1988). During the early stages of precluster formation, Rough and Ato act antagonistically. Rough downregulates the expression of Ato in two of the three cells of the R8 equivalence group, which then differentiate into R2 and R5 (Dokucu et al. 1996). Consistent with this idea, *rough* null mutations in R2/R5 show ectopic Ato expression and therefore fail to differentiate properly. Conversely, misexpression of Rough causes loss of Ato in R8 cells (Dokucu et al. 1996).

4.3 R3/R4 Specification

The last cells to be specified within the precluster are R3 and R4. Specification of these cells requires the activity of the Sevenup (Svp) transcription factor. The *svp* gene encodes two orphan nuclear receptors differing in their putative ligand binding domains that show 75% identity to the human COUP protein (Mlodzik et al. 1990). Svp is expressed in R3/R4 in the precluster, and later in R1/R6 in the mature cluster, and is required for proper differentiation of these cell types (Fig. 2E). Loss of Svp protein from these cells converts them into R7 type, while its ectopic expression causes complex cell fate changes that depend on the timing of ectopic activation (Hiromi et al. 1993).

At the point of recruitment, the two cells of the R3/R4 pair appear equivalent. However, in a more mature ommatidium, the R3/R4 photoreceptors adopt asymmetric positions creating a chirality for the cluster (Strutt and Mlodzik 1995; Fanto et al. 1998). This differentiation between R3/R4 requires Notch signaling. Loss of Notch function converts the R3/R4 pair into R3/R3. Conversely, gain of function mutations in Notch cause the R3/R4 pair to assume R4/R4 fate (Cooper and Bray 1999; Fanto and Mlodzik 1999).

5 The Second Wave of Morphogenesis

The cells generated in the second round of morphogenesis can also be identified by a number of cell-specific transcription factors (reviewed in Kumar and Moses 1997). Among such proteins are Bar, required in R1/R6, Svp, required in R1/R6 (and R3/R4 of the precluster), DPax-2 required in the four cone cells, and Prospero (Pros), required in R7 and the cone cells (Fig. 2; Mlodzik et al. 1990; Higashijima et al. 1992; Kauffmann et al. 1996; Fu and Noll 1997). It is important to realize that by the time a cell expresses either Bar, Svp, DPax-2, or Pros, it is already different from its neighbor by the very virtue of its expression of such a cell-specific factor. The pattern of differentially expressed transcription factors must be established by a protein that is widely expressed in the precursor population. Lz is such a protein, as it is expressed in all of the undifferentiated cells posterior to the furrow (Fig. 1B), and its function is required in this pool of pluripotent cells. Lz regulates all of the known transcription factors required in the cells arising from the second wave of morphogenesis (Daga et al. 1996; Flores et al. 2000; Xu et al. 2000). Loss of Lz function results in the loss of expression of Bar in R1/R6, loss of Pros in R7 and cone cells, and loss of DPax-2 in cone cells (Fig. 3), indicating that the wild-type role of Lz is to activate these targets. Also, in *lz* null eyes, Svp is ectopically expressed in cone cells, suggesting that Lz either directly or indirectly causes repression of Svp in cone cells (Fig. 3; Daga et al. 1996). By both positively and negatively regulating different genes in the pool of undifferentiated cells, Lz prepatterns the cells and allows them to assume distinct fates. Lz function may only be needed for a short period in the undifferentiated cells to

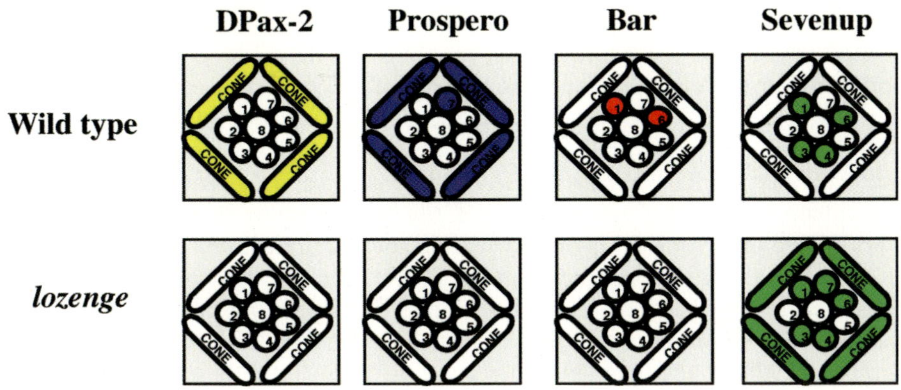

Fig. 3. Lozenge regulates the expression of several cell-specific transcription factors. Schematics depict the expression of proteins in wild-type and *lz* mutant ommatidia. The expression of DPax-2, Prospero, and Bar is lost in *lz* mutants, and Sevenup is expressed ectopically in the four cone cells

program the transcriptional status of critical genes required for a cell's specification. In doing so, Lz makes these cells competent to respond to signaling pathways and adopt appropriate developmental programs. This scenario is somewhat similar to the role of Swi5 in the transcriptional activation of the *ho* gene in yeast (Cosma et al. 1999).

5.1 R1/R6 Specification

Little is known about the signaling pathways which regulate the cell fate of the R1/R6 photoreceptors. BarH1 and BarH2 homeodomain proteins are expressed specifically in the R1/R6 cells (Fig. 2B) and are essential for their differentiation (Higashijima et al. 1992). Bar expression in these cells is dependent on Lz function in their precursors. In *lz* mutants, the expression of Bar is lost in R1 and R6 (Fig. 3). Svp is also expressed in these cells, and mutations in *svp* lead to the conversion of these cells to the R7 fate (Mlodzik et al. 1990; Daga et al. 1996). Phyllopod (Phyl) is another nuclear protein that is expressed at high levels in R1/R6. Expression of Phyl is dependent on the Ras/Raf pathway. In loss of function mutations of *phyl*, R1/R6 and R7 cells are not specified properly (Chang et al. 1995; Dickson et al. 1995).

5.2 R7 Specification

R7 is the last photoreceptor to be inducted into the ommatidial cluster and is the most widely studied cell type in the eye. Analysis of R7 cell fate determination dates back to the identification of the *sevenless* (*sev*) mutation in screens for mutants defective in phototaxis (Harris et al. 1976). Later studies demonstrated that in the *sev* mutation the presumptive R7 cell is converted into an equatorial cone cell fate (Tomlinson and Ready 1987). The *sev* gene encodes a receptor tyrosine kinase (RTK) expressed in, but not limited to, the R7 cell (Banerjee et al. 1987; Hafen et al. 1987). In spite of its broad expression pattern, only the R7 cell fate is affected in a *sev* mutant (Banerjee et al. 1987; Tomlinson et al. 1987). Activation of the Sev receptor is mediated by the Boss (Bride of Sevenless) protein which is specifically expressed in the R8 cell (Kramer et al. 1991; Van Vactor et al. 1991). Like Sev, loss of Boss function results in loss of R7 cells. However, unlike Sev, Boss functions in a non-cell autonomous manner in the R8 cell to influence R7 cell fate (Reinke and Zipursky 1988). The *sev* pathway has been the basis of many sensitized genetic screens which have been critical in relating the Ras pathway to the activation of tyrosine kinase receptors (reviewed in Daga and Banerjee 1994; Simon 1994; Zipursky and Rubin 1994; Dickson 1995).

Several lines of evidence established that the RTK signal mediated by Sev is a trigger that initiates cell fate determination of the R7 cell but is not the sole determinant of its fate. In swapping experiments, Sev and EGFR functions were found to be interchangeable (Freeman 1996). Also, if Rough is expressed in the R7 precursor, this cell assumes an outer cell fate but remains dependent on the Sev signal for its specification (Basler et al. 1990; Kimmel et al. 1990). Why it is necessary to have both Sev and EGFR function (Freeman 1996; Tio and Moses 1997) in R7 development is currently unclear.

Some mechanistic details of the nuclear events within the R7 cell following the activation of the RTKs have been uncovered. High levels of RTK signaling result in transcription of *phyl* in the R7 precursor (Chang et al. 1995; Dickson et al. 1995). Phyl then binds Sina (Seven in absentia), which is essential for R7 fate specification (Carthew and Rubin 1990). The Sina/Phyl complex is essential for the degradation of Ttk88, which normally represses the transcription of many neuron-specific genes including *pros*, which is expressed at high levels in R7 (Kauffmann et al. 1996). Mutational analysis of the *pros* eye-specific enhancer has shown that expression of this gene is also controlled by direct binding of Lz and the Pointed (Pnt) protein which is activated by RTK signaling (Xu et al. 2000).

Notch signaling is also required for the specification of R7 cells. Loss of the Notch ligand Delta in R1 and R6 causes the R7 cells to assume an R1/R6 cell type (Cooper and Bray 2000; Tomlinson and Struhl 2001). Furthermore, ectopic expression of activated Notch in R1 and R6 converts them into R7-like cells (Kauffmann et al. 1996; Tomlinson and Struhl 2001). Taken together, these studies suggest that a combination of the Sev signal from R8 and the Notch signal from R1/R6 is required for proper R7 specification (Tomlinson and Struhl 2001).

5.3 Cone Cell Specification

The four lens-secreting cone cells are recruited to the ommatidial cluster after all of the photoreceptors have joined. Cone cells are non-neuronal and develop from a group of cells that belong to the R7 equivalence group. This is based on the observation that overactivation of the Ras pathway in cone cells will cause them to adopt an R7 fate (Gaul et al. 1992; Lai and Rubin 1992; Rogge et al. 1992). Like the photoreceptors, cone cells also express a unique set of transcription factors, including DPax-2, Pros, and Cut (Kauffmann et al. 1996; Fu and Noll 1997). Mechanisms controlling the activation of the target genes *DPax-2* and *pros* have been studied in some detail (Flores et al. 2000; Xu et al. 2000; Nagaraj and Banerjee, unpubl.). The expression of DPax-2 and Pros specifically in the cone cells depends upon the function of the Lz protein as well as downstream effectors of the Notch and EGFR pathways (Fig. 4A). In loss of function mutations in *lz*, the expression of DPax-2 and Pros in cone cells is lost. Furthermore, Lz binding sites were identified in the eye-specific enhancers of *DPax-2* and *pros*, and site-directed mutations in these sites result in complete loss of their expression, establishing that their control by Lz is through direct transcriptional regulation. The cone cell-specific expression of DPax-2 is essential for the control of the Cut protein which is required for cone cell differentiation (Flores et al. 2000; Xu et al. 2000).

5.4 Pigment Cell Specification

Primary pigment cells express the transcription factors Bar and DPax-2. Pigment cells are not properly specified in the absence of BarH1, and misexpression of BarH1 can convert cone cells into primary pigment cells. Therefore, BarH1 is critical for the differentiation of pigment cells (Hayashi et al. 1998). Induction and patterning of these cells appear to be regulated by a combination of signaling inputs from cone cells, as well as from within the pigment cell equivalence group (Miller and Cagan 1998). Loss of cone cell development leads to the loss of primary pigment cell differentiation suggesting that cone cells provide essential signals for the differentiation of these cell types. Notch and the EGFR signaling pathways have also been shown to be involved in the cell fate specification of these cells. In certain mutant alleles of *Notch*, there is a loss of the primary pigment cells suggesting Notch is normally required for the specification of the primary pigment cell fate within the equivalence group (Cagan and Ready 1989). Loss of EGFR signal also results in the loss of pigment cell specification whereas activation of the EGFR signal results in the overspecification of these cells (Freeman 1996).

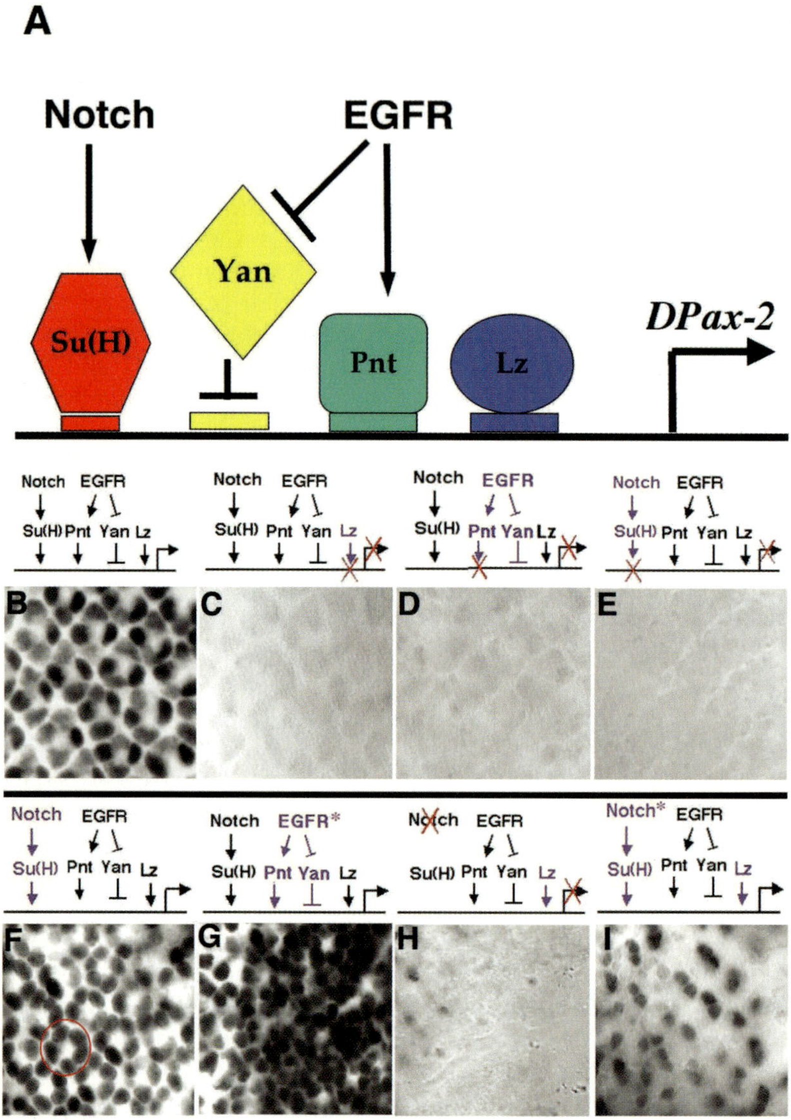

Fig. 4. Combinatorial regulation of *DPax*-2. **A** Model of tripartite control of *DPax*-2 by Notch, EGFR, and Lz. **B–I** The schematic above each panel represents the combination of inputs on the *DPax*-2 eye-specific enhancer resulting in its expression below. *Blue lettering* indicates altered signaling pathways. **B** Wild-type *DPax*-2 expression in cone cells. **C–E** Mutation of binding sites for Lozenge **C**, Pnt/Yan **D**, and Su(H) **E** on the *DPax*-2 enhancer leads to complete loss of DPax-2 expression in cone cells. **F** Activation of Notch in the R7 cell leads to ectopic expression of DPax-2 in this cell (*five cells circled* (four cones and R7)). **G** Activation of EGFR in undifferentiated cells is sufficient to cause expression of DPax-2 in these cells. **H** Lz expression alone does not result in ectopic expression of DPax-2 in R3/R4 cells, whereas in **I** coexpression of Lozenge and activated Notch causes ectopic activation of DPax-2 in R3/R4 cells

6 Generating and Testing Combinatorial Models

A central theme emerging from studies of cell fate specification in the eye undertaken by many different laboratories is that only a small number of signaling pathways is involved in this process. Notch and an RTK (EGFR or Sev) are the only two signaling pathways that have been shown to generate the complex pattern of cellular diversity away from the furrow. Boss/Sev signaling as a dedicated system devoted to the development of a single cell type seems to be more of an exception than a rule. Instead, it is the timely mix and match of the multifunctional Notch and EGFR signals that controls a multitude of cellular processes in the eye. Investigators in the field have used many sophisticated genetic schemes to get around the problem of pleiotropic function of the pathways. Still, it is difficult, in mutant studies, to separate one function of these pathways from the others. For example, EGFR function is essential for cell proliferation and viability as well as specification of fate. Suppose a gene fails to be transcribed in a specific cell in an EGFR mutant background. It is not easy to decide whether the EGFR pathway controls the expression of this gene or whether the cell has undergone a fate change and as a secondary consequence, lost expression of the marker. Another similar problem arises from the fact that the transcription factors that control the cell-specific markers are, of necessity, widely expressed. Once again, phenotypes resulting from elimination of say Yan or Suppressor of Hairless (Su(H)) function are extremely complex and it is difficult to establish definitively that they are involved in the transcriptional control of a cell-fate specifying transcription factor. One approach to overcome this difficulty is to identify eye-specific enhancers of such target genes, mutate binding sites on the enhancer for activated transcription factors and transform a reporter gene into flies under the control of this mutated enhancer. The cell-specific expression due to the enhancer can then be assessed in a completely wild-type background. Such an approach has been used for the *pros* and *DPax-2* enhancers (Flores et al. 2000; Xu et al. 2000).

For *DPax-2*, a minimal enhancer that controls the eye-specific expression in the four cone cells of each ommatidium (Fu et al. 1998) includes three Lz binding sites, six Yan binding sites (of which two also bind Pnt) and eight Su(H) binding sites (Flores et al. 2000). When the binding sites for any one of these factors are eliminated, the enhancer can no longer support cone cell expression (Fig. 4C–E). Thus, DPax-2 expression is limited to cone cells because cone cell precursors express Lz and receive enough EGFR and Notch signals at the time of their development. Presumably, at least one of these three activation mechanisms is not sufficiently functional at the correct time in cells that do not express DPax-2. Such a combinatorial model is easily tested by further genetic analysis.

Undifferentiated cells do not receive sufficient RTK signal in wild-type flies. However, DPax-2 expression can be achieved in these cells by ectopically activating EGFR (Fig. 4G). Similarly, the R7 precursors express Lz and receive RTK signals, yet they do not express DPax-2. However, expression of an activated

	Lz	EGFR	Notch	*D-Pax2*
R3/R4	-	+	-	-
R7	+	+	-	-
undiff	+	-	+	-
Cone	+	+	+	+

Fig. 5. A combinatorial code for cell fate specification based on expression of DPax-2

form of Notch will cause ectopic DPax-2 expression in R7 (Fig. 4F). Finally, the R3/R4 cells can also be made to express DPax-2 if both Lz and activated Notch are coexpressed in their precursors (Fig. 4H, I).

The nuclear effectors of the EGFR and Notch signal transduction pathways, Yan, PntP2, and Su(H), and the transcriptional regulator Lz are all ubiquitously expressed. Lz can bind the enhancers of a large number of genes and prepare them to be activated. Whether these genes are in fact expressed will depend upon their position and the timely activation of transcription factors that function downstream of incoming signals. Each enhancer will be different in its combinatorial logic of activation and repression. The model shown in Fig. 5 is based on the expression pattern of a single gene and therefore does not take into account signaling levels and timing differences that may be sufficient to activate one set of target genes but not another. For example, Notch signal from R1/R6 is important for specification of R7 fate (Cooper and Bray 2000; Tomlinson and Struhl 2001) but this signal is either too late or too weak to activate DPax-2 in the R7 cell. To generate the necessary diversity, signals will combine in all possible ways. For example, recent studies indicate that the expression of the Notch ligand Delta in photoreceptor cells is controlled by the EGFR pathway (Nagaraj, Tsuda, Zipursky and Banerjee, unpubl.). Thus, EGFR first functions serially and then in parallel with Notch to determine cone cell fate. In principle, if controlling regions of all the cell-specific genes were available and were subjected to a similar mutational analysis, then the temporal and spatial coding for ommatidial assembly would be fully understood. We predict that the number of signaling pathways involved will remain very small, even when the entire code is revealed.

References

Baker NE, Yu SY (2001) The EGF receptor defines domains of cell cycle progression and survival to regulate cell number in the developing *Drosophila* eye. Cell 104:699–708

Baker NE, Zitron AE (1995) *Drosophila* eye development: Notch and Delta amplify a neurogenic pattern conferred on the morphogenetic furrow by scabrous. Mech Dev 49:173–189

Baker NE, Yu S, Han D (1996) Evolution of proneural atonal expression during distinct regulatory phases in the developing *Drosophila* eye. Curr Biol 6:1290–1301

Banerjee U, Renfranz PJ, Pollock JA, Benzer S (1987) Molecular characterization and expression of sevenless, a gene involved in neuronal pattern formation in the *Drosophila* eye. Cell 49: 281–291

Basler K, Yen D, Tomlinson A, Hafen E (1990) Reprogramming cell fate in the developing *Drosophila* retina: transformation of R7 cells by ectopic expression of rough. Genes Dev 4:728–739

Bonini NM, Leiserson WM, Benzer S (1993) The eyes absent gene: genetic control of cell survival and differentiation in the developing *Drosophila* eye. Cell 72:379–395

Cagan RL, Ready DF (1989) Notch is required for successive cell decisions in the developing *Drosophila* retina. Genes Dev 3:1099–1112

Carthew RW, Rubin GM (1990) seven in absentia, a gene required for specification of R7 cell fate in the *Drosophila* eye. Cell 63:561–577

Chang HC, Solomon NM, Wassarman DA, Karim FD, Therrien M, Rubin GM, Wolff T (1995) phyllopod functions in the fate determination of a subset of photoreceptors in *Drosophila*. Cell 80:463–472

Cheyette BN, Green PJ, Martin K, Garren H, Hartenstein V, Zipursky SL (1994) The *Drosophila* sine oculis locus encodes a homeodomain-containing protein required for the development of the entire visual system. Neuron 12:977–996

Cho KO, Choi KW (1998) Fringe is essential for mirror symmetry and morphogenesis in the *Drosophila* eye. Nature 396:272–276

Cooper MT, Bray SJ (1999) Frizzled regulation of Notch signalling polarizes cell fate in the *Drosophila* eye. Nature 397:526–530

Cooper MT, Bray SJ (2000) R7 photoreceptor specification requires Notch activity. Curr Biol 10:1507–1510

Cosma MP, Tanaka T, Nasmyth K (1999) Ordered recruitment of transcription and chromatin remodeling factors to a cell cycle- and developmentally regulated promoter. Cell 97:299–311

Czerny T, Halder G, Kloter U, Souabni A, Gehring WJ, Busslinger M (1999) twin of eyeless, a second Pax-6 gene of *Drosophila*, acts upstream of eyeless in the control of eye development. Mol Cell 3:297–307

Daga A, Banerjee U (1994) Resolving the sevenless pathway using sensitized genetic backgrounds. Cell Mol Biol Res 40:245–251

Daga A, Karlovich CA, Dumstrei K, Banerjee U (1996) Patterning of cells in the *Drosophila* eye by Lozenge, which shares homologous domains with AML1. Genes Dev 10:1194–1205

De Nooij JC, Hariharan IK (1995) Uncoupling cell fate determination from patterned cell division in the *Drosophila* eye. Science 270:983–985

Desplan C (1997) Eye development: governed by a dictator or a junta? Cell 91:861–864

Dickson B (1995) Nuclear factors in sevenless signalling. Trends Genet 11:106–111

Dickson BJ, Dominguez M, van der Straten A, Hafen E (1995) Control of *Drosophila* photoreceptor cell fates by phyllopod, a novel nuclear protein acting downstream of the Raf kinase. Cell 80:453–462

Dokucu ME, Zipursky SL, Cagan RL (1996) Atonal, rough and the resolution of proneural clusters in the developing *Drosophila* retina. Development 122:4139–4147

Dominguez M, Hafen E (1997) Hedgehog directly controls initiation and propagation of retinal differentiation in the *Drosophila* eye. Genes Dev 11:3254–3264

Dominguez M, Wasserman JD, Freeman M (1998) Multiple functions of the EGF receptor in *Drosophila* eye development. Curr Biol 8:1039–1048

Fanto M, Mlodzik M (1999) Asymmetric Notch activation specifies photoreceptors R3 and R4 and planar polarity in the *Drosophila* eye. Nature 397:523–526

Fanto M, Mayes CA, Mlodzik M (1998) Linking cell-fate specification to planar polarity: determination of the R3/R4 photoreceptors is a prerequisite for the interpretation of the Frizzled mediated polarity signal. Mech Dev 74:51–58

Flores GV, Daga A, Kalhor HR, Banerjee U (1998) Lozenge is expressed in pluripotent precursor cells and patterns multiple cell types in the *Drosophila* eye through the control of cell-specific transcription factors. Development 125:3681–3687

Flores GV, Duan H, Yan H, Nagaraj R, Fu W, Zou Y, Noll M, Banerjee U (2000) Combinatorial signaling in the specification of unique cell fates. Cell 103:75–85

Freeman M (1996) Reiterative use of the EGF receptor triggers differentiation of all cell types in the *Drosophila* eye. Cell 87:651–660

Fu W, Noll M (1997) The Pax2 homolog sparkling is required for development of cone and pigment cells in the *Drosophila* eye. Genes Dev 11:2066–2078

Fu W, Duan H, Frei E, Noll M (1998) shaven and sparkling are mutations in separate enhancers of the *Drosophila* Pax2 homolog. Development 125:2943–2950

Gaul U, Mardon G, Rubin GM (1992) A putative Ras GTPase activating protein acts as a negative regulator of signaling by the Sevenless receptor tyrosine kinase. Cell 68:1007–1019

Gehring WJ, Ikeo K (1999) Pax 6: mastering eye morphogenesis and eye evolution. Trends Genet 15:371–377

Gibson MC, Schubiger G (2000) Peripodial cells regulate proliferation and patterning of *Drosophila* imaginal discs. Cell 103:343–350

Greenwood S, Struhl G (1999) Progression of the morphogenetic furrow in the *Drosophila* eye: the roles of Hedgehog, Decapentaplegic and the Raf pathway. Development 126:5795–5808

Hafen E, Basler K, Edstroem JE, Rubin GM (1987) Sevenless, a cell-specific homeotic gene of *Drosophila*, encodes a putative transmembrane receptor with a tyrosine kinase domain. Science 236:55–63

Halder G, Callaerts P, Gehring WJ (1995) Induction of ectopic eyes by targeted expression of the eyeless gene in *Drosophila*. Science 267:1788–1792

Harris WA, Stark WS, Walker JA (1976) Genetic dissection of the photoreceptor system in the compound eye of *Drosophila melanogaster*. J Physiol 256:415–439

Hayashi T, Kojima T, Saigo K (1998) Specification of primary pigment cell and outer photoreceptor fates by BarH1 homeobox gene in the developing *Drosophila* eye. Dev Biol 200:131–145

Heberlein U, Treisman JE (2000) Early retinal development in *Drosophila*. Results Probl Cell Differ 31:37–50

Higashijima S, Kojima T, Michiue T, Ishimaru S, Emori Y, Saigo K (1992) Dual Bar homeo box genes of *Drosophila* required in two photoreceptor cells, R1 and R6, and primary pigment cells for normal eye development. Genes Dev 6:50–60

Hiromi Y, Mlodzik M, West SR, Rubin GM, Goodman CS (1993) Ectopic expression of seven-up causes cell fate changes during ommatidial assembly. Development 118:1123–1135

Irvine KD (1999) Fringe, Notch and making developmental boundaries. Curr Opin Genet Dev 9:434–441

Jarman AP, Grau Y, Jan LY, Jan YN (1993) atonal is a proneural gene that directs chordotonal organ formation in the *Drosophila* peripheral nervous system. Cell 73:1307–1321

Jarman AP, Grell EH, Ackerman L, Jan LY, Jan YN (1994) atonal is the proneural gene for *Drosophila* photoreceptors. Nature 369:398–400

Jun S, Wallen RV, Goriely A, Kalionis B, Desplan C (1998) Lune/eye gone, a Pax-like protein, uses a partial paired domain and a homeodomain for DNA recognition Proc Natl Acad Sci USA 95:13720–13725

Kauffmann RC, Li S, Gallagher PA, Zhang J, Carthew RW (1996) Ras1 signaling and transcriptional competence in the R7 cell of *Drosophila*. Genes Dev 10:2167–2178

Kimmel BE, Heberlein U, Rubin GM (1990) The homeo domain protein rough is expressed in a subset of cells in the developing *Drosophila* eye where it can specify photoreceptor cell subtype. Genes Dev 4:712–727

Kramer H, Cagan RL, Zipursky SL (1991) Interaction of bride of sevenless membrane-bound ligand and the sevenless tyrosine-kinase receptor. Nature 352:207–212

Kumar J, Moses K (1997) Transcription factors in eye development: a gorgeous mosaic? Genes Dev 11:2023–2028

Kumar JP, Moses K (2001) EGF receptor and Notch signaling act upstream of Eyeless/Pax6 to control eye specification. Cell 104:687–697

Kumar JP, Tio M, Hsiung F, Akopyan S, Gabay L, Seger R, Shilo BZ, Moses K (1998) Dissecting the roles of the *Drosophila* EGF receptor in eye development and MAP kinase activation. Development 125:3875–3885

Lai ZC, Rubin GM (1992) Negative control of photoreceptor development in *Drosophila* by the product of the yan gene, an ETS domain protein Cell 70:609–620

Lai ZC, Harrison SD, Karim F, Li Y, Rubin GM (1996) Loss of tramtrack gene activity results in ectopic R7 cell formation, even in a sina mutant background. Proc Natl Acad Sci USA 93:5025–5030

Ma C, Moses K (1995) Wingless and patched are negative regulators of the morphogenetic furrow and can affect tissue polarity in the developing *Drosophila* compound eye. Development 121:2279–2289

Miller DT, Cagan RL (1998) Local induction of patterning and programmed cell death in the developing *Drosophila* retina. Development 125:2327–2335

Mlodzik M, Hiromi Y, Weber U, Goodman CS, Rubin GM (1990) The *Drosophila* seven-up gene, a member of the steroid receptor gene superfamily, controls photoreceptor cell fates. Cell 60:211–224

Papayannopoulos V, Tomlinson A, Panin VM, Rauskolb C, Irvine KD (1998) Dorsal-ventral signaling in the *Drosophila* eye. Science 281:2031–2034

Penton A, Selleck SB, Hoffmann FM (1997) Regulation of cell cycle synchronization by decapentaplegic during *Drosophila* eye development. Science 275:203–206

Pichaud F, Treisman J, Desplan C (2001) Reinventing a common strategy for patterning the eye. Cell 105:9–12

Reinke R, Zipursky SL (1988) Cell-cell interaction in the *Drosophila* retina: the bride of sevenless gene is required in photoreceptor cell R8 for R7 cell development Cell 55:321–330

Rogge R, Cagan R, Majumdar A, Dulaney T, Banerjee U (1992) Neuronal development in the *Drosophila* retina: the sextra gene defines an inhibitory component in the developmental pathway of R7 photoreceptor cells. Proc Natl Acad Sci USA 89:5271–5275

Royet J, Finkelstein R (1997) Establishing primordia in the *Drosophila* eye-antennal imaginal disc: the roles of decapentaplegic, wingless and hedgehog. Development 124:4793–4800

Saint R, Kalionis B, Lockett TJ, Elizur A (1988) Pattern formation in the developing eye of *Drosophila melanogaster* is regulated by the homeo-box gene, rough. Nature 334:151–154

Seimiya M, Gehring WJ (2000) The *Drosophila* homeobox gene optix is capable of inducing ectopic eyes by an eyeless-independent mechanism. Development 127:1879–1886

Shen W, Mardon G (1997) Ectopic eye development in *Drosophila* induced by directed dachshund expression. Development 124:45–52

Simon MA (1994) Signal transduction during the development of the *Drosophila* R7 photoreceptor. Dev Biol 166:431–442

Spencer SA, Powell PA, Miller DT, Cagan RL (1998) Regulation of EGF receptor signaling establishes pattern across the developing *Drosophila* retina. Development 125:4777–4790

Strutt DI, Mlodzik M (1995) Ommatidial polarity in the *Drosophila* eye is determined by the direction of furrow progression and local interactions. Development 121:4247–4256

Strutt H, Strutt D (1999) Polarity determination in the *Drosophila* eye. Curr Opin Genet Dev 9:442–446

Thomas BJ, Gunning DA, Cho J, Zipursky L (1994) Cell cycle progression in the developing *Drosophila* eye: roughex encodes a novel protein required for the establishment of G1. Cell 77:1003–1014

Tio M, Moses K (1997) The *Drosophila* TGF alpha homolog Spitz acts in photoreceptor recruitment in the developing retina. Development 124:343–351

Tomlinson A, Struhl G (2001) Delta/Notch and Boss/Sevenless signals act combinatorially to specify the *Drosophila* R7 photoreceptor. Mol Cell 7:487–495

Tomlinson A, Bowtell DD, Hafen E, Rubin GM (1987) Localization of the sevenless protein, a putative receptor for positional information, in the eye imaginal disc of *Drosophila*. Cell 51:143–150

Tomlinson A, Kimmel BE, Rubin GM (1988) rough, a *Drosophila* homeobox gene required in photoreceptors R2 and R5 for inductive interactions in the developing eye. Cell 55:771–784

Treisman JE, Heberlein U (1998) Eye development in *Drosophila*: formation of the eye field and control of differentiation. Curr Top Dev Biol 39:119–158

Treisman JE, Rubin GM (1995) wingless inhibits morphogenetic furrow movement in the *Drosophila* eye disc. Development 121:3519–3527

Van Vactor DL Jr, Cagan RL, Kramer H, Zipursky SL (1991) Induction in the developing compound eye of *Drosophila*: multiple mechanisms restrict R7 induction to a single retinal precursor cell. Cell 67:1145–1155

Xiong WC, Montell C (1993) tramtrack is a transcriptional repressor required for cell fate determination in the *Drosophila* eye. Genes Dev 7:1085–1096

Xu C, Kauffmann RC, Zhang J, Kladny S, Carthew RW (2000) Overlapping activators and repressors delimit transcriptional response to receptor tyrosine kinase signals in the *Drosophila* eye. Cell 103:87–97

Younossi-Hartenstein A, Nassif C, Green P, Hartenstein V (1996) Early neurogenesis of the *Drosophila* brain. J Comp Neurol 370:313–329

Zipursky SL, Rubin GM (1994) Determination of neuronal cell fate: lessons from the R7 neuron of *Drosophila*. Annu Rev Neurosci 17:373–397

Tissue Polarity in the Retina

Marek Mlodzik[1]

1 Tissue Polarization in Development

1.1 What is Tissue Polarity?

In multicellular organisms, most tissues derived from epithelial cell sheets form highly organized structures that are not only polarized in the apical-basolateral axis but also display a polarization within the plane of the epithelium (Fig. 1). The function of many organs or tissues requires this additional axis of polarity within the epithelium, namely a uniform polarity of single cells or multicellular units within the plane of the epithelium. This type of polarization of cells is usually referred to as epithelial planar polarity, planar cell polarity, or as mostly used in *Drosophila*, tissue polarity. Such polarization is evident in most epidermal structures (e.g., the ordered appearance of scales in fish or feathers in birds), in neuroepithelia (e.g., the inner ear epithelium, where the stereocilia bundles are aligned for normal sensitivity to sound) as well as in internal organs (e.g., in the oviduct, with the cilia allowing directional transport of an egg). Another very good example is the exocuticle in insects, as all the respective tissues are derived from the single cell layer epithelial imaginal discs (Adler 1992; Gubb 1993; Eaton 1997). Similarly, the *Drosophila* retina, as it is derived from a single cell layer epithelium, the eye imaginal disc, displays a tissue or planar polarity (Dietrich 1909). It is an intriguing problem how cells that are hundreds of cell diameters apart adopt the same polarity in the plane.

1.2 Why Is the Retina Polarized?

As the function of the visual system is to receive and transmit information to the brain to form images, precise retinotopic projections of retinal photoreceptors onto the optic lobes in the brain are an important prerequisite. The *Drosophila* eye, like all insect eyes, is a compound eye containing several hundred ommatidia, each representing a separate unit. Each point in space is perceived by only one ommatidium. As therefore, only the composite input of

[1] Mount Sinai School of Medicine, Department of Molecular, cell and Developmental Biology, Annenberg Bldg. 18-92, Box 1007, One Gustave L. Levy Place, New York, New York 10029, USA

Results and Problems in Cell Differentiation, Vol. 37
K. Moses (Ed.): Drosophila Eye Development

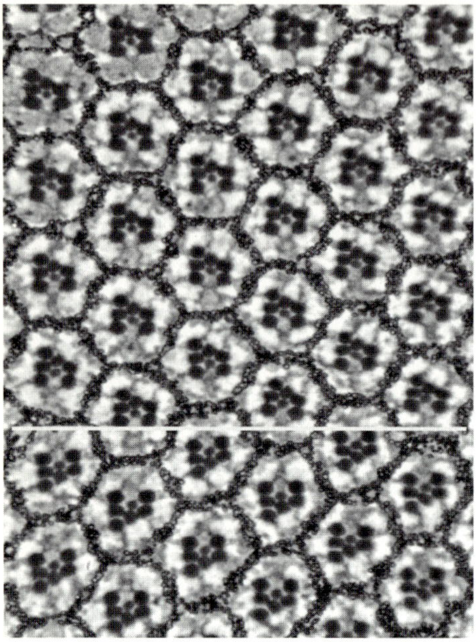

Fig. 1. Tangential section through the retina, illustrating the arrangement of the ommatidia with respect to the antero-posterior and dorso-ventral axes. The equator (mirror image symmetry line) is indicated by a *white line* in the section. The coordinates and relative orientation of the retinal field are indicated on the *right*

several ommatidia can create an image, the alignment of a single ommatidium needs to be very precise with respect to its neighbors and the whole eye field. Thus, a whole aspect of *Drosophila* eye development and patterning relates to the precise organization of the ommatidia with respect to both their neighboring ommatidia and their position within the eye. This aspect is referred to as tissue or planar polarity within the retina. This polarization is not only necessary for the correct arrangement of single ommatidia within the eye, but also for the proper arrangement of photoreceptors within each ommatidium itself. Both aspects are critical for the correct innervation and neuronal connectivity in the optic lobes (Clandinin and Zipursky 2000) and thus for image formation and vision in general.

2 The Arrangement of the Ommatidia Within the Plane of the Retina

2.1 Establishment of Ommatidial Polarity During Development

The *Drosophila* eye is polarized in a spectacular way. Polarity is reflected in the mirror image arrangement of ommatidia (or facets) of opposite chiral forms across the dorso-ventral (D/V) midline. The ommatidia are arranged within

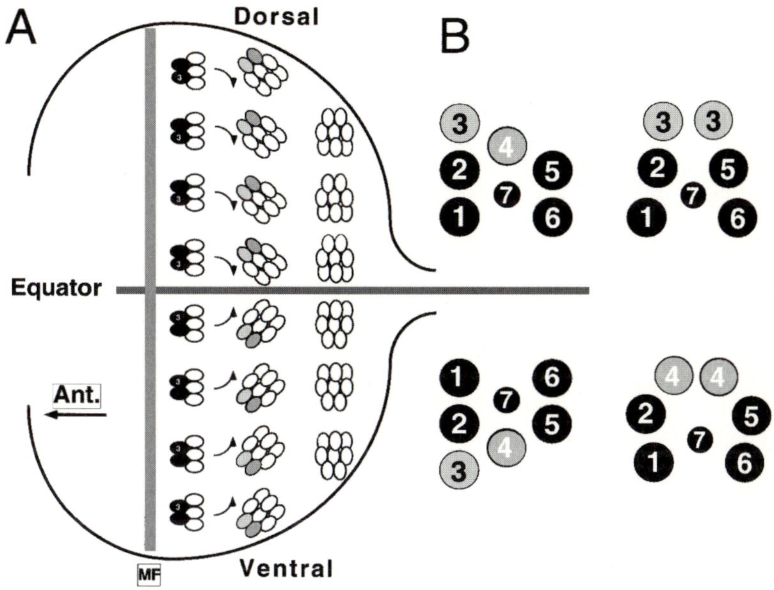

Fig. 2. Schematic presentation of the establishment of planar polarity in the retina. **A** Cartoon of the logic of polarity generation during eye development. Initially, the ommatidial preclusters are organized in the A/P axis and are symmetrical. The morphogenetic furrow is indicated by a *gray vertical bar*. Subsequently, the ommatidial clusters rotate 90° towards the equator and at the end their symmetry is broken and chirality is evident by the difference in the R3/R4 cells. **B** Schematic presentation of chiral organization of mature ommatidia (cf. Fig. 1) In addition to the two chiral forms, symmetrical clusters with R3/R3 or R4/R4 pairs can be found in mutants affecting this process. The R3 (*light gray with black number*) and R4 (*dark gray with white number*) are indicated; all other R-cells are shown in *white* in **A** or in *black with white numbers* in **B**

the epithelial plane with respect to both the antero-posterior (A/P) and the dorso-ventral (D/V) axes (Fig. 1; reviewed in (Blair 1999; Mlodzik 1999; Reifegerste and Moses 1999; Strutt and Strutt 1999). The antero-posterior arrangement is established by the direction of the progression of the morphogenetic furrow (MF, reviewed in (Heberlein and Moses 1995; Treisman and Heberlein 1998); see also Lee and Treisman, this Vol.), whereas the D/V alignment is generated in response to a polarizing signal that organizes the ommatidia around the dorso-ventral midline, the so-called equator (Eq). The critical signaling events that directly govern the D/V organization of the single ommatidia occur posterior to the MF, before or at the stage of the five-cell precluster (see Lee and Treisman, Chap. 2, this Vol.; Zheng et al. 1995; Strutt et al. 1997; Tomlinson et al. 1997; Wehrli and Tomlinson 1998).

As the MF advances anteriorly, ommatidial preclusters emerge, having a single axis of symmetry and facing in the same direction. Within the following 6–8 h the clusters rotate by 45° away from the A/P axis (Figs. 2, 3). They maintain this angle for about 8 h, before rotating a further 45°, bringing them

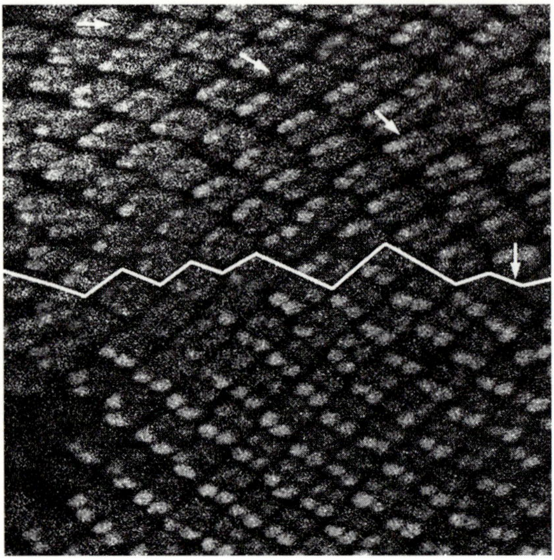

Fig. 3. Partial view of a developing eye imaginal disc demonstrating the regularity of polarity establishment. Ommatidial clusters are marked with anti-Elav (*dark grainy gray*; labeling all photoreceptors as they join the cluster) and the *seven-up* expression in R3/R4 (*light gray*) Seven-up is expressed initially in the R3/R4 pair (later also in R1/6, not shown) Anterior is left. The morphogenetic furrow is on the left side just outside the shown field. Orientation of some ommatidial clusters is highlighted with *white arrows* in upper, dorsal half; the equator is marked by *white zigzag line*

to a final 90° from their original position in the A/P axis. Subsequently, the ommatidial clusters remain in this orientation. Ommatidia in the dorsal and ventral halves of the eye imaginal disc rotate in opposite directions, thus creating a line of mirror-image symmetry, the equator (running along the D/V midline). Chirality becomes evident in mature ommatidia, where the R3 and R4 photoreceptors are asymmetrically positioned at the tip of the ommatidial trapezoids (Figs. 1, 2).

Initially, in developing ommatidial preclusters, the R3/R4 precursor pair is symmetrically arranged in the eye imaginal discs, with the R3 precursor being closer to the equator than R4 (Fig. 2). Concomitant with their rotation, ommatidial clusters also lose their symmetry. Opposite chiral forms are established in the dorsal and ventral eye halves, respectively, as R3 and R4 become asymmetrically positioned within the ommatidial cluster (Fig. 2; Dietrich 1909; Tomlinson 1988; Tomlinson and Ready 1987; Wolff and Ready 1991, 1993; Blair 1999; Mlodzik 1999; Reifegerste and Moses 1999; Strutt and Strutt 1999).

The polarity features mentioned above have suggested that the polarizing signal during eye development originates at the D/V midline, the equator. How the D/V midline (equator) is established, which is also critical for the definition of the point where the furrow initiates, is discussed in detail in Lee and

Treisman (this Vol.). In brief, recent work has implicated several signaling molecules and pathways in setting up the dorso-ventral midline in the 2nd larval instar eye disc, which serves later in the third instar larvae as the equator. These include *pannier*, the Wg pathway, the homeo domain genes of the *Iroquois* complex and the *hopscotch*/JAK-STAT pathway. Ultimately, the interplay of these signals leads to the expression *fringe* in the ventral half of the disc and thus to Notch activation at the boundary of Fringe expressing and non-expressing cells (reviewed in Blair 1999; Strutt and Strutt 1999; see Lee and Treisman, this Vol.). The polarizing signal, often referred to as factor X, is thought to emanate from the equator later in the third instar eye disc, polarizing the eye field and the developing ommatidial clusters (Mlodzik 1999; Reifegerste and Moses 1999; Strutt and Strutt 1999).

2.2 The Role of the R3 and R4 Photoreceptors in Ommatidial Polarization

Genetic analysis of mutants affecting planar polarity has indicated that the R3/R4 photoreceptor pair is critical for the establishment of ommatidial polarity (Zheng et al. 1995). First, it is important that the R3/R4 photoreceptor subtype is correctly induced by the nuclear receptor Seven-up prior to (or at) the time of tissue polarity signaling (Fanto et al. 1998). Second, following this subtype specification, the correct cell fate within the R3 /R4 photoreceptor pair with respect to one another determines the polarity of the ommatidium. It is thought that the polarizing signal is received first, or at higher levels, by the R3/R4 precursor that is closer to the source (of the presumed signal), the equator, leading to its specification as a R3 cell and determining the direction of rotation. In mutants affecting tissue polarity (see below) both the chirality and the direction and degree of rotation become random (Gubb 1993; Theisen et al. 1994; Zheng et al. 1995). An example of a retinal tissue polarity mutant phenotype is shown in Fig. 4 for *dishevelled*. Confirming the importance of the R3/R4 pair in this context, in tissue polarity mutants some ommatidia remain symmetrical, giving rise to either V or U-shaped adult ommatidia with non-chiral R3/R3 or R4/R4 photoreceptor pairs (Fig. 2).

3 Genetic Control of Retinal Polarization

3.1 The "Tissue Polarity" or Planar Polarity Genes

Many of the components required for planar polarity establishment in the retina have been identified in genetic screens and are also required for planar polarity in most (if not all) other tissues (e.g., wing, legs and body wall). These are thus referred to as "primary polarity genes" (Adler 1992; Gubb 1993). They comprise the following group: *frizzled* (*fz*), *dishevelled* (*dsh*), *prickle-spiny legs*

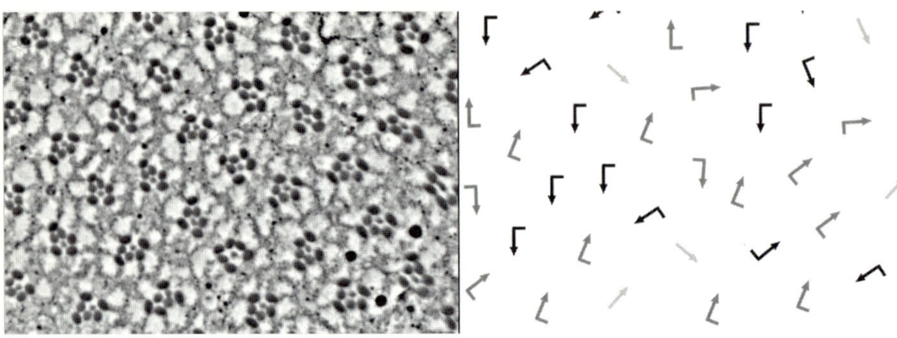

Fig. 4. A typical retinal polarity phenotype as represented by the *dsh* mutant. Compare to Fig. 1 for wild-type arrangement. Right side is a schematic representation of the field shown on the *left*. Note that the two chiral forms, represented by *black* and *gray arrows with "flag"* are randomly intermixed and that the arrows point in random directions. Ommatidia that have lost the chirality are shown as *light gray arrows without "flag"*

(*pk-sple*), *rhoA*, *mishapen* (*msn*), *strabismus* (also called *Van Gogh*) and *flamingo* (also called *starry night*) (Adler 1992; Gubb 1993; Theisen et al. 1994; Zheng et al. 1995; Strutt et al. 1997; Taylor et al. 1998; Wolff and Rubin 1998; Chae et al. 1999; Gubb et al. 1999; Paricio et al. 1999; Usui et al. 1999; see Table 1 for details).

In addition to these primary polarity genes, several genes have been identified that are required for polarity generation in specific tissues. For example, *fuzzy*, *inturned* and *multiple wing hairs* (*mwh*) only affect cell polarity in the wing (Adler et al. 1994), whereas mutations in *nemo* and *roulette* appear only to affect ommatidial polarity (Choi and Benzer 1994; see Sect. 5). Based on these observations, it was suggested that the genes affecting polarity in all tissues might be components of a common signaling pathway, responsible for reading and relaying a common polarization signal, whereas genes like *fuzzy* or *nemo* act as tissue-specific effectors of this signaling pathway. Whereas all the known primary polarity genes affect both chirality and direction and degree of rotation, the eye-specific polarity genes appear only to affect the rotation of the ommatidia. They do not affect the R3/R4 cell fate decision and thus the chirality generation (Choi and Benzer 1994; see Sect. 6.5).

3.2 Frizzled and Dishevelled

The most prominent tissue polarity gene is *frizzled* (*fz*). Elegant work by Paul Adler and colleagues has led to a detailed understanding of its phenotypic and molecular features (Vinson and Adler 1987; Vinson et al. 1989; Adler et al. 1990; Krasnow and Adler 1994; Wang et al. 1994; Jones et al. 1996). Among the planar polarity genes *fz* displays a unique feature. Whereas most of the other genes

Table 1. Planar polarity genes in retinal tissue polarity

Gene	Phenotypic defects in mutants	Molecular features
Primary planar polarity genes:		
frizzled (*fz*)	Random chirality and rotation	Seven-pass transmembrane receptor, binds Wg/Wnt ligands
dishevelled (*dsh*)	Random chirality and rotation (see Fig. 4 for example)	Cytoplasmic protein with three conserved domains (DIX, PDZ, DEP), recruited to membrane via Fz-signaling
prickle (*pk*) (a.k.a. *prickle-spiny legs*)	Random chirality and rotation in *sple* and null alleles	Cytoplasmic protein with three LIM domains and PET domain
strabismus (stbm) (a.k.a. *Van Gogh*)	Random chirality and rotation	Putative transmembrane protein, no functional homology
flamingo (*fmi*) (a.k.a. *starry night*)	Not described	Seven-pass transmembrane protein with Cadherin domains in extracellular part
RhoA	Random chirality and rotation	Small GTPase
misshapen (*msn*)	Random chirality and rotation	STE20-like S/T protein kinase
Secondary polarity genes:		
nemo (*nmo*)	Rotation stops after the first 45°	Kinase, MAPK family member
roulette (*rlt*)	Rotation does not stop at 90°, when completed	?
lamininA	Random degree of rotation	Cell adhesion molecule

(with the exception of *stbm/Vang*; see Sect. 3.3) are required autonomously in the mutant cells, *fz* tissue appears also to affect neighboring wild-type cells on the distal (wing) or polar (eye) side of the mutant clone (Vinson and Adler 1987; Zheng et al. 1995). This observation has been interpreted such that, in addition to the cell-autonomous requirement of *fz*, the signal cannot travel through an *fz* patch of cells, implying that Fz is not only necessary for reading the signal, but also for its propagation. Molecular cloning of *fz* suggested that it might function as a receptor: its primary sequence predicted a seven-pass transmembrane receptor-like molecule (Vinson et al. 1989). Recently, the Frizzled family of transmembrane receptors has been shown to act as the receptors for the Wnt family of secreted growth factors (Bhanot et al. 1996). In other developmental contexts, Fz can act as the receptor for Wg itself. This function is, however, redundant with Fz2 (Bhat 1998; Kennerdell and Carthew 1998; Bhanot et al. 1999; Chen and Struhl 1999; Mueller et al. 1999). Thus, although the planar polarity-specific Fz ligand has not yet been identified, it is likely that it is another member of the Wnt family. This missing ligand is often referred to as factor X (Wehrli and Tomlinson 1998).

Of the other planar polarity genes, *dishevelled* (*dsh*) is the best characterized (Klingensmith et al. 1994; Krasnow et al. 1995; Theisen et al. 1994). Genetic epistasis analysis has placed *dsh* downstream of *fz* in planar polarity signaling (Krasnow et al. 1995; Strutt et al. 1997). The *dsh* gene encodes a 70-kDa cytoplasmic protein, and although it has no similarities to proteins with known biochemical functions, it contains three conserved domains (Klingensmith et al. 1994; Theisen et al. 1994; Yanagawa et al. 1995). Homologues of Dsh have been identified in many organisms, ranging from nematodes to humans. All Dsh proteins identified share three highly conserved domains: a DIX domain, a central PDZ domain and a C-terminal DEP domain (reviewed in Boutros and Mlodzik 1999). All three domains have been implicated as protein-protein interaction modules, and thus Dsh might serve as an adapter molecule. Although Dsh acts downstream of Fz, it remains unclear how these proteins are molecularly linked.

In the context of polarity in the retina and R3/R4 specification, Fz and Dsh are thought to generate a bias within the early symmetrical R3/R4 pair (Zheng et al. 1995). More specifically, they specify the R3 fate (Fanto and Mlodzik 1999; Tomlinson and Struhl 1999). The particular roles of several other primary polarity genes that appear to function in the context of Frizzled signaling are discussed below.

3.3 A Frizzled Mediated Planar Polarity Signaling Pathway Is Emerging

Several of the primary planar polarity genes and other signaling components have been found to interact genetically with *fz* and *dsh*. Recently, a combination of genetic and biochemical studies has demonstrated that in the eye the planar polarity pathway downstream of Fz and Dsh consists of the small GTPase RhoA and the STE20-like kinase Misshapen (Msn) (Strutt et al. 1997; Boutros et al. 1998; Paricio et al. 1999). As with *fz* and *dsh*, both *RhoA* and *msn* are required for the generation of planar polarity in all tissues analyzed (Strutt et al. 1997; Paricio et al. 1999). In addition, genetic interactions in the eye and biochemical experiments also indicate that a Jun N-terminal Kinase (JNK)-type mitogen-activated protein kinase (MAPK) module and the AP-1 transcription factor act downstream of Dsh and Msn in this context (Boutros et al. 1998; Paricio et al. 1999; Weber et al. 2000). The small GTPase Rac (represented by *Drac1* and *Drac2* in *Drosophila*) also appears to be involved in this pathway regulating tissue polarity in the retina. Based on dominant negative and gain-of-function studies and supported by genetic interactions with the respective *Drac1* and *Drac2* deficiencies, Rac is thought to mediate the activation of Msn and the JNK-cascade downstream of Dsh (Boutros et al. 1998; Fanto et al. 2000). These observations indicate that in the eye Fz signals to the nucleus via a Dsh/Rac/Msn pathway that leads to the activation of JNK-(and p38) type kinase cascades and Jun (Fanto et al. 2000; Weber et al. 2000; summarized in Fig. 5). These factors acting downstream of Dsh are not appreciably involved in Wg signaling, and thus a signaling cascade that is distinct from the canon-

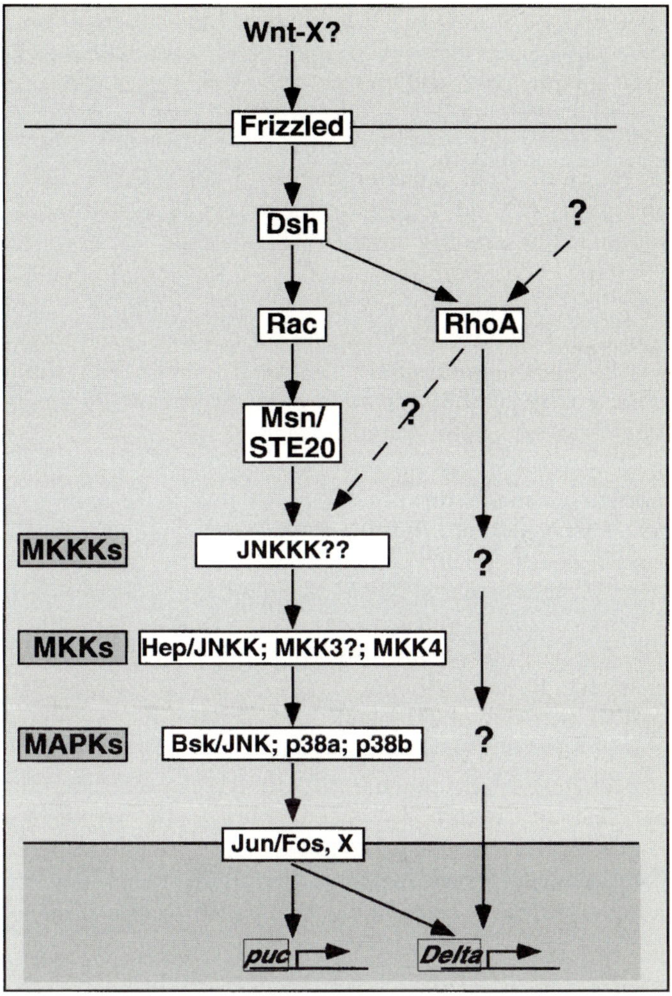

Fig. 5. Schematic model of the Frizzled signaling pathway required for R3/R4 chirality decision and direction of ommatidial rotation. See text for details

ical Wg/Wnt pathway has emerged for tissue polarity in the eye (and probably planar polarity in general).

The involvement and requirements of the kinases acting downstream of Msn as determined by genetic interactions and biochemical studies remain, however, unclear. Both *Drosophila* Dsh and its human homologues have been shown to act as potent activators of JNK signaling (Boutros et al. 1998; Li et al. 1999). Although mutations in the JNK-cascade dominantly suppress gain-of-function genotypes of *fz*, *dsh*, *msn* and *Rac*, they do not show significant planar polarity phenotypes in simple loss-of-function analyses, suggesting that there is redundancy at this level in the cascade (Strutt et al. 1997; Boutros et al. 1998; Paricio et al. 1999; Fanto et al. 2000; Weber et al. 2000). Recent genetic experiments have suggested that the observed redundancy of the JNK module might be due to the action of the related p38 type MAPK module (Paricio et al. 1999; Weber et al. 2000). As components of either kinase cascade are able to cross-phosphorylate the respective recipients in either module, this possibility is supported by existing biochemical analysis.

Despite the identification of these new components of Fz/planar polarity signaling, it remains unclear how these components are molecularly linked. Although Fz recruits Dsh to the membrane (Axelrod et al. 1998), there is no direct molecular interaction between Fz and Dsh. Similarly, it is not known what factor(s) Dsh binds to, in order to activate its downstream planar polarity effectors Rac, RhoA and/or Msn/STE20.

3.4 How Is Frizzled Activity Regulated?

Despite the increasing knowledge about the role of Frizzled in planar polarity generation and the insights of how the D/V-midline, the equator, is established (see Lee and Treisman, this Vol.), the ligand for Fz in this context is still unknown. Based on the fact that Frizzled family receptors generally bind Wnt ligands (Bhanot et al. 1996), a candidate for the planar polarity Fz ligand should be a Wnt gene. Nevertheless, so far none of the Wnt genes analyzed in sufficient detail appears to be the right candidate. Thus, one might speculate that the "search" has been going in the wrong direction and that the Fz planar polarity ligand is not a Wnt. This could be envisioned as either a "heterologous" ligand binding directly to Fz, or that Fz is part of a receptor complex where it signals, but a co-receptor contributes to ligand binding in analogy to the Patched/Smoothened receptor complex in Hedgehog signal transduction. Here, Patched binds Hedgehog, whereas Smoothened is the signaling part of the receptor complex (reviewed in Ingham 1998; Johnson and Scott 1998).

Recently, the analysis of *four jointed* (*fj*) has revealed that it might be a factor involved in this context (Zeidler et al. 1999). Interestingly, *fj* encodes a type II transmembrane protein that is expressed in a D/V gradient in the eye imaginal disc. This expression is regulated by the Notch, JAK/STAT and Wingless pathways consistent with the idea that it mediates their effects in D/V patterning and retinal polarity (Zeidler et al. 1999). Loss-of-function clones and ectopic expression analyses of *fj* have revealed non-autonomous defects in ommatidial polarity within the D/V axis. However, complete removal of *fj* function resulted in only very mild polarity defects (Zeidler et al. 1999). These observations suggest that Fj participates redundantly in D/V polarization, possibly by modulating either the activity or stability of a secreted factor (factor X), or more directly the activity of the Fz receptor as part of factor X itself.

Biochemical and genetic experiments to identify the genes and proteins Fj interacts with will be necessary to resolve these issues.

3.5 Other Primary Polarity Genes Involved in Retinal Polarity

In addition, three other primary polarity genes have been identified that do not (yet) appear to have a clear link to *fz* (Table 1). The first that has been around for some time is *prickle-spiny legs*, now called *prickle (pk)* (Gubb et al. 1999). It encodes several protein isoforms containing three Lin-11, Isl-1 Mec-3 (LIM) domains and a Prickle, Espinas and Testin (PET) domain and might thus be involved in protein-protein interactions. It has been proposed that it might be serving a scaffolding function (Gubb et al. 1999). The genetics of the *pk* locus are complex, as there are several protein isoforms, with isoform-specific alleles corresponding to the original genetic *prickle* and *spiny legs (sple)* complementation groups. Whereas the null alleles affect all isoforms and show a phenotype in all tissues, mutant alleles that are either *pk-prickle* or *pk-sple*-specific affect only subsets of tissues; e.g., *pk-prickle* affects the wing but not the eye, whereas *pk-sple* affects the eye and not the wing. Strikingly, the tissue-specific phenotypes of the isoform-specific alleles are stronger than the respective phenotype of the null alleles (Gubb et al. 1999), suggesting that a balance of the respective isoforms is critical for the function of Prickle. The role(s) of *pk* and its specific isoforms in planar polarity establishment, and their relationship to Fz signaling remains, however, obscure due to the lack of molecular or genetic interactions (Strutt et al. 1997; Gubb et al. 1999; Mlodzik 2000).

The second, *strabismus* (*stbm*; also known as *Van Gogh/Vang*) shows the same phenotypic features as *fz* and *dsh* (Taylor et al. 1998; Wolff and Rubin 1998). On close inspection, however, its requirement in the eye is opposite to that of *fz* and *dsh*: *stbm* is required in the R4 and not in the R3 cell (Wolff and Rubin 1998). Molecular features of *stbm* are not informative, as it shares no functional homology with other proteins. Nevertheless, its sequence predicts a multi-pass transmembrane protein (Wolff and Rubin 1998). Its requirement in R4 might indicate that the role of Stbm is to antagonize Fz signaling. This is further supported by the observation that the non-autonomous "shadow" of *stbm* (*Vang*) clones, as reported in the wing, points in the opposite direction as compared to *fz*- clones (Taylor et al. 1998).

Most recently, a new primary polarity gene, *flamingo* (*fmi*; also known as *starry night*), has been identified (Chae et al. 1999; Usui et al. 1999), which encodes a Cadherin superfamily member with features of a seven-pass transmembrane receptor-like protein. Since Fmi is differentially localized at cell-cell boundaries in the wing in a *fz* and *dsh*-dependent manner, it is likely to act downstream of Frizzled signaling, as a general effector or modulator of the Fz signal (Usui et al. 1999). Although Usui et al. (1999) mention that *fmi* also displays a retinal polarity phenotype, this has not been described in any significant detail and thus the role of *fmi* in the generation of ommatidial polarity and its potential involvement in R3/R4 specification remain unclear.

4 How Is the Polarity Signal Interpreted Within a Single Ommatidium?

How does Fz/planar polarity signaling generate polarity within a group of cells, like the ommatidial cluster in the eye? Genetic manipulation has shown that relative Fz activity in the R3/R4 photoreceptor pair is critical for polarity establishment. In particular, the cell that has higher Fz activity (or might acquire it first in a wild-type background) will become the R3 cell, giving an ommatidium the respective chirality (Zheng et al. 1995; Fanto and Mlodzik 1999; Tomlinson and Struhl 1999). However, R3 and R4 precursors are direct neighbors within the five-cell precluster and can be over a hundred cells away from the presumptive source of the signal (for ommatidia on the polar sides of the disc). Consequently, the difference in Fz activity in these two cells should be very small. It is thus difficult to imagine how a 100% reliable read-out is generated between the R3/R4 cells. In addition, in *fz* and *dsh* mutants, most ommatidia still remain chiral with R3/R4 being specified in these clusters (albeit at random).

These observations suggested that there could be a secondary signal acting downstream of Fz that is involved in R3/R4 specification. Recently, the Notch pathway has been implicated in the generation of chirality, and Notch signaling was shown to specify R4 (Cooper and Bray 1999; Fanto and Mlodzik 1999; Tomlinson and Struhl 1999; summarized in Fig. 6). Expression analysis of the membrane-associated Notch ligand, Delta, demonstrates that *Delta* is a transcriptional target of Fz signaling in the R3 precursor (Cooper and Bray 1999; Fanto and Mlodzik 1999). Fz and several downstream components of Fz, including Dsh, Rac, RhoA, Hep/JNKK and the transcription factor dJun have been shown to regulate Delta expression within the R3/R4 pair (Cooper and Bray 1999; Fanto and Mlodzik 1999; Fanto et al. 2000; Weber et al. 2000).

Following its transcriptional Fz-signaling-mediated upregulation, Delta in turn amplifies the Fz signal by activating Notch signaling in the neighboring R4 precursor, locking the binary R3/R4 cell fate and chirality decision in place (Cooper and Bray 1999; Fanto and Mlodzik 1999; Tomlinson and Struhl 1999). The signal amplification is even enhanced by the fact that activated Notch signaling represses *Delta* expression in R4 (summarized in Fig. 6). This two-tiered mechanism explains how a small initial difference in Fz signaling between the R3 and R4 precursor cells reliably generates the correct decision. This observation also explains why chirality, albeit stochastically, is still generated in *fz* and *dsh* null mutants.

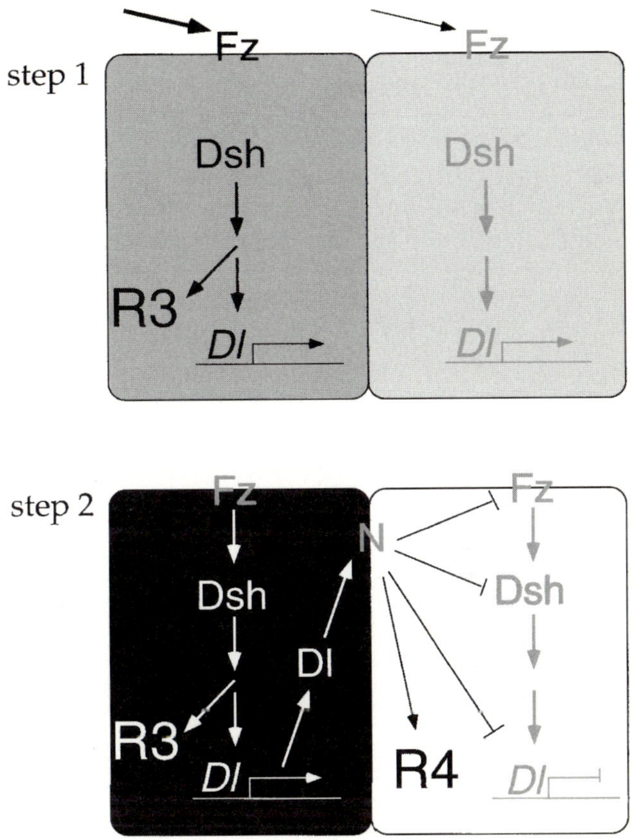

Fig. 6. Two tiered mechanism of chirality generation within the R3/R4 pair. An initial small difference in Fz signaling levels is amplified by the transcriptional upregulation of *Delta* and the subsequent activation of Notch signaling. Notch in turn inhibits Delta transcription, creating a solid binary cell fate decision. In Fz pathway mutants the asymmetry of *Delta* upregulation is lost and a stochastic Delta/Notch interaction generates randomly a R3/R4 difference and chirality. See text for details.

5 Ommatidial Rotation

Despite the recent insights into the Fz signaling pathway and its role in establishing ommatidial chirality and tissue polarity in the retina, very little is known about the subsequent ommatidial rotation. The rotation is an important aspect of polarity in the retina, and a fascinating biological problem, as groups of cells (the ommatidial preclusters) have to rotate as a whole within the epithelium. Strikingly, rotation initiates at a point when some of the photoreceptor precursors (R8, R2/R5) have already initiated axonal outgrowth towards the optic lobes in the brain.

In all the mutants of primary polarity genes, rotation is still taking place, even though at random, indicating that the two processes, determination of polarity/chirality and subsequent rotation, are (at least in part) independent. There are very few genes that appear to affect only the rotation aspect of the process. These are mainly *roulette* (*rlt*) and *nemo* (Choi and Benzer 1994). Interestingly, *rlt* and *nemo* have opposite effects on the rotation process. Whereas in *nemo* mutants ommatidia only rotate the first 45°, in the *rlt* mutant ommatidia fail to stop at 90° (Choi and Benzer 1994), suggesting that *nemo* is positively required in the process, whereas *rlt* acts negatively. The *nemo*, *roulette* double mutant displays the same defect as *nemo* alone and, thus, it was postulated that *nemo* acts upstream of *rlt* in a genetic cascade regulating the rotation. Whereas *nemo* encodes a distant member of the MAPK superfamily, the molecular nature of the *rlt* gene is not yet known (Choi and Benzer 1994).

Are these two genes linked to *fz* or any other of the primary polarity genes? Although there is no evidence for a clear link between the Fz pathway and *nemo* or *rlt*, it has been shown that *fz*, *nemo* double mutants produce a more severe phenotype than either single mutant. Specifically, in a *fz*, *nemo* double mutant many ommatidia do not rotate at all (Zheng et al. 1995), suggesting that *fz* has some positive effect on Nemo activity. However, as in *fz* or *dsh* mutants rotation is not much affected, except for its direction and degree becoming random, the role of the primary polarity genes in rotation remains unclear.

As the rotation process per se implies that the ommatidial precursor cells have to change their adhesive properties in the process, it is an interesting question whether and what cell adhesion features and molecules are involved in the context. Interestingly, the only other gene with a specific defect in rotation, in addition to *nemo* and *rlt*, is the cell adhesion molecule *Laminin* (Henchcliffe et al. 1993). In *LamininA* mutants the rotation appears random, whereas the chirality is not affected. However, the function of Laminin has not been linked to any other gene/molecule in this context and so this remains just an interesting observation at this point.

6 General Conclusions

What are the major lessons and evolutionary generalities that we can learn from the establishment of tissue polarity in the retina? The recent discovery of an alternative Wnt/Fz pathway (see Fig. 5) in the generation polarity in the eye is an important feature. As it appears now, this type of Fz signaling pathway is conserved in evolution and, for example, also used in the coordination of cellular migrations in vertebrate convergent extension during gastrulation (Heisenberg et al. 2000; Wallingford et al. 2000). Similarly, Wnt/Fz signaling has been implicated in many contexts throughout the animal kingdom, ranging from embryonic patterning to proliferation control. Thus, it is likely that studies of retinal polarity and patterning could turn out relevant models for human disease (see also Bonini and Fortini, this Vol.).

Moreover, the two tiered Fz/Notch activation to polarize groups of cells might be a general mechanism for polarity establishment in multicellular units. The asymmetric activation of Notch-signaling in response to a graded signal is possibly also used in the generation of polarity in the multicellular feather bud precursors in vertebrates, where polarized expression of Notch pathway components has been reported (Chen et al. 1997). This appears reminiscent of the regulation of *Delta* expression by the Fz/planar polarity signaling pathway in the eye. Although, it is not yet known whether Fz signaling is involved in the context of feather bud polarization, the usage of Fz and Notch in the same processes can be observed in several other tissues (e.g., bristle patterning and wing development in *Drosophila*). The potentially antagonistic or synergistic Fz-Notch interactions have led to speculations that Notch is also involved in Wg signal transmission (Couso and Martinez Arias 1994). The observation that Dsh can interact directly with Notch as shown in a yeast two hybrid assay (Axelrod et al. 1996) suggests that the pathway interactions are complex, and possibly involve quenching of pathway components.

Acknowledgements. I am grateful to all the members of the Mlodzik lab and Suzanne Eaton for helpful and stimulating discussions, and Jennifer Curtiss for improving my English and making the chapter readable. This work in the Mlodzik lab is supported by the NIH-NEI 1 RO1 EY13256 grant.

References

Adler PN (1992) The genetic control of tissue polarity in *Drosophila*. BioEssays 14:735–741

Adler PN, Vinson C, Park WJ, Conover S, Klein L (1990) Molecular structure of *frizzled*, a *Drosophila* tissue polarity gene. Genetics 126:401–416

Adler PN, Charlton J, Park WJ (1994) The *Drosophila* tissue polarity gene *inturned* functions prior to wing hair morphogenesis in the regulation of hair polarity and number. Genetics 137:1–8

Axelrod J, Matsuno K, Artavanis-Tsakonas S, Perrimon N (1996) Interaction between Wingless and Notch signaling pathways mediated by Dishevelled. Science 271:1826–1832

Axelrod JD, Miller JR, Shulman JM, Moon RT, Perrimon N (1998) Differential requirement of Dishevelled provides signaling specificity in the Wingless and planar cell polarity signaling pathways. Genes Dev 12:2610–2622

Bhanot P, Brink M, Samos CH, Hsieh J-C, Wang Y, Macke JP, Andrew D, Nathans J, Nusse R (1996) A new member of the *frizzled* family from *Drosophila* functions as a Wingless receptor. Nature 382:225–230

Bhanot P, Fish M, Jemison JA, Nusse R, Nathans J, Cadigan KM (1999) Frizzled and DFrizzled-2 function as redundant receptors for wingless during *Drosophila* embryonic development [In Process Citation]. Development 126:4175–4186

Bhat KM (1998) Frizzled and Frizzled2 play a partially redundant role in Wingless signaling and have similar requirements to Wingless in neurogenesis. Cell 95:1027–1036

Blair S (1999) Eye development: Notch lends a handedness. Curr Biol 9:356–360

Boutros M, Mlodzik M (1999) Dishevelled: at the crossroads of divergent intracellular signaling pathways. Mech Dev 83:27–37

Boutros M, Paricio N, Strutt DI, Mlodzik M (1998) Dishevelled activates JNK and discriminates between JNK pathways in planar polarity and *wingless* signaling. Cell 94:109–118

Chae J, Kim MJ, Goo JH, Collier S, Gubb D, Charlton J, Adler PN, Park WJ (1999) The *Drosophila* tissue polarity gene starry night encodes a member of the protocadherin family. Development 126:5421–5429

Chen C-N, Struhl G (1999) Wingless transduction by the Frizzled and Frizzled2 proteins of *Drosophila.* Development 126:5441–5452

Chen CWJ, Jung HS, Jiang TX, Chuong CM (1997) Asymmetric expression of Notch/Delta/Serrate is associated with the anterior-posterior axis of feather buds. Dev Biol 188:181–188

Choi K-W, Benzer S (1994) Rotation of photoreceptor clusters in the developing *Drosophila* eye requires the *nemo* gene. Cell 78:125–136

Clandinin TR, Zipursky SL (2000) Afferent growth cone interactions control synaptic specificity in the *Drosophila* visual system. Neuron 28:427–436

Cooper MTD, Bray SJ (1999) Frizzled regulation of Notch signalling polarizes cell fate in the *Drosophila* eye. Nature 397:526–529

Couso JP, Martinez Arias A (1994) Notch is required for Wingless signaling in *Drosophila.* Cell 79:259–272

Dietrich W (1909) Die Facettenaugen der Dipteren. Z Wiss Zool 92:465–539

Eaton S (1997) Planar polarity in *Drosophila* and vertebrate epithelia. Curr Opin Cell Biol 9

Fanto M, Mlodzik M (1999) Asymmetric Notch activation specifies photoreceptors R3 and R4 and planar polarity in the *Drosophila* eye. Nature 397:523–526

Fanto M, Mayes CA, Mlodzik M (1998) Linking cell-fate specification to planar polarity: determination of the R3/R4 photoreceptors is a prerequisite for the interpretation of the Frizzled mediated polarity signal. Mech Dev 74:51–58

Fanto M, Weber U, Strutt DI, Mlodzik M (2000) Nuclear signaling by Rac abd Rho GTPases is required in the establishment of epithelial planar polarity in the *Drosophila* eye. Curr Biol 10 (in press)

Gubb D (1993) Genes controlling cellular polarity in *Drosophila.* Development (Suppl) 1993: 269–277

Gubb D, Green C, Huen D, Coulson D, Johnson G, Tree D, Collier S, Roote J (1999) The balance between isoforms of the prickle LIM domain protein is critical for planar polarity in *Drosophila* imaginal discs. Genes Dev 13:2315–2327

Heberlein U, Moses K (1995) Mechanisms of *Drosophila* retinal morphogenesis: the virtues of being progressive. Cell 81:987–990

Heisenberg CP, Tada M, Rauch GJ, Saude L, Concha ML, Geisler R, Stemple DL, Smith JC, Wilson SW (2000) Silberblick/Wnt11 mediates convergent extension movements during zebrafish gastrulation. Nature 405:76–81

Henchcliffe C, Garcia-Alonso L, Tang J, Goodman CS (1993) Genetic analysis of laminin A reveals diverse functions during morphogenesis in *Drosophila.* Development 118:325–337

Ingham PW (1998) Transducing Hedgehog: the story so far. EMBO J 17:3505–3511

Johnson RL, Scott MP (1998) New players and puzzles in the Hedgehog signaling pathway. Curr Opin Genet Dev 8:450–456

Jones KH, Liu J, Adler PN (1996) Molecular analysis of EMS-induced frizzled mutations in *Drosophila melanogaster.* Genetics 142:205–215

Kennerdell JR, Carthew RW (1998) Use of dsRNA-mediated genetic interference to demonstrate that *frizzled* and *frizzled2* act in the *wingless* pathway. Cell 95:1017–1026

Klingensmith J, Nusse R, Perrimon N (1994) The *Drosophila* segment polarity gene *dishevelled* encodes a novel protein required for response to the *wingless* signal. Genes Dev 8:118–130

Krasnow RE, Adler PN (1994) A single *frizzled* protein has a dual role in tissue polarity. Development 120:1883–1893

Krasnow RE, Wong LL, Adler PN (1995) *dishevelled* is a component of the *frizzled* signalling pathway in *Drosophila.* Development 121:4095–4102

Li L, Yuan H, Xie W, Mao J, Caruso AM, McMahon A, Sussman DJ, Wu D (1999) Dishevelled proteins lead to two signalling pathways: regulation of Lef-1 and c-Jun N-terminal kinase in mammalian cells. J Biol Chem 274:129–134

Mlodzik M (1999) Planar polarity in the *Drosophila* eye: a multifaceted view of signaling specificity and cross-talk. EMBO J 24 (in press)

Mlodzik M (2000) Spiny legs and prickled bodies: new insights and complexities in planar polarity establishment. Bioessays 22:311–315

Mueller H, Samanta R, Wieschaus E (1999) Wingless signaling in the *Drosophila* embryo: zygotic requirements and the role of the *frizzled* genes. Development 126:577–586

Paricio N, Feiguin F, Boutros M, Eaton S, Mlodzik M (1999) The *Drosophila* STE20-like kinase Misshapen is required downstream of the Frizzled receptor in planar polarity signaling. EMBO J 18:4669–4678

Reifegerste R, Moses K (1999) The genetics of epithelial polarity and pattern in the *Drosophila* retina. BioEssays 21:275–285

Strutt DI, Weber U, Mlodzik M (1997) The role of RhoA in tissue polarity and Frizzled signalling. Nature 387:292–295

Strutt H, Strutt DI (1999) Polarity determination in the *Drosophila* eye. Curr Opin Genet Dev 9:442–446

Taylor J, Abramova N, Charlton J, Adler PN (1998) Van Gogh: a new *Drosophila* tissue polarity gene. Genetics 150:199–210

Theisen H, Purcell J, Bennett M, Kansagara D, Syed A, Marsh JL (1994) *dishevelled* is required during *wingless* signalling to establish both cell polarity and cell identity. Development 120:347–360

Tomlinson A (1988) Cellular interactions in the developing *Drosophila* eye. Development 104: 183–193

Tomlinson A, Ready DF (1987) Cell fate in the *Drosophila* ommatidium. Dev Biol 123:264–275

Tomlinson A, Struhl G (1999) Decoding vectorial information from a gradient: sequential roles of the receptors Frizzled and Notch in establishing planar polarity in the *Drosophila* eye. Development 126:5725–5738

Tomlinson A, Strapps WR, Heemskerk J (1997) Linking Frizzled and Wnt signaling in *Drosophila* development. Development 124:4515–4521

Treisman JE, Heberlein U (1998) Eye development in *Drosophila*: formation of the eye field and control of differentiation. Curr Top Dev Biol 39:119–158

Usui T, Shima Y, Shimada Y, Hirano S, Burgess RW, Schwarz TL, Takeichi M, Uemura T (1999) Flamingo, a seven-pass transmembrane cadherin, regulates planar cell polarity under the control of Frizzled. Cell 98:585–595

Vinson CR, Adler PN (1987) Directional non-cell autonomy and the transmission of polarity information by the *frizzled* gene of *Drosophila*. Nature 329:549–551

Vinson CR, Conover S, Adler PN (1989) A *Drosophila* tissue polarity locus encodes a protein containing seven potential transmembrane domains. Nature 338:263–264

Wallingford JB, Rowning BA, Vogeli KM, Rothbacher U, Fraser SE, Harland RM (2000) Dishevelled controls cell polarity during *Xenopus* gastrulation. Nature 405:81–85

Wang W-J, Liu J, Adler PN (1994) The *frizzled* gene of *Drosophila* encodes a membrane protein with an odd number of transmembrane domains. Mech Dev 45:127–137

Weber U, Paricio N, Mlodzik M (2000) Jun mediates Frizzled induced R3/R4 cell fate distinction and planar polarity determination in the *Drosophila* eye. Development 127 (in press)

Wehrli M, Tomlinson A (1998) Independent regulation of anterior/posterior and equatorial/polar polarity in the *Drosophila* eye; evidence for the involvement of Wnt signaling in the equatorial/polar axis. Development 125:1421–1432

Wolff T, Ready DF (1991) The beginning of pattern formation in the *Drosophila* compound eye: the morphogenetic furrow and the second mitotic wave. Development 113:841–850

Wolff T, Ready DF (1993) Pattern formation in the *Drosophila* retina. In: Martinez-Arias MBA (ed) The development of *Drosophila melanogaster*. Cold Spring Harbor Press, Cold Spring Harbor, pp 1277–1326

Wolff T, Rubin GM (1998) *strabismus*, a novel gene that regulates tissue polarity and cell fate decisions in *Drosophila*. Development 125:1149–1159

Yanagawa S, van Leeuwen F, Wodarz A, Klingensmith J, Nusse R (1995) The Dishevelled protein is modified by Wingless signalling in *Drosophila.* Genes Dev 9:1087–1097

Zeidler MP, Perrimon N, Strutt DI (1999) The four-jointed gene is required in the *Drosophila* eye for ommatidial polarity specification. Curr Biol 9:1363–1372

Zheng L, Zhang J, Carthew RW (1995) *frizzled* regulates mirror-symmetric pattern formation in the *Drosophila* eye. Development 121:3045–3055

Regulation of Growth and Cell Proliferation During Eye Development

Thomas P. Neufeld[1] Iswar K. Hariharan[2]

1 Introduction

The adult eye is composed of 750–800 ommatidia each consisting of 20 cells. Thus there are approximately 16,000 viable cells in the adult eye. Approximately 2000 of the cells generated during eye development are eliminated by apoptosis (Wolff and Ready 1993). In addition, two of the four cells that are part of the bristle group have degenerated late in development (Perry 1968). Thus the total number of cells generated is close to 20,000. Assessments of the number of cells in the eye primordium in the embryo and newly hatched larva vary considerably (see below) but an estimate of 20 precursor cells is not unreasonable. Thus there is approximately a 1000-fold increase in the number of cells during eye development. Assuming that all cells keep proliferating, such an increase would require ten rounds of cell division.

The increase in cell number is also paralleled by a substantial increase in the size of the organ. This is because cell growth occurs throughout the larval and pupal stages of development. However, cell growth and cell division are only coupled loosely. Cell growth occurs prior to the onset of cell proliferation, during the proliferative phase and continues to occur after cells have stopped dividing.

All cells are not equal with respect to their growth and proliferative properties. The eye-imaginal disc has a size and shape that is different from other discs. Moreover, the patterns of cell division are subject to strict spatiotemporal regulation. This is most apparent in the disc of the third larval instar. Different cell types also differ in the extent of postmitotic growth. It is therefore obvious that patterning mechanisms must eventually influence the processes that regulate growth and cell cycle entry.

In this chapter, we primarily discuss the mechanisms that are known to regulate growth and cell number in the developing *Drosophila* eye. We first provide a brief description of the growth and proliferative properties of the cells that participate in eye development. We then discuss the mechanisms and signaling pathways that regulate the growth of imaginal discs in general. A

[1] Department of Genetics, Cell Biology and Development, University of Minnesota, Minneapolis, Minnesota 55455, USA

[2] Massachusetts General Hospital Cancer Center, Building 149, 13th Street, Charlestown, Massachusetts 02129, USA

Results and Problems in Cell Differentiation, Vol. 37
K. Moses (Ed.): Drosophila Eye Development

wide variety of experimental approaches have been utilized over the years to study imaginal discs, ranging from disc transplantation to flow cytometry. We summarize the conclusions from many of these studies. In addition to the mechanisms that regulate the growth of all discs, several modes of regulation appear to be somewhat unique to the eye-imaginal disc such as the cell cycle arrest in the morphogenetic furrow and the "second mitotic wave". Insights gained from these eye-specific modes of regulation will be presented. Finally, we review the initial links that have been established between patterning mechanisms and the regulators of growth and proliferation.

2 Growth and Cell Proliferation During Eye Development

The development of the *Drosophila* eye has been reviewed in a number of publications (cited in Wolff and Ready 1993). A particularly detailed description of the cellular events that occur during eye development is found in Wolff and Ready (1993). Here, we discuss those aspects of eye development that pertain directly to disc growth and cell proliferation.

The eye-antennal disc develops from a group of 6–23 cells of the blastoderm embryo (García-Bellido and Merriam 1969; Wieschaus and Gehring 1976). Almost no cell division is detected in these cells after 12 h of embryogenesis and they resume their divisions 13–15 h after hatching. Using histological methods, estimates of the number of cells in the eye-antennal disc of a freshly hatched larva range from 70 (Madhavan and Schneiderman 1977) to 95 (Newby and Thelander 1950). The eye portion of the disc has been estimated to consist of about 40 cells (Madhavan and Schneiderman 1977). These numbers are considerably larger than the estimate of 13 cells in 10 h embryos deduced from clonal analysis (Wieschaus and Gehring 1976). X-ray-induced mitotic recombination may have resulted in significant cell death. Alternatively, all the cells observed and directly counted in the disc may not give rise to progeny and might represent an overestimate of actual precursor cells. Thus the two approaches may provide complementary information.

The eye-antennal disc is morphologically recognizable at the end of embryogenesis and cells start dividing again 13–15 h after hatching. Their estimated division time during the second half of the first larval instar is 7 h (Madhavan and Schneiderman 1977). Cell numbers in the eye-antennal disc increase logarithmically and have a doubling time of approximately 6 h during the second instar (Chevais 1944; Becker 1957). The estimated number of cells increases from 130 at the end of the first larval instar to 1300–1600 at the beginning of the third instar (Becker 1957; Wolff and Ready 1993). Approximately three to four divisions may occur anterior to the morphogenetic furrow in the third larval instar. By analogy with the wing disc, the duration of the cell cycle may increase in the third instar eye disc. Future studies utilizing GFP-marked clones should arrive at more precise estimates of division times in the eye disc at various stages of development. It has also been noted that the proliferative capacity of all cells is not the same; an anteriorly placed twin-spot is often larger than its sibling clone that lies immediately posterior to it (Becker 1957; Wolff and Ready 1993).

Interestingly, growth and cell proliferation are only loosely coupled during these larval stages (Madhavan and Schneiderman 1977). In the period that immediately follows hatching, entry into the cell cycle is preceded by 12–14 h of cell growth. During this time, cell volume has increased almost sixfold and the volume of their nuclei has increased almost threefold. From this point, cell size decreases throughout larval development. At the end of the larval stage, disc cells are only twice as large as their ancestors in the newly hatched larva.

During the third larval instar, the morphogenetic furrow begins to sweep across the disc from its posterior edge. The number of cells along the morphogenetic furrow determines the number of ommatidia in the adult eye. Lateral inhibitory mechanisms define regularly spaced preclusters of cells in the morphogenetic furrow (Baker et al. 1990). The preclusters are the founders of the developing ommatidia. Thus, in a larger disc composed of more cells, more preclusters are determined along the furrow. Conversely, in a smaller disc composed of fewer cells, fewer preclusters are determined leading to a smaller adult eye composed of fewer ommatidia. Increased cell proliferation at later stages of development cannot lead to an increase in the number of ommatidia; it can only result in an excess of interommatidial cells that are in most instances eliminated by apoptosis.

The cells in the precluster that differentiate into photoreceptor cells R8, R2, R5, R3 and R4 do not divide again. All other cells enter a synchronous round of cell division after the furrow has passed that is referred to as the second mitotic wave (Ready et al. 1976; Wolff and Ready 1991a). Baker and Yu (2001) found an average of 9.7 cells entered the second mitotic wave per precluster. Thus, approximately a third of the cells have stopped dividing at the furrow and the remaining two thirds enter the second mitotic wave. The cells of the second mitotic wave undergo a highly synchronous S-phase that is visualized as a sharp band of BrdU incorporation immediately posterior to the morphogenetic furrow (Fig. 1A). Mitoses, as visualized by the anti-phospho H3 antibody, mostly occur three to five rows behind the morphogenetic furrow (Fig. 1B; Baker and Yu 2001) but some occur much later. Cells derived from the second mitotic wave differentiate into the remaining photoreceptor cells (R1, R6 and R7), the cone cells, the pigment cells and the cell that divides twice during the pupal stage to generate the four cells of the interommatidial bristle.

During the pupal stage of development, cell fate assignments are completed. Approximately 2,000 interommatidial cells are eliminated by apoptosis (Wolff and Ready 1991b, 1993). The precursor cell of the bristle group divides and its two progeny each divide again to generate the four cell unit that generates the bristle. The S-phases are initiated in the center of the eye approximately 14 h

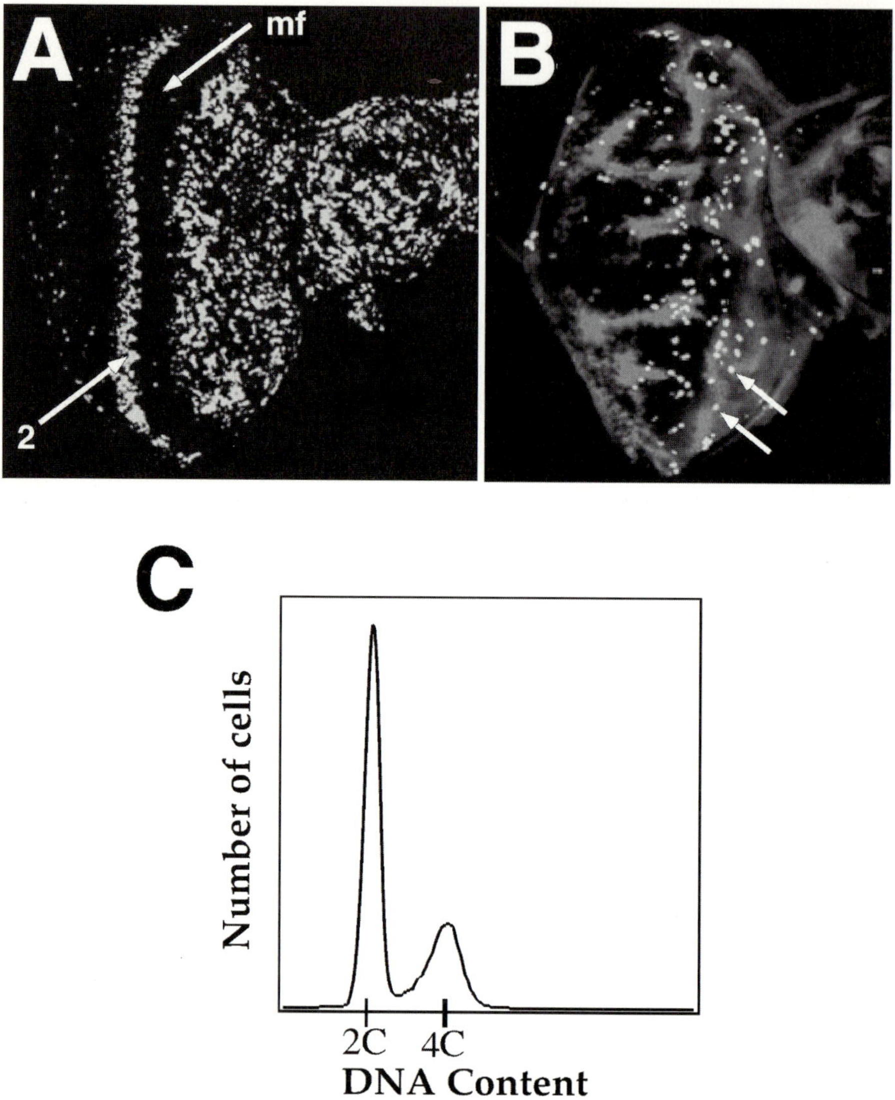

Fig. 1. Methods for visualizing cell proliferation in the eye-imaginal disc. **A** Incorporation of BrdU is used to visualize S-phases in a wild-type third instar eye-antennal disc. *MF* refers to the morphogenetic furrow. *2* refers to the "second mitotic wave". Anterior is to the *right*. **B** Mitoses are visualized using an anti-phospho H3 antibody in a third instar eye-antennal disc. Clones of mutant tissue can also be generated using *eyFLP* and an anti-β-galactosidase antibody that marks wild-type cells to visualize the effects of individual mutations on the patterns of mitoses. In this case a wild-type chromosome was flipped to illustrate the method. Examples of cells in mitosis are *arrowed*. For details see Tapon et al. (2001). **C** Flow cytometry can be used to determine the proportion of cells in different phases of the cell cycle. In this case, wild-type third instar eye-antennal discs were cut along the morphogenetic furrow, cells were dissociated from the posterior portion and analyzed for their DNA content. This profile demonstrates that most cells posterior to the morphogenetic furrow have a 2 C DNA content and are likely to be in G1

after pupariation and spread radially. Mitosis trails S-phases by about 6 h (Wolff and Ready 1993). Two of the four cells of the bristle group, the tormogen and trichogen, degenerate late in development (Perry 1968). The pupal stage is also characterized by dramatic changes in cell morphology as cells elaborate specialized structures such as the rhabdomeres. These changes are concomitant with significant postmitotic cell growth.

3 Mechanisms That Regulate Imaginal Disc Growth

Overall growth of the eye-imaginal disc is achieved through relatively coordinated increases in mass and cell number. In this section, we review work from a large number of groups over the past several decades that has helped to define the mechanisms that regulate the growth of imaginal discs. While many of the experiments described were conducted using wing or leg discs, the mechanisms they have defined pertain to the regulation of growth and proliferation in the eye-imaginal disc.

3.1 Disc Autonomous Mechanisms That Regulate Growth

Several lines of evidence support the idea that the regulation of size is largely intrinsic to the developing imaginal disc. When transplanted into female host abdomens, wing discs from mid-third instar larvae have been found to terminate growth at the proper size and cell number (Bryant and Simpson 1984; Bryant and Levinson 1985). These studies demonstrated that the mechanisms governing disc growth are largely disc-autonomous. Similar conclusions were drawn from experiments in which imaginal disc regeneration, caused by clonal induction of a cell-lethal mutation, was found to delay pupariation for up to several days (see below). During the extended larval period, the intact, non-regenerating discs stopped growing when they reached their normal mature size, despite the clear presence of hormonal conditions supportive of growth in these animals (Simpson et al. 1980).

This local autonomy also apparently extends to the compartment level; growth within a given compartment is for the most part unaffected by significant changes in growth or division rates in cells of the adjacent compartment (Simpson 1976; Weigmann et al. 1997; Neufeld et al. 1998). Such autonomy can be readily observed in wing discs comprised of a wild-type posterior compartment and a Minute anterior compartment. Despite the discrepancy between anterior and posterior compartment growth rates, such discs give rise to adult wings of normal size and proportion (Morata and Ripoll 1975). In such discs, the posterior compartment reaches its normal mature size well before the anterior; it then stops growing and waits while the anterior compartment catches up (T. P. Neufeld, unpubl. observ.). Together, these experiments indicate that the information dictating final disc size is contained within each disc, and that the presence of nutrients and growth factors capable of supporting disc growth is not sufficient to overcome this limit.

3.2 Non-Autonomous Control of Disc Growth

Despite this level of autonomous regulation, growth of the eye and other adult structures is also controlled by factors external to the disc. Hormones, growth factors, and nutrients each have a profound influence on disc growth. For example, larvae cultured on media of poor nutritional quality develop slowly and pupariate at an abnormally small size, resulting in small adults. Interestingly, larvae respond to starvation in one of two ways, depending on whether they have reached a critical point in the mid-third instar stage. Starvation prior to this point results in a reversible, long-term growth arrest, whereas larvae that have developed beyond this point respond to starvation by pupating, often at a smaller than normal size (Simpson 1979). The molecular basis of this switch is unknown, but its existence indicates an intricate interplay between nutrition and hormonal signaling. Moreover, ecdysone appears to be required at low levels for imaginal disc cell growth and proliferation, in addition to its better known role as a molting hormone (Postlethwait and Schneiderman 1970).

Several studies have identified the larval fat body as the source of a secreted factor(s) that supports imaginal disc growth. Synthetic media conditioned by fat body is capable of allowing disc growth in vitro (Davis and Shearn 1977); similarly, in vitro proliferation of quiescent larval neuroblasts can be induced by co-culturing with fat body (Britton and Edgar 1998). Bryant and coworkers have recently identified a family of chitinase-related polypeptides, the imaginal disc growth factors (IDGFs), which are secreted from the fat body and can stimulate growth of imaginal disc cell cultures (Kawamura et al. 1999). In addition, Shearn and colleagues (Martin et al. 2000) have shown that *minidiscs* encodes an amino acid transporter expressed in the fat body, where it is required for imaginal disc growth. Together, these observations, along with the striking changes in fat body morphology induced by starvation, support a model in which the fat body serves as an intermediate between nutrition and imaginal disc growth.

Another level of non-autonomous control of imaginal disc growth involves interactions between the discs themselves. Working with butterfly and beetle larvae in 1998, Nijhout and Emlen found that reducing or eliminating one or more developing imaginal tissues resulted in increased growth of neighboring structures. For example, surgical removal of butterfly hind wing discs caused enlargement of adjacent forewings, and artificial selection of beetles for relatively small horn size led to an increase in the size of the eye. These compensatory changes in growth were found only in neighboring structures, and suggest that developing discs compete with each other for limited nutrients or growth factors.

3.3 Signaling Pathways Regulating Imaginal Disc Growth

Tissue and organ growth ultimately occurs at the cellular level. Therefore, a large effort has gone into identifying and characterizing genes that regulate cell growth in response to external stimuli. In this section, we briefly summarize recent advances in our understanding of these genes. Although some of the studies described have focused in greatest detail on other tissues, particularly the wing imaginal disc, many of the generalities are likely to apply to the eye, especially during the period of apparently unpatterned growth prior to furrow progression.

3.3.1 PI-3 Kinase Pathway

Studies in mammalian cell culture have found that inositol lipids phosphorylated at the D3 position of the inositol ring act to recruit to the cell membrane a number of proteins containing pleckstrin homology domains (Leevers et al. 1999). Membrane localization of such proteins, which include the serine-threonine kinases Akt and Pdk1, participates in their activation and thus initiates signaling along a pathway which ultimately leads to increased cell growth, in part through effects on translation (Fig. 2).

Mutations in a number of genes encoding components of this pathway have been identified in *Drosophila*, and epistasis analyses have largely confirmed the relationships amongst these proteins initially determined in mammalian cells (Chen et al. 1996; Bohni et al. 1999; Goberdhan et al. 1999; Huang et al. 1999; Montagne et al. 1999; Verdu et al. 1999; Weinkove et al. 1999; Gao et al. 2000; Oldham et al. 2000; Zhang et al. 2000; Brogiolo et al. 2001). In addition, analysis of this pathway in *Drosophila* has led to an appreciation of its role in regulating cell, organ and body size. For example, cell size can be increased by overexpression of *PI3K*, *dAkt*, or *dS6K*, or by loss of function mutations in *dPTEN*, a negative regulator of this pathway. Conversely, inactivation of signaling, via mutations in *Inr*, *chico*, *PI3K*, *dAkt*, *dTOR*, or *dS6K*, or overexpression of *dPTEN*, causes a cell autonomous reduction in size of a variety of cell types, including adult retinal, wing epithelial, and bristle cells, as well as imaginal and endoreplicating cells of the larva. In addition to these effects on cell size, several experiments have demonstrated that this pathway also regulates the rate and extent of overall growth of cells, tissues, organs, and of the organism as a whole. First, clones of cells which overexpress *PI3K* or *dAkt*, or which lack *dPTEN* function, grow at a faster rate and to a larger size than wild-type clones. Second, overexpression of *PI3K* or *dAkt* throughout the posterior compartment causes increased growth relative to the anterior compartment. Third, overexpression of *dPTEN* in eye or wing discs causes a reduction in the size of the resulting adult structures, whereas eye-specific inactivation of *dPTEN* results in larger than normal eyes. Finally, viable alleles of *Inr*, *chico*, *dAkt* and *dS6K* produce flies of diminished size.

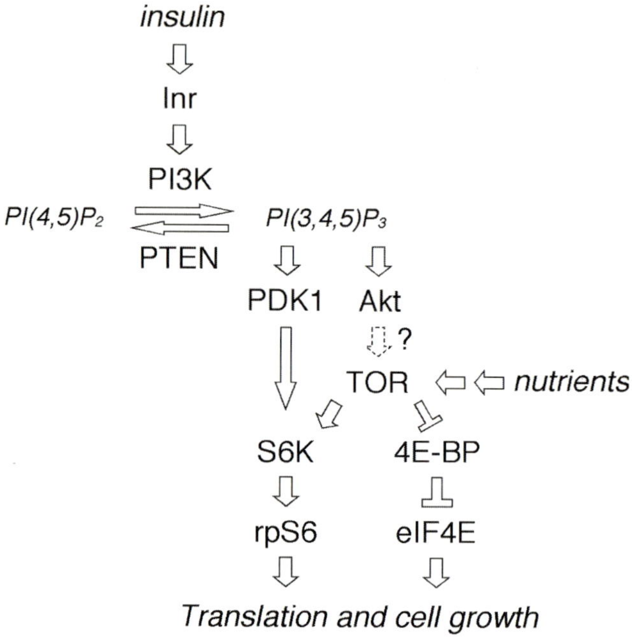

Fig. 2. PI3K signaling pathway. Activation of PI3K by the insulin receptor (*Inr*) and other receptor tyrosine kinases causes an increase in phosphoinositol (3,4,5) phosphate levels; the lipid phosphatase activity of PTEN acts to decrease these levels. Increased PI (3,4,5) phosphate levels causes membrane recruitment and activation of proteins containing pleckstrin homology domains, such as the Akt and PDK1 kinases. Through a TOR-dependent signaling event, Akt and PDK1 are thought to stimulate cellular biosynthesis in two ways: activation of p70 S6 kinase (S6K) allows translation of ribosomal proteins and other translation components, while the activity of eIF4E (cap binding protein) is critical for translation of mRNAs with highly structured 5′ UTRs

Many of the phenotypes caused by mutations in this pathway are similar to those of flies that were deprived of nutrients during larval development: cell and body size are reduced, development is delayed, and patterning is normal. These apparent similarities have led to the proposal that the PI3K pathway may act to modulate rates of growth in accordance with nutrient conditions (Edgar 1999). Such a model would be in keeping with the exquisite sensitivity of PI3K activity to insulin, which in mammals acts as an intermediate between serum nutrient levels and cell growth. In addition, TOR proteins (mTOR/FRAP in mammals, dTOR in flies) are implicated in sensing or responding to nutrients (reviewed in Gingras et al. 2001). Mutations in *dTOR* have been shown mimic nutrient deprivation in a number of ways beyond the overt growth defects: like starvation, loss of *dTOR* causes a similar cell type-specific cell cycle arrest, a reduction in nucleolar size, and morphological changes in the fat body (Oldham et al. 2000; Zhang et al. 2000). Thus, this pathway may have evolved to ensure that cellular growth rates are appropriate for a given diet.

3.3.2 Myc

Members of the proto-oncogene network of Myc/Max/Mad transcription factors are frequently deregulated in tumors, and have been implicated in controlling cell proliferation, differentiation, and death (Amati et al. 1998). Recent work, including identification of a number of Myc's transcriptional targets, has suggested that one of Myc's primary functions is to regulate cellular growth (Coller et al. 2000). In flies, a Myc homologue is encoded by the *diminutive (dm)* locus (Gallant et al. 1996). *dm* mutants share many phenotypes with mutants of the PI3K pathway: loss of *dm* function causes a similar delay in development and reduction in cell and body size, whereas overexpression increases cellular growth and has similar effects on cell cycle phasing (Johnston et al. 1999; see below). Interestingly, despite a potent ability to stimulate cellular growth, overexpression of dMyc was found to be unable to increase compartment size, in apparent contrast to the effects of PI3K. *dm* mutants also differ phenotypically from *Inr*, *chico* and *S6K* mutants in having a disproportionate effect on bristle size, similar to that seen in *Minute* mutants. In flies, the sole Myc target identified thus far is *pitchoune*, which encodes a DEAD box RNA helicase involved in ribosomal RNA processing (Zaffran et al. 1998). Together, these results suggest that Myc may promote growth in part by regulating ribosome biogenesis.

3.3.3 Ras/MAPK

Activation of the Ras pathway has long been associated with cell proliferation and survival in *Drosophila* (Nishida et al. 1988; Clifford and Schupbach 1989). Although important for proliferation, Ras signaling is not essential, since clones of cells lacking Ras can proliferate if given a growth advantage in a Minute background (Diaz-Benjumea and Hafen 1994; Halfar et al. 2001). Such cells that survive to adulthood are usually severely reduced in size and are often misspecified. Expression of activated *Ras1* in the eye using *ey-GAL4* leads to dramatic overgrowth/hyperplasia, as well as extensive cell death (Karim and Rubin 1998). Furthermore, clonal activation of *Ras1* in imaginal discs has been found to cause increased growth rates, cell enlargement, and promotion of S phase entry (Prober and Edgar 2000). These effects could be accounted for in part by post-transcriptional increases in the abundance of dMyc and Cyclin E protein. *Ras1* activation also appears to cause changes in cell affinity, as clones of cells expressing activated *Ras1* "round up", minimizing contact with wild type cells.

The striking effects on cell and organ growth in response to PI3K, Ras or Myc signaling have provided a framework to address the issue of how the cell division cycle responds to changes in growth rate (Fig. 3). Three generalities have emerged from these lines of investigation. First, the increase in growth following activation of these pathways does not lead to increased disc cell pro-

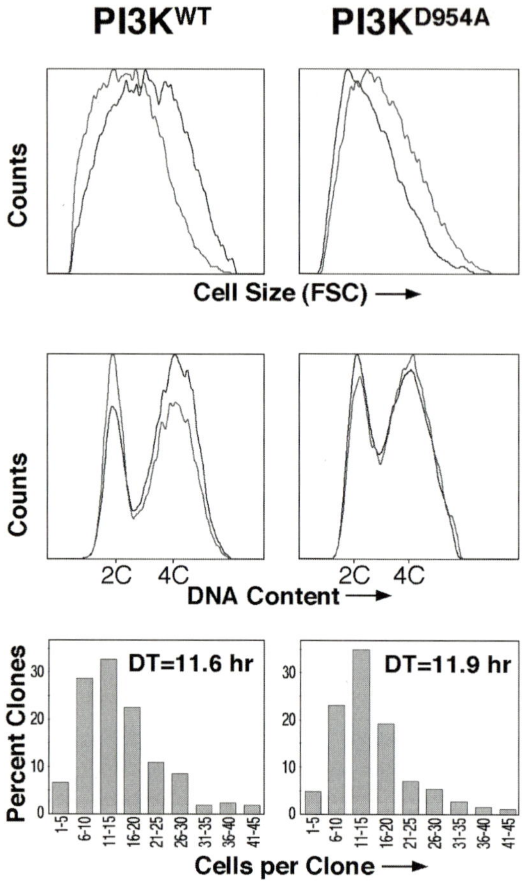

Fig. 3. FACS analysis and cell division rates of PI3K-expressing cells. PI3K regulates cell size (*top panels*) and cell cycle phasing (*middle panels*). Expression of wild type or dominant negative (D945A) PI3K was induced clonally at 72 h AED; 48 h later, wing discs were dissected, dissociated into single cells, and analyzed by flow cytometry. PI3K-expressing cells are indicated by *black traces*; wild-type cells by *gray*. Expression of PI3K increases cell size and reduces the proportion of cells in G1 (2 C DNA content), whereas inhibiting PI3K activity reduces cell size and slightly increases the G1 population. The bottom panel displays the results of doubling-time (*DT*) calculations for $PI3K^{WT}$ and $PI3K^{D954A}$ cells. Clones were induced at 72 h AED, and the number of cells per clone was counted 43 h later. No differences in average cell doubling time were observed between $PI3K^{WT}$, $PI3K^{D954A}$, and wild-type control clones. For details, see Weinkove et al. (1999)

liferation. Careful measurements of clones in which these pathways have been activated have revealed no difference in cell number from controls. Second, despite this lack of an observable increase in cell proliferation rates, FACS analysis of disc cells following activation of these pathways generally reveals a relative increase in the proportion of cells in the S and G2 phases of the cell cycle. This suggests that stimulation of cell growth can promote S phase entry. Presumably, the reason this does not lead to increased proliferation is that the G2/M transition is controlled independently of growth (see below). Third, whereas activation of PI3K, Ras or Myc signaling can promote cell growth but not cell proliferation, their inactivation generally decreases the rate of both cell growth and cell proliferation. The diminished size of such cells indicates that cell growth rates are reduced to a greater degree than rates of cell proliferation, suggesting that the observed proliferation defects are secondary to the reduced rates of growth. Note that elimination of slow-growing cells by competition would tend to increase the apparent reduction in cell doubling rates, thus causing an overestimation of the effects of these mutations on cell proliferation.

Several exceptions to these rules have been observed. For example, in contrast to PI3K activation, loss of *dPTEN* has been found to cause significant increases in both growth and proliferation, suggesting dPTEN may have PI phosphatase-independent functions (Gao et al. 2000). Furthermore, whereas cell clones overexpressing PI3K or Ras proliferate at normal rates, overexpression of these genes more broadly throughout the disc can apparently lead to increases in cell number (Karim and Rubin 1998; Leevers et al. 1999). Finally, dS6K mutants are reduced in size but have a normal number of cells, violating the rule that decreased growth rates result in reduced proliferation. However, these mutants also suffer a severe developmental delay, indicating a decrease in the rate, if not the extent, of cell proliferation.

3.3.4 Cyclin D/cdk4

Extensive studies in mammalian cell culture have identified Cyclin D and its associated kinase, cdk4, as key cell cycle regulators, which together promote entry into S phase by phosphorylating the retinoblastoma protein, (pRB; Sherr 1995). A major tenet of this model is that cdk4 activity increases in response to serum growth factors, and thus connects cell cycle progression to cellular growth rates. Recent experiments in *Drosophila* have supported an alternative view that Cyclin D and cdk4 act to promote cell growth, rather than to sense and respond to it (Datar et al. 2000; Meyer et al. 2000). Unlike the G1/S regulator Cyclin E, whose overexpression leads to induction of S phase without affecting growth (Neufeld et al. 1998), overexpression of Cyclin D and cdk4 in eye and wing imaginal disc cells was found to cause a coordinated increase in cell growth and proliferation rates. Interestingly, all phases of the cell cycle were equally accelerated (Datar et al. 2000). Expression of Cyclin D and cdk4 also stimulated the growth of post-mitotic cells. These cellular effects are clearly distinct from the results of PI3K, Myc or Ras activation, in which G1/S progression is accelerated but cell doubling rates are not. Furthermore, loss of cdk4 leads to a ~20% reduction in adult body size, but unlike these other pathways, has no effect on developmental timing or cell size (Meyer et al. 2000).

These exciting results suggest that Cyclin D and cdk4 regulate growth and proliferation in novel ways.

3.3.5 Tsc1 and Tsc2

The inherited human disease tuberous sclerosis, characterized by benign tumors, is caused by mutations in either the *TSC1* or *TSC2* genes. The *TSC1* gene encodes a protein with a hydrophobic stretch and two coiled-coil domains. The TSC2 protein has a small region which displays sequence similarity to the GTPase activating protein for Rap1. TSC1 and TSC2 are found in a complex and their precise function is unknown.

Loss of function mutations in either *Tsc1* or *Tsc2* (also known as *gigas*) lead to discs of increased size characterized by large cells (Ferrus and Garcia-Bellido 1976; Ito and Rubin 1999) of normal ploidy (Gao and Pan 2001; Potter et al. 2001; Tapon et al. 2001). The growth rate of cells (rate of mass accumulation) is increased and their doubling times are only marginally shorter than wild-type cells. Mutant cells appear to have a shorter G1 phase of the cell cycle. Conversely, combined overexpression of both *Tsc1* and *Tsc2* together results in reduced growth and a slowing down of the cell cycle. It is unclear how exactly these genes function in the context of known signaling pathways. However, the mutant phenotypes are influenced by changes in the levels of cyclins as well as components of the PI3K pathway.

3.3.6 *Drosophila* "Tumor-Suppressor" Genes

A number of genes have been identified over the years which, when mutated, results in overgrowth of imaginal discs (reviewed by Gateff 1994; Watson et al. 1994). The structures generated by these mutant discs are usually not patterned normally. Such genes include *warts/LATS* (Justice et al. 1995; Xu et al. 1995), *discs large* (Woods and Bryant 1991), *scribble* (Bilder et al. 2000) and many others (Gateff 1994; Watson et al. 1994). We will learn much more over the coming years about the precise mechanism by which these proteins influence cell growth and cell cycle progression.

3.3.7 Non-Autonomous Regulators of Growth

Mutations in the *Drosophila* homologue of the neurofibromatosis type 1 (NF1) gene, result in reduced growth resulting in flies that have smaller wings and eyes (The et al. 1997). Clonal analysis in cells of the wing disc indicates that the requirement for NF1 is non-cell autonomous. This raises the possibility that NF1 may function in a neuroendocrine pathway that regulates growth.

Nitric oxide (NO) is a small molecule that can diffuse freely between adjacent cells. Cells of developing imaginal discs express NO synthase. Inhibition of NO synthase leads to increased cell proliferation and an increase in the size of imaginal discs and an overgrowth of the adult organs that develop from those discs (Kuzin et al. 1996). NO might function to co-ordinate the proliferation of clusters of adjacent cells as has been observed in the developing wing disc.

4 Mechanisms That Regulate Growth and Cell Proliferation in the Eye-Imaginal Disc

In this section, we review mechanisms that have been implicated in the regulation of growth and cell proliferation in the eye-imaginal disc.

4.1 Cell Proliferation in First and Second Larval Instar Discs

We still know very little about what drives cell proliferation in the eye-imaginal disc. Both Dpp and Hh are likely to have a role in promoting cell proliferation in the early stages of eye development. Mutations that either reduce Dpp expression or that reduce Dpp-mediated signaling result in reduced cell proliferation and smaller eyes (Masucci et al. 1990; St Johnston et al. 1990; Blackman et al. 1991; Heberlein et al. 1993; Burke and Basler 1996). Expression of Dpp in the early stages of eye development requires Hh (Royet and Finkelstein 1997). Clones of the EGF receptor (*egfr*), or components of the receptor tyrosine kinase (RTK) signaling pathway also show greatly reduced cell proliferation, indicating that this pathway functions in promoting cell proliferation in the eye disc (Baker and Rubin 1989; Xu and Rubin 1993; Halfar et al. 2001).

Notch has been shown to be an important regulator of the extent of cell proliferation in eye discs (Cho and Choi 1998; Dominguez and de Celis 1998; Papayannopoulos et al. 1998). Overexpression of activated forms of the Notch receptor or its ligand, Delta, using an *eyeless-GAL4* driver line leads to an eye of increased size containing increased numbers of ommatidia. This is a reflection of increased cell numbers in the eye imaginal disc at the time of precluster and R8 specification. In larger discs containing increased numbers of cells, more ommatidia are specified. Notch activation, as measured by the activity of the *E(spl)mβ-CD2* reporter gene is visualized in the second instar disc along the dorsoventral midline. Curiously, Notch is activated along the "equatorial groove" that was originally described by Ready, Hanson and Benzer (Ready et al. 1976). The activation of Notch along the equatorial groove occurs because of a juxtaposition of cells that express the glycosyltransferase Fringe with those that do not. Expression of the Iroquois genes (*mirror*, *araucan* and *caupolican*) occurs in the dorsal half of the disc. These genes repress *fringe* expression.

Ventral cells express *fringe*. As in the wing disc, Notch activation occurs at the boundary that separates these two groups of cells. We currently do not understand how Notch activation along the dorsoventral midline promotes cell growth and proliferation uniformly throughout the eye-imaginal disc. However, given that increasing Notch activity can make the eye much bigger than normal, it is conceivable that the level of Notch activation is an important determinant of the size of the imaginal disc and hence of the adult eye.

As expected, a reduction in the function of genes that are required for normal growth and proliferation reduces the size of the imaginal disc and hence the size of the adult eye. For instance, a reduction of Cyclin E levels in the eye disc leads to a detectable decrease in the amount of BrdU incorporation observed and consequently to an adult eye of diminished size (Secombe et al. 1998).

4.2 Cell Proliferation in the Anterior Domain of the Third-Instar Discs

Cell proliferation in the anterior portion of the eye-imaginal disc is still poorly characterized. It is generally assumed that these cells are proliferating asynchronously and that their proliferation is not patterned in any way. This notion is based on observations of BrdU incorporation and expression of Cyclin E, Cyclin A and Cyclin B which appear to be randomly distributed (Horsfield et al. 1998). However, these cell divisions have not been examined as carefully as those in the more easily observable wing disc where clusters of cells appear to undergo S-phase together and other groups of cells are synchronized with respect to their mitotic divisions.

4.3 Synchronization of Cells in the Morphogenetic Furrow

While cells are cycling in an apparently random fashion anterior to the morphogenetic furrow, they become arrested in the furrow itself and are thought to be arrested in G1. In wild-type imaginal discs, no BrdU incorporation is observed in the furrow and an increased number of cells is observed to be undergoing mitosis immediately anterior to the furrow (Thomas et al. 1994). Also consistent with cells in the furrow being in G1 is the observation that they do not express Cyclin B, a marker for cells that have passed G1/S and have not yet divided (Baker and Yu 2001). High levels of RNA for the *string* gene are observed immediately anterior to the furrow (Thomas et al. 1994; Heberlein et al. 1995). Since *string* encodes a phosphatase that is capable of dephosphorylating and activating the cdc2 protein kinase, it has been suggested that high levels of String drive cells through mitosis and this leads to their accumulation in G1 (Thomas et al. 1994). This hypothesis has been strengthened by more recent studies that show that string can shorten the G2 phase of imaginal disc cell cycles (Neufeld et al. 1998). However, although *string* RNA levels are high, elevated levels of String protein are only observed in a small subset of mitotic cells anterior to the morphogenetic furrow (Horsfield et al. 1998).

Two pathways have been directly implicated in maintaining the G1 arrest in the morphogenetic furrow. The first pathway involves the *roughex* (*rux*) gene (Thomas et al. 1994). Mutations in *rux* abolish G1 arrest in the furrow completely. *rux* encodes a protein that has been shown to bind to cyclin A, facilitate its nuclear translocation and degradation (Sprenger et al. 1997; Avedisov et al. 2000). Cyclin E/cdk2 activity is capable of phosphorylating and inactivating Rux (Thomas et al. 1997). Thus the absence of CyclinE/cdk2 activity together with the inhibition of Cyclin A activity due to *rux* function is thought to block S-phase entry or progression in the furrow. Posterior to the furrow, elevated Cyclin E levels in subsets of cells could inactivate Rux and allow their entry into the S-phase of the second mitotic wave.

A second pathway required for arrest in the furrow is signaling mediated via the Dpp receptor (Horsfield et al. 1998). Clones of cells mutant for the Dpp receptor fail to arrest in the anterior half of the morphogenetic furrow and continue to proliferate asynchronously. However, these cells do arrest in the posterior half of the furrow and are also delayed with respect to their entry into the S-phase of the second mitotic wave. The expression of the cdk inhibitor Dacapo is not altered in Dpp receptor mutant clones indicating that this mechanism of arrest is not analogous to the induction of the cdk inhibitor p27 in response to TGFβ in mammalian cells. Moreover, a reduction in BrdU incorporation caused by overexpression of Dpp is still observed in *rux* mutants indicating that Dpp is likely to function independently of *rux*.

It has been suggested that the synchronous cell cycle arrest in the morphogenetic furrow facilitates cell fate assignments via intercellular communication. Indeed the adult eyes of *rux* mutant flies show disorganized ommatidia with greatly reduced numbers of R cells (Thomas et al. 1994). Certain other explanations are also tenable. The morphogenetic furrow can also be thought of as a moving A/P compartment boundary in the eye-antennal disc. Cell cycle arrest is also observed in certain other compartment boundaries such as the zone of non-proliferating cells found in the dorso-ventral boundary of the wing disc (O'Brochta and Bryant 1985). Mitotically quiescent cells may be necessary to perform some function of a compartment boundary that we still do not understand.

4.4 Regulation of The Second Mitotic Wave

Cells excluded from the precluster enter a synchronous cell cycle referred to as the second mitotic wave (Ready et al. 1976; Wolff and Ready 1991a). The S-phase of the second mitotic wave occurs immediately posterior to the morphogenetic furrow and is visualized as a tight band of BrdU incorporation across the disc. While cells are highly synchronized during S-phase, the remainder of the cycle is less synchronous. Cells undergoing mitoses are mostly seen in rows three to five posterior to the morphogenetic furrow. However, some cells divide much later and a small number appear to remain arrested in G2, as evidenced by elevated levels of cyclin B and may never go through a mitotic division (Wolff and Ready 1991a; Baker and Yu 2001). Cells generated by the second mitotic wave are recruited to become photoreceptor cells R1, R6 and R7, the cone cells, the pigment cells and the precursor cells for the groups of cells that produce the interommatidial bristle. However, none of these fates are contingent upon the extra division. Blocking the second mitotic wave by expression of the human cdk inhibitor p21 in cells posterior to the morphogenetic furrow reduces the size of the precursor cell pool but does not prevent the specification of any specific cell fates (de Nooij and Hariharan 1995).

At present, we know very little about the signals that trigger the second mitotic wave. Cells that have already been recruited to the five-cell clusters are blocked from entering the second mitotic wave. When *egfr* function is disrupted, all cells other than R8 re-enter the cell cycle (Baker and Yu 2001). Thus, an *egfr* mediated pathway must function in blocking cell cycle progression in these cells. This effect is not dependent on the recruitment of these cells to the photoreceptor cell fate by *spitz*-mediated signaling. In *spitz* mutant clones, only R8 cells are assigned a photoreceptor fate, but the remaining cells of the precluster do not re-enter the cell cycle. Thus cell cycle exit of the precluster cells is contingent upon a *spitz*-independent function of *egfr*. Two other genes have been implicated as positive regulators of the second mitotic wave. In imaginal discs homozygous for the *suppressor of roughex 2B* mutations, the stripe of BrdU incorporation characteristic of the second mitotic wave is absent (Dong et al. 1997). In clones of cells mutant for *daughterless (da)*, no mitotic divisions are observed posterior to the furrow (Brown et al. 1996). Since the absence of S-phases have not been demonstrated by analyzing BrdU incorporation, we can only conclude that *da* function is required at some stage prior to mitosis.

G2-M progression in the second mitotic wave is activated by a signal from differentiating preclusters (Baker and Yu 2001). This signal appears to act via *egfr*. Overexpression of *string* can bypass this requirement for *egfr*. A likely explanation therefore, is that *egfr* activation leads to the accumulation of String protein that facilitates the G2-M transition by dephosphorylating and activating the cdc2 kinase.

Why is there a second mitotic wave? One possibility is that an evolutionary ancient compound eye was composed of ommatidia that had fewer photoreceptors and accessory cells. In that situation, the divisions in front of the furrow may have been sufficient to generate enough cells. As compound eyes became more sophisticated, more photoreceptor and accessory cells were added per ommatidium. Increasing the number of cell divisions anterior to the furrow could no longer generate these additional cells, since increasing the number of the cells ahead of the furrow would lead to the specification of more preclusters and hence the whole exercise would be self-defeating. Thus an additional wave of division was added *after* the preclusters were determined, so as to increase the number of cells per ommatidium. This speculative model would be vindicated if one were able to find some unusual creature with ommatidia consisting of fewer cells and an eye disc where cell divisions occurred only anterior to the morphogenetic furrow.

4.5 Exit from the Cell Cycle

Posterior to the second mitotic wave, most cells have completed mitosis and are arrested in G1. Evidence that they are in G1 comes from the observation that most cells that have gone through the S-phase of the second mitotic wave also appear to complete mitosis in the larval eye disc. More recently, experiments using flow cytometry, have confirmed that most cells posterior to the morphogenetic furrow have a 2 C DNA content (Fig. 1C).

The mechanisms that regulate exit from the cell cycle in the eye are likely to be similar to those that have been shown to operate in the embryonic epidermis and peripheral nervous system. In those situations, a precise and timely exit from the cell cycle is achieved by the down-regulation of the activity of CyclinE-cdk2 activity together with increased expression of the cyclin-dependent kinase inhibitor Dacapo (Knoblich et al. 1994; de Nooij et al. 1996; Lane et al. 1996). In the eye imaginal disc, the expression patterns of Cyclin E and *dacapo* are consistent with their likely functions (Richardson et al. 1995; de Nooij et al. 1996, 2000). The E2F1 transcription factor, together with its dimerization partner DP1 appears necessary for the expression of several genes that are required for S-phase including PCNA and *ribonucleotide reductase 2.* Overexpression of E2F posterior to the furrow induces the entry of quiescent cells into S-phase (Du et al. 1996b). *Rbf* functions as a repressor of E2F-inducible genes and is likely to function in maintaining G1 arrest. Mutations in *Rbf* lead to a breakdown of G1 arrest in the embryo (Du and Dyson 1999). Other genes that function in maintaining post-mitotic arrest include the tuberous sclerosis homologues *Tsc1* and *Tsc2* (Gao and Pan 2001; Potter et al. 2001; Tapon et al. 2001).

Interestingly, arrest in the G1 phase of the cell cycle does not appear to be necessary for the specification of most cell fates. Overexpression of either the Roughex protein (Thomas et al. 1997) or the retinoblastoma homologue (Rbf; Du et al. 1996a) in cells posterior to the morphogenetic furrow blocks the second mitotic wave either during or after S-phase (i.e., in G2). However, in both situations, most cell fates can be assigned indicating that arrest in G1 is unnecessary. These experiments do not exclude the possibility that arrest at some stage in the cell cycle is indeed necessary; actively cycling cells may not be receptive to signals that specify fate.

4.6 Post-Mitotic Growth of the Eye

The final size of the adult eye is determined in significant part by the extent of post-mitotic growth of the cells of the imaginal disc. This occurs during the pupal phase of eye development. Several pathways have been shown to regulate cell size in the adult eye. These include PI3K-mediated signaling, the activity of the Cyclin D/cdk4 kinase and the homologues of the Tuberous sclerosis complex (Tsc1 and 2). Each of these pathways appears to regulate growth in cycling cells as well as in post-mitotic cells.

Altering the activity of components of the insulin receptor-mediated signaling pathway in post-mitotic cells increases their size. Using *sev-GAL4* which is expressed exclusively in post-mitotic cells or *GMR-GAL4* which is expressed mostly in post-mitotic cells, the expression of PI3-kinase (Leevers et al. 1996) and Akt (Verdu et al. 1999) have been shown to result in an increase in the size of retinal cells. Consistent with a role for this pathway in promoting growth, inactivating mutations in PTEN, the lipid phosphatase that antagonizes PI3-kinase function, also increases cell size (Goberdhan et al. 1999; Huang et al. 1999; Gao et al. 2000). As discussed earlier, the complex of Cyclin D and its effector kinase cdk4 promote post-mitotic growth (Datar et al. 2000). In contrast, a complex that includes the Tsc1 and Tsc2 proteins restricts post-mitotic growth (Gao and Pan 2001; Potter et al. 2001; Tapon et al. 2001).

Our understanding of the mechanisms that regulate post-mitotic growth is in its infancy. Screens being conducted in a number of laboratories are likely to identify key regulators.

5 Connections Between Disc Patterning and Growth

The characteristic size and shape of the eye disc, as well as that of other imaginal discs and their resulting adult structures, indicates that disc growth must be regulated in a patterned fashion. The rate and/or duration of growth and cell proliferation must vary amongst different regions of the disc. The gene networks involved in establishing the anterior/posterior (A/P) and dorsal/ventral (D/V) axes of the imaginal discs are strongly implicated in such regulation.

5.1 Lessons from the Wing Disc

The links between patterning and growth have been most thoroughly investigated in the wing imaginal disc (Blair 1995). Outgrowth of the wing is directed by organizing centers established along the A/P and D/V compartment boundaries. Expression of the transcription factors Engrailed (En) and Apterous (Ap) is limited to the posterior and dorsal compartments, respectively, where they act as selector genes to regulate compartmentalization and wing morphogenesis. En-expressing cells express Hedgehog (Hh), which induces Dpp expression in a narrow band of adjacent anterior cells. Dpp then acts as a morphogen across the A/P axis of the wing disc, regulating expression of target genes such as *optomotor-blind*, *spält*, and *brinker* in a dosage-sensitive fashion. In addition to patterning the disc, Dpp also promotes disc growth, as evidenced by the following observations: (1) loss of *dpp* in the wing disc leads to a severe reduction in wing size (Spencer et al. 1982); (2) clones of cells mutant for the Dpp receptor Tkv are out-competed by surrounding wild-type cells, whereas clones expressing an activated version of Tkv overproliferate (Burke and Basler 1996; Nellen et al. 1996); (3) ectopic expression of Dpp can cause pattern duplications involving a large increase in wing size (Zecca et al. 1995); (4) increased levels of Dpp in its normal expression domain leads to disc overgrowth, primarily in the A/P direction (Nellen et al. 1996). A similar mechanism is at work along the D/V axis. Ap induces expression of Fringe (Fng) and Serrate (Ser) in dorsal cells, which leads to activation of Notch in cells along the D/V compartment boundary (Fleming et al. 1997; Panin et al. 1997). Notch signaling in these cells results in expression of Wingless (Wg), which, like Dpp, serves to pattern and promote growth of the wing (Diaz-Benjumea and Hafen 1994; de Celis et al. 1996). Thus, regulation of growth and patterning by a common network of genes allows coordination of these processes during wing development, resulting in discs and adult appendages of proper size and shape.

5.2 Patterned Growth in the Eye Disc; Notch and Morphogen Gradients

Is growth of the eye controlled by a similar compartmental mechanism? As discussed earlier, Notch activation along the dorsal/ventral midline of the eye imaginal disc has been found to be crucial for both growth and patterning of the developing eye. A group of homeodomain proteins, Mirror, Aruacan, and Caupolican, are expressed in the dorsal half of the eye, where they act to restrict expression of Fng to the ventral region (Cho and Choi 1998; Dominguez and de Celis 1998; Papayannopoulos et al. 1998). As in the wing, borders between Fng-expressing and non-expressing cells are crucial for Notch activation and subsequent patterning and outgrowth, as either loss of Fng or its uniform over-expression causes a dramatic reduction in eye size, which can be rescued by expression of an activated form of Notch (Cho and Choi 1998; Dominguez and de Celis 1998; Papayannopoulos et al. 1998). Presumably, Notch activation along the disc midline leads to production of a secreted ligand, perhaps Wg-related (Reifegerste and Moses 1999), that directs growth and patterning along the D/V axis of the eye.

The eye disc is not divided into anterior and posterior compartments, and thus growth along this axis is not regulated by a static midline organizer. Nonetheless, Dpp and Hh, expressed initially along the posterior disc margin, are critical for promoting growth during early eye development. Mutations that decrease Dpp expression in the eye primordia result in severely reduced eyes (Masucci et al. 1990; St Johnston et al. 1990; Blackman et al. 1991; Heberlein et al. 1993), and clones of Tkv mutant cells proliferate poorly (Burke and Basler 1996). Furthermore, ectopic Hh expression leads to enlarged eyes (Heberlein et al. 1995). In both wings and eyes, organizing centers are located where D/V (Notch, Wg) and A/P (Dpp, Hh) are juxtaposed, either naturally or experimentally.

Ironically, despite being driven by patterning genes, imaginal disc cell proliferation is largely unpatterned, particularly during the rapid periods of growth in the second and early third instars. During these stages, patterns of BrdU incorporation or phospho-H3 expression in the eye and wing disc are uniform, and do not reflect the localized expression patterns of Dpp or Wg (Adler and MacQueen 1984). Although the differences in growth rates sufficient to affect tissue shape are likely to be subtle and may therefore be missed by such assays, these results clearly indicate that Dpp and Wg do not act as simple mitogens to drive proliferation in a dosage-sensitive manner. In fact, the first observable proliferation pattern in the wing disc occurs as a region of the disc known as the zone of non-proliferating cells (ZNC) exits the cell cycle in response to Wg signaling (O'Brochta and Bryant 1985; Johnston and Edgar 1998). During the late third instar, Wg actually has an inhibitory effect on cell proliferation. In a similar vein, overexpression of Dpp in the third-instar eye disc results in a reduction of BrdU incorporation and Dpp appears to be necessary for the cycle arrest in the morphogenetic furrow (Horsfield et al. 1998). Thus both Wg and Dpp appear to have context-dependent effects on cell proliferation.

If patterning molecules do not have direct mitogenic effects, how do they promote imaginal disc growth? One attractive model posits that the driving force for growth in the disc is the slope of the concentration gradient of morphogens such as Dpp, rather than their absolute levels (French et al. 1976; Day and Lawrence 2000). In this view, growth is promoted by discontinuities in the levels of signaling activity, and proceeds until these differences are eliminated. This type of model is compelling in that it offers a satisfying explanation for the activation of cell proliferation following surgical removal of disc fragments, as well as the phenomenon of wound healing in general. However, such models are countered by a number of experimental observations, most significantly that clonal expression of a uniformly activated receptor leads to hyperproliferation throughout the clone, rather than at the clone margins as these models would predict.

Interestingly, there may be some pattern on a local level; adjacent, nonclonal, clusters of approximately five cells have been found to advance together through different phases of the cell cycle (Adler and MacQueen 1981; Mathi and Larsen 1988; Milan et al. 1996). The significance of such clustering of cells is unknown, but it suggests the existence of a novel mode of short range interaction or cell coupling which remains to be fully addressed. Since nitric oxide (NO) is a readily diffusible molecule with anti-proliferative properties, it has been suggested that NO might function to co-ordinate the proliferation of small groups of cells (Kuzin et al. 1996).

5.3 Growth Control by Other Patterning Factors

A group of seven genes has been implicated in the earliest steps of eye specification. In general, ectopic expression of any member of this group is sufficient to induce ectopic eyes, whereas loss of function alleles prevent eye formation (reviewed in Heberlein and Treisman 2000). Interestingly, at least two members of this group have a role in proliferation control which is at least partially independent of their effects on eye specification. Clones of *sine oculis* or *eyes absent* induced early during eye development were found to cause overproliferation (Pignoni et al. 1997). Although the mutant cells were unable to develop into photoreceptors, the observed hyperproliferation was not simply a result of a failure to arrest in G1 and differentiate, since overproliferation was also seen in cells prior to passage of the morphogenetic furrow. Furthermore, these cells continued to express *eyeless*, indicating that overproliferation was not due to a gross change in cell identity (Pignoni et al. 1997). In addition, mutant clones of *dachsund* have been found to overproliferate, although only in clones which included the eye margin (Mardon et al. 1994).

5.4 Influence of the Peripodial Membrane

Recent observations have revealed that the peripodial membrane, a layer of squamous epithelial cells overlying the disc proper, has previously unexpected effects on disc growth and patterning. Gibson and Schubiger (2000) and Choi and colleagues (Cho et al. 2000) observed that peripodial cells of the eye and wing imaginal discs extend microtubule-based processes through the disc lumen, and that these extensions contact the apical surfaces of the columnar epithelium below. Surgical removal or genetic ablation of the peripodial membrane was shown to dramatically reduce disc growth. Furthermore, modulation of Fng, Ser or Hh expression in the peripodium also caused a reduction in eye size. Ectopic peripodial Hh expression was also found to induce Ser expression in the disc proper. These studies underscore the need to incorporate a three dimensional view to our thinking of patterning and growth in the eye disc.

5.5 Interactions Between Patterning Networks and Growth Pathways

While activation of growth effectors such as PI3K, cyclin D, Myc, and Ras are required and sufficient to promote growth in some contexts, whether (and if so, how) these pathways are regionally modulated by the signals that direct patterned growth in the eye remains poorly understood. Given the large number of genes and diverse signaling pathways that have been observed to influence growth of the eye, it would appear daunting to formulate a unifying model that can incorporate these myriad observations. Aiding in this effort will be experiments in which growth or patterning is modified by manipulations of one gene network, followed by examination of the effects on other growth pathways, through epistasis tests, genetic interactions, and activity assays. Such a cross-disciplinary approach is needed to untangle the relationships between patterning and growth.

The few described examples of this type of experiment appear promising. For example, cell cycle exit in the ZNC of the wing disc has long been known to depend on Wg signaling. More recently, it has been found that Wg achieves this by inhibiting expression of dMyc (Johnston et al. 1999). Similar connections between Dpp-type ligands and S6K signaling have recently been reported in other systems (Petritsch et al. 2000).

6 Concluding Remarks

The development of the *Drosophila* eye provides an ideal opportunity to study the mechanisms that regulate growth and cell proliferation in the context of a developing organ. Recent studies have begun to identify pathways that regulate the growth and proliferation of individual cells in response to patterning cues and extracellular signals. The challenge for the future is to understand how multiple inputs can together modulate the cell cycle machinery and biosynthetic pathways in groups of cells so as to coordinate their growth and proliferation in order to build a structure of the desired size and shape such as the *Drosophila* eye.

Acknowledgements. We thank Kelly Tseng and Nicolas Tapon for the images used in Fig. 1, Tanya Wolff and Barbara Thomas for answers to last minute questions, and Kieran Harvey for comments on the manuscript. I.H. is grateful to Drs. M.D. Ramanathan and Semmangudi Srininvasier for continued inspiration.

References

Adler PN, MacQueen M (1981) Partial coupling of the cell cycles of neighboring imaginal disc cells. Exp Cell Res 133:452–456

Adler PN, MacQueen M (1984) Cell proliferation and DNA replication in the imaginal wing disc of *Drosophila melanogaster*. Dev Biol 103:28–37

Amati B, Alevizopoulos K,Vlach J (1998) Myc and the cell cycle. Front Biosci 3:D250–268

Avedisov SN, Krasnoselskaya I, Mortin M, Thomas BJ (2000) Roughex mediates G(1) arrest through a physical association with cyclin A. Mol Cell Biol 20:8220–8229

Baker NE, Rubin GM (1989) Effect on eye development of dominant mutations in *Drosophila* homologue of the EGF receptor. Nature 340:150–153

Baker NE, Yu SY (2001) The EGF receptor defines domains of cell cycle progression and survival to regulate cell number in the developing *Drosophila* eye. Cell 104:699–708

Baker NE, Mlodzik M, Rubin GM (1990) Spacing differentiation in the developing *Drosophila* eye: a fibrinogen-related lateral inhibitor encoded by scabrous. Science 250:1370–1377

Becker HJ (1957) Über Röntgenmosaikflecken und Defektmutationen am Auge von *Drosophila* und die Entwicklungsphysiologie des Auges. Z Indukt Abstamm Verbungsl 88:333–373

Bilder D, Li M, Perrimon N (2000) Cooperative regulation of cell polarity and growth by *Drosophila* tumor suppressors. Science 289:113–116

Blackman RK, Sanicola M, Raftery LA, Gillevet T, Gelbart WM (1991) An extensive 3′ cis-regulatory region directs the imaginal disk expression of decapentaplegic, a member of the TGF-beta family in *Drosophila*. Development 111:657–666

Blair SS (1995) Compartments and appendage development in *Drosophila*. Bioessays 17:299–309

Bohni R, Riesgo-Escovar J, Oldham S, Brogiolo W, Stocker H, Andruss BF, Beckingham K, Hafen E (1999) Autonomous control of cell and organ size by CHICO, a *Drosophila* homolog of vertebrate IRS1–4. Cell 97:865–875

Britton JS, Edgar BA (1998) Environmental control of the cell cycle in *Drosophila*: nutrition activates mitotic and endoreplicative cells by distinct mechanisms. Development 125:2149–2158

Brogiolo W, Stocker H, Ikeya T, Rintelen F, Fernandez R, Hafen E (2001) An evolutionarily conserved function of the *Drosophila* insulin receptor and insulin-like peptides in growth control. Curr Biol 11:213–221

Brown NL, Paddock SW, Sattler CA, Cronmiller C, Thomas BJ, Carroll SB (1996) daughterless is required for *Drosophila* photoreceptor cell determination, eye morphogenesis, and cell cycle progression. Dev Biol 179:65–78

Bryant PJ, Levinson P (1985) Intrinsic growth control in the imaginal primordia of *Drosophila*, and the autonomous action of a lethal mutation causing overgrowth. Dev Biol 107:355–363

Bryant PJ, Simpson P (1984) Intrinsic and extrinsic control of growth in developing organs. Q Rev Biol 59:387–415

Burke R, Basler K (1996) Dpp receptors are autonomously required for cell proliferation in the entire developing *Drosophila* wing. Development 122:2261–2269

Chen C, Jack J, Garofalo RS (1996) The *Drosophila* insulin receptor is required for normal growth. Endocrinology 137:846–856

Chevais S (1944) Determinisme de la taille de l'ocil chez le mutant bar de la *Drosophile*. Intervention d'une substance diffusible specifique. Bull Biol Fr Belg 78:71–110

Cho KO, Choi KW (1998) Fringe is essential for mirror symmetry and morphogenesis in the *Drosophila* eye. Nature 396:272–276

Cho KO, Chern J, Izaddoost S, Choi KW (2000) Novel signaling from the peripodial membrane is essential for eye disc patterning in *Drosophila*. Cell 103:331–342

Clifford RJ, Schupbach T (1989) Coordinately and differentially mutable activities of torpedo, the *Drosophila melanogaster* homolog of the vertebrate EGF receptor gene. Genetics 123:771–787

Coller HA, Grandori C, Tamayo P, Colbert T, Lander ES, Eisenman RN, Golub TR (2000) Expression analysis with oligonucleotide microarrays reveals that MYC regulates genes involved in growth, cell cycle, signaling, and adhesion. Proc Natl Acad Sci USA 97:3260–3265

Datar SA, Jacobs HW, de la Cruz AF, Lehner CF, Edgar BA (2000) The *Drosophila* cyclin D-Cdk4 complex promotes cellular growth. EMBO J 19:4543–4554

Davis KT, Shearn A (1977) In vitro growth of imaginal disks from *Drosophila melanogaster*. Science 196:438–440

Day SJ, Lawrence PA (2000) Measuring dimensions: the regulation of size and shape. Development 127:2977–2987

de Celis JF, Garcia-Bellido A, Bray SJ (1996) Activation and function of Notch at the dorsal-ventral boundary of the wing imaginal disc. Development 122:359–369

de Nooij JC, Hariharan IK (1995) Uncoupling cell fate determination from patterned cell division in the *Drosophila* eye. Science 270:983–985

de Nooij JC, Letendre MA, Hariharan IK (1996) A cyclin-dependent kinase inhibitor, Dacapo, is necessary for timely exit from the cell cycle during *Drosophila* embryogenesis. Cell 87:1237–1247

de Nooij JC, Graber KH, Hariharan IK (2000) Expression of the cyclin-dependent kinase inhibitor Dacapo is regulated by cyclin E. Mech Dev 97:73–83

Diaz-Benjumea FJ, Hafen E (1994) The sevenless signalling cassette mediates *Drosophila* EGF receptor function during epidermal development. Development 120:569–578

Dominguez M, de Celis JF (1998) A dorsal/ventral boundary established by Notch controls growth and polarity in the *Drosophila* eye. Nature 396:276–278

Dong X, Zavitz KH, Thomas BJ, Lin M, Campbell S, Zipursky SL (1997) Control of G1 in the developing *Drosophila* eye: rca1 regulates Cyclin A. Genes Dev 11:94–105

Du W, Dyson N (1999) The role of RBF in the introduction of G1 regulation during *Drosophila* embryogenesis. EMBO J 18:916–925

Du W, Vidal M, Xie JE, Dyson N (1996a) RBF, a novel RB-related gene that regulates E2F activity and interacts with cyclin E in *Drosophila*. Genes Dev 10:1206–1218

Du W, Xie JE, Dyson N (1996b) Ectopic expression of dE2F and dDP induces cell proliferation and death in the *Drosophila* eye. EMBO J 15:3684–3692

Edgar BA (1999) From small flies come big discoveries about size control. Nat Cell Biol 1:E191–193

Ferrus A, Garcia-Bellido A (1976) Morphogenetic mutants detected in mitotic recombination clones. Nature 260:425–426

Fleming RJ, Gu Y, Hukriede NA (1997) Serrate-mediated activation of Notch is specifically blocked by the product of the gene fringe in the dorsal compartment of the *Drosophila* wing imaginal disc. Development 124:2973–2981

French V, Bryant PJ, Bryant SV (1976) Pattern regulation in epimorphic fields. Science 193:969–981

Gallant P, Shiio Y, Cheng PF, Parkhurst SM, Eisenman RN (1996) Myc and Max homologs in *Drosophila*. Science 274:1523–1527

Gao X, Neufeld TP, Pan D (2000) *Drosophila* PTEN regulates cell growth and proliferation through PI3K-dependent and -independent pathways. Dev Biol 221:404–418

Gao X, Pan D (2001) TSC1 and TSC2 tumor suppressors antagonize insulin signaling in cell growth. Genes Dev 15:1383–1392

García-Bellido A, Merriam JR (1969) Cell lineage of the imaginal discs in *Drosophila* gynandromorphs. J Exp Zool 170:61–76

Gateff E (1994) Tumor suppressor and overgrowth suppressor genes of *Drosophila melanogaster*. Developmental aspects. Int J Dev Biol 38:565–590

Gibson MC, Schubiger G (2000) Peripodial cells regulate proliferation and patterning of *Drosophila* imaginal discs. Cell 103:343–350

Gingras AC, Raught B, Sonenberg N (2001) Regulation of translation initiation by FRAP/mTOR. Genes Dev 15:807–826

Goberdhan DC, Paricio N, Goodman EC, Mlodzik M, Wilson C (1999) *Drosophila* tumor suppressor PTEN controls cell size and number by antagonizing the Chico/PI3-kinase signaling pathway. Genes Dev 13:3244–3258

Halfar K, Rommel C, Stocker H, Hafen E (2001) Ras controls growth, survival and differentiation in the *Drosophila* eye by different thresholds of MAP kinase activity. Development 128:1687–1696

Heberlein U, Treisman JE (2000) Early retinal development in *Drosophila*. Results Problems Cell Differ 31:37–50

Heberlein U, Wolff T, Rubin GM (1993) The TGF beta homolog dpp and the segment polarity gene hedgehog are required for propagation of a morphogenetic wave in the *Drosophila* retina. Cell 75:913–926

Heberlein U, Singh CM, Luk AY, Donohoe TJ (1995) Growth and differentiation in the *Drosophila* eye coordinated by hedgehog. Nature 373:709–711

Horsfield J, Penton A, Secombe J, Hoffman FM, Richardson H (1998) decapentaplegic is required for arrest in G1 phase during *Drosophila* eye development. Development 125:5069–5078

Huang H, Potter CJ, Tao W, Li DM, Brogiolo W, Hafen E, Sun H, Xu T (1999) PTEN affects cell size, cell proliferation and apoptosis during *Drosophila* eye development. Development 126:5365–5372

Ito N, Rubin GM (1999) gigas, a *Drosophila* homolog of tuberous sclerosis gene product-2, regulates the cell cycle. Cell 96:529–539

Johnston LA, Edgar BA (1998) Wingless and Notch regulate cell-cycle arrest in the developing *Drosophila* wing. Nature 394:82–84

Johnston LA, Prober DA, Edgar BA, Eisenman RN, Gallant P (1999) *Drosophila* myc regulates cellular growth during development. Cell 98:779–790

Justice RW, Zilian O, Woods DF, Noll M, Bryant PJ (1995) The *Drosophila* tumor suppressor gene warts encodes a homolog of human myotonic dystrophy kinase and is required for the control of cell shape and proliferation. Genes Dev 9:534–546

Karim FD, Rubin GM (1998) Ectopic expression of activated Ras1 induces hyperplastic growth and increased cell death in *Drosophila* imaginal tissues. Development 125:1–9

Kawamura K, Shibata T, Saget O, Peel D, Bryant PJ (1999) A new family of growth factors produced by the fat body and active on *Drosophila* imaginal disc cells. Development 126:211–219

Knoblich JA, Sauer K, Jones L, Richardson H, Saint R, Lehner CF (1994) Cyclin E controls S phase progression and its down-regulation during *Drosophila* embryogenesis is required for the arrest of cell proliferation. Cell 77:107–120

Kuzin B, Roberts I, Peunova N, Enikolopov G (1996) Nitric oxide regulates cell proliferation during *Drosophila* development. Cell 87:639–649

Lane ME, Sauer K, Wallace K, Jan YN, Lehner CF, Vaessin H (1996) Dacapo, a cyclin-dependent kinase inhibitor, stops cell proliferation during *Drosophila* development. Cell 87:1225–1235

Leevers SJ, Weinkove D, MacDougall LK, Hafen E, Waterfield MD (1996) The *Drosophila* phosphoinositide 3-kinase Dp110 promotes cell growth. EMBO J 15:6584–6594

Leevers SJ, Vanhaesebroeck B, Waterfield MD (1999) Signalling through phosphoinositide 3-kinases: the lipids take centre stage. Curr Opin Cell Biol 11:219–225

Madhavan M, Schneiderman HA (1977) Histological analysis of the dynamics of growth of imaginal discs and histoblast nests during the larval development of *Drosophila melanogaster*. Wilhelm Roux's Arch 183:269–305

Mardon G, Solomon NM, Rubin GM (1994) dachshund encodes a nuclear protein required for normal eye and leg development in *Drosophila*. Development 120:3473–3486

Martin JF, Hersperger E, Simcox A, Shearn A (2000) minidiscs encodes a putative amino acid transporter subunit required non-autonomously for imaginal cell proliferation. Mech Dev 92:155–167

Masucci JD, Miltenberger RJ, Hoffmann FM (1990) Pattern-specific expression of the *Drosophila* decapentaplegic gene in imaginal disks is regulated by 3′ cis-regulatory elements. Genes Dev 4:2011–2023

Mathi SK, Larsen E (1988) Patterns of cell division in imaginal discs of *Drosophila*. Tissue Cell 20:461–472

Meyer CA, Jacobs HW, Datar SA, Du W, Edgar BA, Lehner CF (2000) *Drosophila* Cdk4 is required for normal growth and is dispensable for cell cycle progression. EMBO J 19:4533–4542

Milan M, Campuzano S, Garcia-Bellido A (1996) Cell cycling and patterned cell proliferation in the wing primordium of *Drosophila*. Proc Natl Acad Sci USA 93:640–645

Montagne J, Stewart MJ, Stocker H, Hafen E, Kozma SC, Thomas G (1999) *Drosophila* S6 kinase: a regulator of cell size. Science 285:2126–2129

Morata G, Ripoll P (1975) Minutes: mutants of *Drosophila* autonomously affecting cell division rate. Dev Biol 42:211–221

Nellen D, Burke R, Struhl G, Basler K (1996) Direct and long-range action of a DPP morphogen gradient. Cell 85:357–368

Neufeld TP, de la Cruz AF, Johnston LA, Edgar BA (1998) Coordination of growth and cell division in the *Drosophila* wing. Cell 93:1183–1193

Newby WW, Thelander RP (1950) Early development of the head in normal and tumorous head *D. melanogaster*. Drosoph Inform Serv 24:89–90

Nijhout HF, Emlen DJ (1998) Competition among body parts in the development and evolution of insect morphology. Proc Natl Acad Sci USA 95:3685–3689

Nishida Y, Hata M, Ayaki T, Ryo H, Yamagata M, Shimizu K, Nishizuka Y (1988) Proliferation of both somatic and germ cells is affected in the *Drosophila* mutants of raf proto-oncogene. EMBO J 7:775–781

O'Brochta DA, Bryant PJ (1985) A zone of non-proliferating cells at a lineage restriction boundary in *Drosophila*. Nature 313:138–141

Oldham S, Montagne J, Radimerski T, Thomas G, Hafen E (2000) Genetic and biochemical characterization of dTOR, the *Drosophila* homolog of the target of rapamycin. Genes Dev 14: 2689–2694

Panin VM, Papayannopoulos V, Wilson R, Irvine KD (1997) Fringe modulates Notch-ligand interactions. Nature 387:908–912

Papayannopoulos V, Tomlinson A, Panin VM, Rauskolb C, Irvine KD (1998) Dorsal-ventral signaling in the *Drosophila* eye. Science 281:2031–2034

Perry MM (1968) Further studies on the development of the eye of *Drosophila melanogaster*. II: The inter-ommatidial bristles. J Morphol 124:249–262

Petritsch C, Beug H, Balmain A, Oft M (2000) TGF-beta inhibits p70 S6 kinase via protein phosphatase 2A to induce G1 arrest. Genes Dev 14:3093–3101

Pignoni F, Hu B, Zavitz KH, Xiao J, Garrity PA, Zipursky SL (1997) The eye-specification proteins So and Eya form a complex and regulate multiple steps in *Drosophila* eye development. Cell 91:881–891

Postlethwait JH, Schneiderman HA (1970) Induction of metamorphosis by ecdysone analogues. *Drosophila* imaginal discs cultured in vivo. Biol Bull 138:47–55

Potter CJ, Huang H, Xu T (2001) *Drosophila* Tsc1 functions with Tsc2/gigas to antagonize insulin signaling in the regulation of cell growth, cell proliferation, and organ size. Cell 105:357–368

Prober DA, Edgar BA (2000) Ras1 promotes cellular growth in the *Drosophila* wing. Cell 100: 435–446

Ready DF, Hanson TE, Benzer S (1976) Development of the *Drosophila* retina, a neurocrystalline lattice. Dev Biol 53:217–240

Reifegerste R, Moses K (1999) Genetics of epithelial polarity and pattern in the *Drosophila* retina. Bioessays 21:275–285

Richardson H, O'Keefe LV, Marty T, Saint R (1995) Ectopic cyclin E expression induces premature entry into S phase and disrupts pattern formation in the *Drosophila* eye imaginal disc. Development 121:3371–3379

Royet J, Finkelstein R (1997) Establishing primordia in the *Drosophila* eye-antennal imaginal disc: the roles of decapentaplegic, wingless and hedgehog. Development 124:4793–4800

Secombe J, Pispa J, Saint R, Richardson H (1998) Analysis of a *Drosophila* cyclin E hypomorphic mutation suggests a novel role for cyclin E in cell proliferation control during eye imaginal disc development. Genetics 149:1867–1882

Sherr CJ (1995) D-type cyclins. Trends Biochem Sci 20:187–190

Simpson P (1976) Analysis of the compartments of the wing of *Drosophila melanogaster* mosaic for a temperature-sensitive mutation that reduces mitotic rate. Dev Biol 54:100–115

Simpson P (1979) Parameters of cell competition in the compartments of the wing disc of *Drosophila*. Dev Biol 69:182–193

Simpson P, Berreur P, Berreur-Bonnenfant J (1980) The initiation of pupariation in *Drosophila*: dependence on growth of the imaginal discs. J Embryol Exp Morphol 57:155–165

Spencer FA, Hoffmann FM, Gelbart WM (1982) Decapentaplegic: a gene complex affecting morphogenesis in *Drosophila melanogaster*. Cell 28:451–461

Sprenger F, Yakubovich N, O'Farrell PH (1997) S-phase function of *Drosophila* cyclin A and its downregulation in G1 phase. Curr Biol 7:488–499

St Johnston RD, Hoffmann FM, Blackman RK, Segal D, Grimaila R, Padgett RW, Irick HA, Gelbart WM (1990) Molecular organization of the decapentaplegic gene in *Drosophila melanogaster*. Genes Dev 4:1114–1127

Tapon N, Ito N, Dickson BJ, Treisman JE, Hariharan IK (2001) The *Drosophila* tuberous sclerosis complex gene homologues restrict cell growth and cell proliferation. Cell 105:345–355

The I, Hannigan GE, Cowley GS, Reginald S, Zhong Y, Gusella JF, Hariharan IK, Bernards A (1997) Rescue of a *Drosophila* NF1 mutant phenotype by protein kinase A. Science 276:791–794

Thomas BJ, Gunning DA, Cho J, Zipursky L (1994) Cell cycle progression in the developing *Drosophila* eye: roughex encodes a novel protein required for the establishment of G1. Cell 77:1003–1014

Thomas BJ, Zavitz KH, Dong X, Lane ME, Weigmann K, Finley RL Jr, Brent R, Lehner CF, Zipursky SL (1997) roughex down-regulates G2 cyclins in G1. Genes Dev 11:1289–1298

Verdu J, Buratovich MA, Wilder EL, Birnbaum MJ (1999) Cell-autonomous regulation of cell and organ growth in *Drosophila* by Akt/PKB. Nat Cell Biol 1:500–506

Watson KL, Justice RW, Bryant PJ (1994) *Drosophila* in cancer research: the first fifty tumor suppressor genes. J Cell Sci (Suppl) 18:19–33

Weigmann K, Cohen SM, Lehner CF (1997) Cell cycle progression, growth and patterning in imaginal discs despite inhibition of cell division after inactivation of *Drosophila* Cdc2 kinase. Development 124:3555–3563

Weinkove D, Neufeld TP, Twardzik T, Waterfield MD, Leevers SJ (1999) Regulation of imaginal disc cell size, cell number and organ size by *Drosophila* class I(A) phosphoinositide 3-kinase and its adaptor. Curr Biol 9:1019–1029

Wieschaus E, Gehring W (1976) Clonal analysis of primordial disc cells in the early embryo of *Drosophila melanogaster*. Dev Biol 50:249–263

Wolff T, Ready DF (1991a) The beginning of pattern formation in the *Drosophila* compound eye: the morphogenetic furrow and the second mitotic wave. Development 113:841–850

Wolff T, Ready DF (1991b) Cell death in normal and rough eye mutants of *Drosophila*. Development 113:825–839

Wolff T, Ready DF (1993) Pattern formation in the *Drosophila* retina. In: Bate M, Martinez Arias A (eds) The development of *Drosophila melanogaster*. Cold Spring Harbor Laboratory Press, Plainview, NY, pp 1277–1325

Woods DF, Bryant PJ (1991) The discs-large tumor suppressor gene of *Drosophila* encodes a guanylate kinase homolog localized at septate junctions. Cell 66:451–464

Xu T, Rubin GM (1993) Analysis of genetic mosaics in developing and adult *Drosophila* tissues. Development 117:1223–1237

Xu T, Wang W, Zhang S, Stewart RA, Yu W (1995) Identifying tumor suppressors in genetic mosaics: the *Drosophila* lats gene encodes a putative protein kinase. Development 121:1053–1063

Zaffran S, Chartier A, Gallant P, Astier M, Arquier N, Doherty D, Gratecos D, Semeriva M (1998) A *Drosophila* RNA helicase gene, pitchoune, is required for cell growth and proliferation and is a potential target of d-Myc. Development 125:3571–3584

Zecca M, Basler K, Struhl G (1995) Sequential organizing activities of engrailed, hedgehog and decapentaplegic in the *Drosophila* wing. Development 121:2265–2278

Zhang H, Stallock JP, Ng JC, Reinhard C, Neufeld TP (2000) Regulation of cellular growth by the *Drosophila* target of rapamycin dTOR. Genes Dev 14:2712–2724

Evolution of Color Vision

Franck Pichaud and Claude Desplan[1]

1 The Retinal Mosaic of the Compound Eye

Charles Darwin and Ramon y Cajal were fascinated by the complexity and apparent perfection of eyes throughout the animal kingdom. Considering the vast repertoire in their design, they proposed that, during evolution, eyes have appeared more than 40 times independently in different *phyla* (Salvini-Plawen and Mayr 1977). Recent data from molecular genetics offer an opposite view of this evolutionary problem since the "master regulatory gene" for eye development, *Pax-6*, is highly conserved in all these species (Halder et al. 1995a,b; Gehring and Ikeo 1999). Thus, although the compound eye of Diptera and the camera-like vertebrate eye present many apparent differences in their design, they seem to have evolved from a common and simple ancestral photoreceptive module (Zuker 1994; Halder et al. 1995b; Desplan 1997; Sheng et al. 1997; Gehring and Ikeo 1999).

The *Drosophila* visual system comprises the retina of the compound eyes and the ocelli. The retina is composed of approximately 800 ommatidia. Each ommatidium contains 8 photoreceptor neurons (PRs) and 12 accessory cells (Ready et al. 1976; Wolff and Ready 1991). The six outer PRs (R1–R6) send their axon projections to the first layer of the optic lobe, the lamina. The two inner PRs (R7 and R8) project to a deeper layer of the optic lobe, the medulla (Dietrich 1909; Trujillo-Cenoz 1969; Campos-Ortega and Strausfeld 1972; Braitenberg and Strausfeld 1973). Accordingly, the R1 to R6 axons are designated short visual fibers while those of R7 and R8 are long visual fibers.

Although the compound eyes share the same general organization, they come with many specialized variations and differ in the organization of their light gathering membranes, the rhabdomeres. For instance, in some groups of lower Diptera, as well as in bees and butterflies, all the rhabdomeres of each ommatidium are fused into a central structure called a rhabdom. The subretinal organization of the photoreceptor axons is such that an image of the outside world is projected point by point onto the lamina and medulla through a principle of apposition. In contrast, the higher Diptera have evolved an open rhabdomere structure such that each photoreceptor within an ommatidium

[1] Laboratory of Molecular Genetics, Department of Biology, New York University, 1009 Main Building, 100 Washington Square East, New York, New York 10003-6688, USA

Results and Problems in Cell Differentiation, Vol. 37
K. Moses (Ed.): Drosophila Eye Development

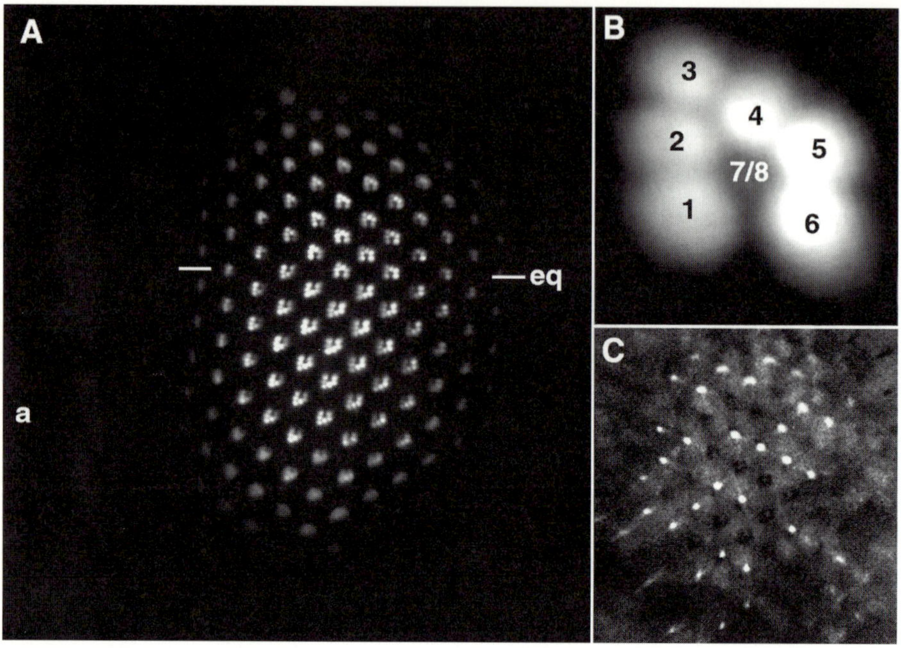

Fig. 1. **A** Imaging of a transgene *prh1-eGFP* after neutralization of the cornea, in the eye of a living *Drosophila*. **B** Magnification of an ommatidium using *prh1-eGFP*; the location of the inner PRs R7 and R8 is shown (*7/8*) **C** Mosaic of *rh3/rh4* expression in inner R7 photoreceptors expressing *prh3-Rh3-eGFP* (***pale*** subtype)

has an independent rhabdomere and a different optical axis from its neighbor. This system, together with a sub-retinal organization referred as neural superposition (Kirschfeld 1967; see Sect. 1.2), allows flies (and mosquitoes) to increase their absolute dim light sensitivity and resolution.

1.1 Rhabdomere and Photoreception

Each of the 800 ommatidia is capped by a convergent dioptrical apparatus called the facet lens, whose ensemble forms the cornea. The incident light is focused by the lens on the tip of the PRs light gathering membranes, the rhabdomeres. These photoreceptive membranes are organized as arrays of microvilli 50–100 nm in diameter whose major protein component is the visual pigment rhodopsin (~70% of total membrane proteins). The outer PR rhabdomeres are about 1.5–2 μm in diameter and extend throughout the depth of the retina (~100 μm in length). The inner PR rhabdomeres are placed one on top of the other in the same optical path and their diameter is smaller, around 1 μm (Figs. 1, 2). The relatively small diameter of the rhabdomeres and the values of refraction index of its surrounding media allow the propagation of light as in an optical waveguide (as in an optic fiber; Lipson et al. 1995). The capture of light by the rhabdomere is optimum for an acceptance angle of 30° and incident light outside this value is excluded (Franceschini and Kirschfeld 1971). Fast migration of pigments (within a few seconds) protects PRs against over-stimulation by excluding some of the incident light, thus acting like "intracellular pupils" that operate within individual cells.

The rhabdomere microvilli are arranged perpendicularly to the axis of the incident light collected by the cornea. The cohesion and fine organization of the tightly packed microvilli depends on the function of several genes including Chaoptin, NinaC and Actin that form the "core-filament". Upon photon reception, the chromophore is changed from *11-cis* to *all trans* hydroxyretinal, leading to a change in conformation and activation of the protein moiety (opsin). In turn, the opsin conformational change activates a G-protein (Gq), the entry point into the phototransduction pathway. Insects have adopted a phosphoinositide-based phototransduction pathway, in contrast to the cyclic GMP-mediated signaling cascade found in vertebrates. A fast and tightly controlled light-response is ensured by a postsynaptic density protein, disc-large, Zo-1 (PDZ) domain containing adapter protein, inactivation no-after potential (INAD) that stabilizes a protein complex localized in the rhabdomere composed of the transient receptor potential/transient receptor potential-like (TRP/TRPL) ion channels, protein kinase C (PKC) and phospholipase C (PLC) enzymes (Chevesich et al. 1997; Tsunoda et al. 1997). Activated Gq activates PLC, leading to inositol trisphosphate (IP3) production via the phosphoinositide-cascade. IP3 in turn causes the opening of the membrane ion channels TRP and TRPL (Niemeyer et al. 1996).

1.2 Image Formation and Neural Superposition

A *Drosophila* eye views the surrounding environment in 800 different directions, thus offering a panoramic visual field. While the spatial sampling achieved by a human eye is about 130 million pixels, the *Drosophila* eye allows a much lower spatial resolution with its 800 pixels. Nevertheless, fly vision is reliable and efficient, probably because flies have evolved effective and computationally efficient solutions such as neural superposition. Serial electron microscopy (Meinertzhagen and O'Neil 1991), confocal microscopy (Hiesinger et al. 2001) and optical staining (Wilcox and Franceschini 1984; Picaud et al. 1990; Clandinin and Zipursky 2000) have helped unravel the detailed organization of the neural cartridges in the lamina, where the outer PRs terminate their axonal projection. In very elegant experiments Picaud and colleagues applied extracellular fluorescent dyes such as Lucifer yellow to the fly retina. Illumination of a single ommatidium resulted in the selective uptake of the dye in the illuminated PRs only. Subsequent anatomical analysis of the lamina demonstrated that the axons of the illuminated PRs were distributed according to the neural superposition principle of higher Diptera (Kirschfeld 1967):

all PR axons forming a neural cartridge, although coming from six different ommatidia, share the same visual axis. In this setup, the signal to noise ratio is increased by a factor of $\sqrt{6}$ as compared to a system with only one input per cartridge or transmission channel (van Hateren 1987). Together with the presence of a broad-spectrum pigment Rh1 (see Sect. 2.2) and a high light quantum catch capacity, the "neural superposition" of outer PR projections is a beautiful adaptation for the achievement of optimal contrast sensitivity in dim light.

1.3 Ocelli

In contrast to the compound eye, each of the three dorsal ocelli is a single lens optical apparatus. Such a lens presents a very high optical aperture and a focal plane that is found far behind the PRs (Schuppe and Hengstenberg 1993). As a consequence, the neuronal image formed by the ocelli is heavily blurred and of very low spatial resolution. While the exact function of ocelli is not completely clear, they likely serve as a system to detect variation of light intensity. All ocellar PRs project to only a few second-order neurons in the brain, called L-neurons (Toh and Tateda 1991; Goodman 1981), and such a high degree of convergence is probably optimum for the detection of a small variation of light intensity (Wison 1978). The ocellar PRs all contain the same opsin, Rh2 (Mismer et al. 1988; Pollock and Benzer 1988), a violet-sensitive opsin whose sensitivity represents an optimum for horizon detection (It is in the UV that the contrast between bright light from the sky and reflected light from the ground is maximum). Studies in bees have shown that the ocellar light response (detection/transmission) is significantly faster than that of the compound eye (for review see Mizunami 1999). This again points to the ocelli as a system to allow fast response for flight control. The measurement by the flying insect of parameters such as roll deviation (rotation along the A/P axis of the body) might be achieved through differences in light response between the central and the two lateral ocelli (Wison 1978).

2 Photoreceptors and Visual Pigments

The exclusive expression of a single rhodopsin in each photoreceptor is a paradigm for a recurrent phenomenon observed in many sensory systems, where the general rule of one receptor molecule per receptor cell applies. This is crucial in order to avoid overlap of sensory inputs that would lead to an ambiguous output. Sensory receptors often belong to large families (e.g., olfactory receptors), and, during development, individual cells make a stochastic choice to express a given receptor molecule and to exclude all others (Chou et al. 1996; Mombaerts et al. 1996; Papatsenko et al. 1997; Clyne et al. 1999, 2000; Vosshall et al. 1999). However, the butterfly PRs appear to have diverged from this general theme, as simultaneous expression of two rhodopsins in the same cell has been reported (Kitamoto et al. 1998). Such variation might represent an evolutionary strategy for broadening the spectrum of light sensitivity through the accumulation of several rhodopsins within single photoreceptors dedicated to black and white vision.

2.1 Color Vision

Color vision can be viewed as the perception by the brain of sensations produced by different light vibrations or wavelengths. Certain conditions have to be met to support true color vision: Color contrast should be detectable independently of light intensity, implying that dimensions such as hue (blueness, greenness, redness) saturation (variation from a given neutral hue with constant brightness) and brightness (from dark to bright) are processed as independent variables. Although a mosaic of pigments presenting different wavelength sensitivity undoubtedly constitute the pre-requisite "hardware" to perform color vision, the subsequent neural wiring determines whether the organism has wavelength-specific behavior or true color vision. Direct synapsis between photoreceptor axons, or indirect connection through interneurons, which allows the comparison of inputs between PRs of different absorption maxima, is a strong indication of color vision. Ultimately, tests based on associative learning strategies are used to demonstrate color vision. By contrast, a wavelength-specific behavior response refers to lack of flexibility in the color/task association that cannot be altered by training (for review, see: Menzel 1979; Goldsmith 1990) e.g., UV-tropism of flies.

Color vision has not been studied in great detail in flies. However, the organization of the fly retina and recent results on the expression of rhodopsin genes strongly suggest that color vision is supported by the long visual fibers photoreceptors R7 and R8, which contain different opsins (Fukushi 1994; see Sect. 3). In flies, inputs from the long fibers appear to be compared in the medulla, allowing discrimination of colored and polarized light (Troje 1993).

2.2 Evolution and Properties of Rhodopsins

Most animal species use a G-coupled seven transmembrane receptor (opsin) linked to a retinal chromophore as their photosensitive molecules. The spectral sensitivity of photo-pigments is, in a large part, determined by the apoprotein opsin, and much work has deciphered the amino acid residues that allow spectral tuning of each molecule (Asenjo et al. 1994; Lin et al. 1998). Opsins can be classified into four groups based on their absorption spectra: UV, blue, green and red. Within vertebrates, opsins can be further subdivided into either "cone opsins" or the "rod-specific opsin". The cone pigments arose first and diversified into the several spectral classes. The rod pigment is proposed to have evolved subsequently from a green cone pigment ancestor (Okano et al.

1992). In this context the phylogeny of opsins is very informative. Although the opsins have diversified after the separation between invertebrates and vertebrates, there are striking similarities in the evolution between these two branches, which probably represents convergent evolution for spectral tuning.

Rod opsin in vertebrates and the outer photoreceptor pigment Rh1 in flies are both used for "black and white" vision, particularly in dim light, and have both evolved from the green branch. This might reflect the need for a broad-spectrum pigment centered around the middle range of wavelengths. Insects other than higher Diptera achieve a broad spectrum of detection by accumulating several opsins within their single light gathering structure, the rhabdom. In flies, Rh1 has a maximum absorption (λmax) centered on the green. A second accessory chromophore moiety attached outside Rh1 further expands the spectrum of absorption towards the UV by energy transfer to the main chromophore (Hamdorf et al. 1973; Kirschfeld and Franceschini 1977). This setup broadens the light spectrum that is detected by the outer PRs from near UV to green.

Interestingly, insects and vertebrates share the same classes of UV, blue and green pigments. The red branch is missing in most insects, but there are a few scattered occurrences of red pigments in several genera (Bernard 1979; Yang and Osorio 1996). These might reflect independent duplication events of the green pigment, as in humans. An extreme case is represented by the butterfly *Papilio* with five spectral classes of receptors (Arikawa et al. 1987), due to as many as six opsins (Briscoe 1998; Kitamoto et al. 1998), which enable this butterfly to have a broad vision spectrum ranging from UV to red. The most heterogeneous retina in terms of wavelength sensitivity is certainly that of the mantis shrimp with its ten types of spectral receptors (Cronin and Marshall 1989).

2.3 Polarized Light Vision

Light can be described as either particle (photon) or electromagnetic (oscillating wave). The electromagnetic wave oscillates perpendicularly to the direction of light travel, and polarized light is defined by a fixed plane of oscillation. Depending on the position of the sun, the polarization of the reflected light varies. Unlike most vertebrates, arthropods are endowed with the ability to detect the plane of polarization of light (for a recent review see Eguchi 1999). The dorsal most part of the fly retina (dorsal margin), which faces the sky, consists of ommatidia with a specialized set of highly dichroic PRs. In the row of ommatidia that flank the head capsule, the orientation of the microvilli of the R7 rhabdomere is orthogonal to that of the R8 microvilli. Both R7 and R8 contain the same UV-sensitive opsin, Rh3 (Zuker et al. 1987), and have larger rhabdomeres than regular R7 and R8 cells (Labhart and Meyer 1999). The development of the dorsal margin is poorly documented. However, a likely prediction is that a dorso-ventral asymmetry is created during eye development so that the dorsal margin PRs are only produced in the dorsal edge of the eye epithelium and not in the ventral one, probably in response to the *wingless* signaling pathway.

The inner PRs in the dorsal margin give a maximal light response when the direction of one of the rhabdomere microvilli coincides with the plane of symmetry of the sky polarization pattern (Rossel 1988) while the other is orthogonal and thus provides a minimal signal. Thus, when the animal is positioned so that the difference between the output of R7 and R8 in the dorsal margin is maximum, it is aligned with the sun and therefore probably in a position to set an appropriate flight course.

3 Spectral Organization of the Fly Retina

Fly vision probably does not rely on building a complete representation of the surroundings. Nevertheless, their aptitudes in flight and their capacities achieved in processes such as tracking or pursuit attest to a perfect adaptation of their visual system to the biotope. Most likely, flies use motion detection to acquire the visual cues needed to achieve these particular tasks (Lehrer and Srinivasan 1994; Hengstenberg et al. 1997). While the fly is moving, it is subject to a pattern of motion across its retina from which it eventually evaluates parameters such as relative distances in space (Srinivasan 1993). The optic lobe (lobula) contains motion-sensitive neurons showing clear directional preferences in their response to mobile stimuli (Hausen 1984; Franceschini 1985). At the level of the retina, it appears that the outer PRs R1–6, with their broad spectral sensitivity, are responsible for motion detection.

3.1 Inner Photoreceptors and Color Vision

The organization in tandem of inner PRs, with R7 in the distal half of the retina and R8 below occupying the remaining proximal half, represents the appropriate hardware to perform color vision (Fig. 2). Since the two rhabdomeres share the same visual axis, the different spectral sensitivities of R7 and R8 allow an analysis of the wavelength content of the light (Braitenberg and Strausfeld 1973; Hardie 1985). The incident light travels through R7 and then R8, these two cells exerting their maximal response in different parts of the light spectrum so that subsequent opponent interactions are possible. Although no point in space is seen simultaneously by four different inner PRs, scanning movements during flight are likely to allow the fly to have a precise idea of its color environment. Interestingly, two muscles are found in the head capsule of Diptera, including *Drosophila*, that are most likely involved in modifying the orientation of the PRs (for review see Franceschini 1997).

R7 and R8 axons bypass the lamina and project to two different layers of the medulla, the R8 axons stopping just before those of R7, but R7 and R8

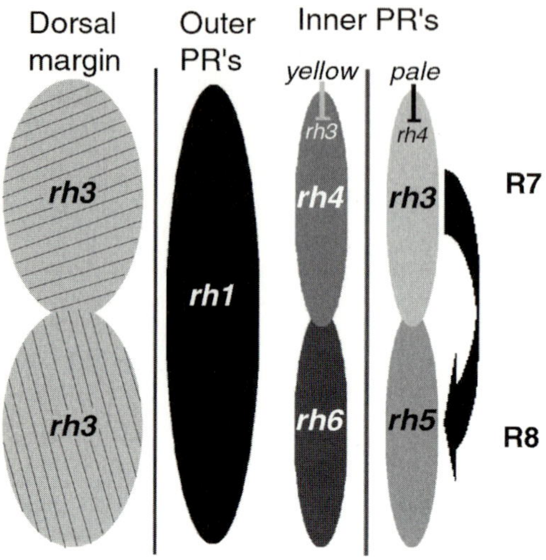

Fig. 2. *Left* to *right* In the dorsal margin of the eye, both R7 and R8 contain Rh3 and are involved in the detection of the vector of polarized light. The outer PRs (R1–R6) form a homogeneous population of PRs that all contain the same opsin, Rh1. Two types of ommatidia form the *y* and *p* subtypes. In the *y* default state, R7 contains Rh4 while R8 contains Rh6. In the *p* acquired state, R7 contains Rh3 while R8 contains Rh5

inputs are likely to be compared in this part of the optic lobe (Strausfeld and Campos-Ortega 1977; Strausfeld 1989). Although the most probable model involves only the long visual fibers R7/R8 as a color opponent system (Troje 1993; Fukushi 1994), the lamina neurons might also be involved in color processing. Indeed, R1–R6 synapse in the lamina on the L3 mononuclear cells that enter the medulla together with the two long visual fibers (R7–R8; Strausfeld 1989). Consistent with this hypothesis, the long visual fibers in some lower Diptera and in bees also have synaptic involvement in the lamina (Ribi 1981; Armett-Kibel and Meinertzhagen 1985).

3.2 Yellow and Pale Ommatidia

Flies have the ability to detect and be strongly attracted by UV light. *Drosophila* possesses two UV-sensitive opsins in R7 (Rh3: R7***pale*** and Rh4: R7***yellow***) and *sevenless* mutants fail to respond to UV light. Therefore, although the fly eye appears to be composed of identical facets, there are clear physiological differences between different ommatidia. The main part of the retina is formed of a mosaic of two stochastically distributed types of ommatidia: 70% of the ommatidia (***yellow*** type) present an R7 cell that contains Rh4 while the underlying R8 contains Rh6. The remaining 30% (***pale*** type; see Fig. 1C) contain Rh3 in R7 and Rh5 in R8 (Hardie 1979; Franceschini et al. 1981a; Montell et al. 1987; Zuker et al. 1987; Chou et al. 1996; Huber et al. 1997; Papatsenko et al. 1997). Although the R7 opsins (Rh3 and Rh4) are present in two non-overlapping subsets of R7, they have only slightly different spectra of absorption in the UV. However, a blue filtering pigment sharpens the absorption of Rh4 in the UV.

Thus, the fly color visual system is tetra-chromatic. In addition, the dorsal margin ommatidia contain Rh3 in both R7 and R8 (see Sect. 2.3).

Finally, although the general organization of the larger fly *Musca domestica* is very similar to that of *Drosophila* (the R7***y/p*** ratio is conserved, as well as the dorsal margin), the male has sacrificed color vision in one third of its retina where the R7 PRs contain Rh1 (instead of Rh3 or Rh4) and project to the lamina instead of the medulla (Hardie 1985). This additional outer-like PR increases the signal to noise ratio by a factor of $\sqrt{7}$ instead of $\sqrt{6}$ for the rest of the retina. This peculiar region of the male retina, baptized the "love spot", is optimized for the detection of the flying female that is viewed through seven rather than six outer PRs (Franceschini et al. 1981b; Hardie et al. 1981).

The complex mosaic organization and diversity encountered in R7/R8 cells in terms of *rh* expression, could not be predicted even by the deep understanding of eye development during morphogenetic furrow progression in the larval imaginal disc (see preceding chapters of this Volume), and it has remained poorly understood. It is important to stress that there are no clonal populations or cell lineage restrictions involved in eye development or in the distribution of the ***yellow*** vs. ***pale*** ommatidia. Clones of cells induced early during eye development have a normal *rh3*/*rh4* ratio in the adult eye (Ready et al. 1976; Pichaud and Desplan 2001). It is likely that the choice of rhodopsin expression in R7 and R8 is the product of fairly late differentiation events that occur during mid-pupation, i.e., days after initial photoreceptor determination.

Several groups have described the heterogeneity of PR light sensitivity using electroretinograms (ERG) to evaluate the spectral properties of visual pigments, or with rhodopsin antibodies (Hardie 1979; Zuker et al. 1988; Chou et al. 1999). In addition, non-invasive microscopy in living flies has been successfully used to unravel the organization of the fly eye mosaic (Franceschini et al. 1981; Pichaud and Desplan 2001; Fig. 1). Such studies have established that, although the distribution of R7***p*** and R7***y*** is stochastic, the ratio is constant (30 vs. 70%). Furthermore, the mutually exclusive expression of rhodopsin genes is absolute.

Recently, the use of different mutant backgrounds has allowed the investigation of the mechanisms involved in setting up the spectral mosaic of the retina. For instance, all R8 cells express *rh6* in a *sevenless* background where no R7 are formed (Papatsenko et al. 1997; Chou et al. 1999). Furthermore, although R8 is the first cell to differentiate in the developing ommatidium during the third instar larval stages, it is not required for R7 to make its choice of *rh* expression (Chou et al. 1999). This strongly suggests that it is R7 that makes the active decision to express a given opsin (*rh3* vs. *rh4*). R7 then communicates the choice to the underlying R8 through a pathway that remains to be discovered; an R7 cell expressing *rh3* is required to force the underlying R8 to express *rh5*. In this model, the ***y*** subtype represents a default state for *rh* expression while the ***p*** subtype is the acquired state. Although expression of *rh* starts late during pupation, it is not clear yet when this decision is made and enforced in R7 and R8.

An exception is found in the case of the dorsal margin (see Sect. 2.3) where the expression of *rh3* in R8 does not seem to require any input from R7: in a *sevenless* background, the dorsal margin R8 still contain Rh3, and not Rh6 as is the case for all other R8. Thus the inner PRs at the dorsal margin are a rather peculiar sub-population, clearly distinct from the ***yellow*** and ***pale*** subtypes.

3.3 Regulation of Rhodopsin Expression

rhodopsin expression is controlled almost entirely at the transcriptional level. Very short promoter fragments can reproduce precise *rh* expression (Fortini and Rubin 1990; Pichaud and Desplan 2001; Tahayato et al., in prep.). Several studies (Fortini and Rubin 1990; Papatsenko et al. 1997, 2001; Tahayato et al., in prep.) strongly suggest that the *rh* promoters present a bi-partite organization. The proximal part of the promoters contains the generic TATA-box element together with a rhodopsin common sequence (RCSI). This sequence is found in all opsin promoters in flies, but is also conserved and required in vertebrate opsin genes (Yu et al. 1996). RCSI is almost identical to the P3 palindromic binding sites defined for Paired-type homeodomains (Wilson et al. 1995, 1996). In fact, RCSI is bound in vivo by a dimer of the Pax6 homeodomain protein (Sheng et al. 1997). Pax6 is considered the "master regulator of eye development" because of its critical role in eye formation in species as different as flies and vertebrates (for a recent review see: Gehring and Ikeo 1999). Furthermore, ectopic expression of Pax6 is able to induce ectopic eyes in both flies and frogs (Halder et al. 1995a,b; Chow et al. 1999), again supporting a role for Pax6 as a gene sitting at the top of the hierarchy of eye determination genes. This conserved role is surprising in view of the independent evolution of eyes as diverse as the single lens eye of vertebrates or the compound eye of insects. Therefore, it was suggested that the conservation of Pax6 might reflect an earlier role of Pax6 as a photoreceptor-specific gene that was later recruited for making the complex eyes of different species: Pax6 might thus control eye development through different routes in different species. Its ancestral role as a regulator of rhodopsin expression, which has been maintained in flies, is therefore fully consistent with its evolution (Sheng et al. 1997). There are two genes in flies that encode a Pax6 protein, *toy* (Czerny et al. 1999) and *eyeless* (Quiring et al. 1994), which have the potential to regulate rhodopsins. Pax6 provides a necessary sub-threshold activation of all rhodopsin genes, which needs to be augmented by other elements that confer photoreceptor subtype specificity. Consistent with this, a multimerized version (3×) of the RCSI site drives expression of a reporter gene in all PRs, and this artificial element appears to be universal in insects (Sheng et al. 1997; Berghammer et al. 1999). Thus, the proximal part of all *rh* promoters provides PR cell identity mediated by Pax6, i.e., by a homeodimer of the product of *ey*, or *toy* or of a heterodimer of the two. A recent investigation has revealed that all RCS-1 sites are not completely equivalent. Subtle differences that are conserved in evolution exist in the RCS-1 sequences of different *rh* promoters (Papatsenko et al. 2001). The presence of binding sites bound by homeoproteins that present a lysine at the position 50 of their HD (K_{50} sites) overlapping RCSI in *rh3* and *rh6* strongly suggests that a heterodimer composed of Pax-6 and a K_{50} homeoprotein (likely to be the product of *orthodenticle*) binds to this sequence in *rh3* and *rh6* (Tahayato et al., in prep.).

The photoreceptor subtype specificity is mediated by the distal part of the rhodopsin promoters. Contrary to the proximal region, the distal elements are very different from one gene to another, although their sequence is conserved throughout evolution (even in distant species of flies). Of particular interest is the fact that the promoters of *rh3* and *rh4*, which are both expressed in R7 cells, share almost no sequence identity. Similarly, *rh5* and *rh6*, which are both expressed in R8 have very limited similarities. This likely reflects the fact that these genes are regulated by different mechanisms: *rh4* and *rh6* represent a "ground state" in R7***y*** and R8***y***, respectively, and therefore are likely to be essentially controlled by negative regulation in other PRs. The promoters of *rh3* and *rh5* represent the "acquired state" in R7***p*** and R8***p***, respectively, and are likely to be controlled by positive regulation. Deletion and site-directed mutagenesis experiments have identified regions of the different promoters that are required to achieve subtype specificity, and the results are generally consistent with this model. Our laboratory is actively looking for the trans-activating factors controlling *rh* expression and has already identified several regulators including the homeodomain protein Orthodenticle (Tahayato et al., in prep.).

The expression of *rh5* in R8 is dictated by R7 since an R7 expressing *rh3* is required to specify an R8 cell expressing *rh5* (see Sect. 3). *rh6* expression is therefore the default state in R8 and its expression is expanded in genetic backgrounds that affect R7 (e.g., *sevenless*) or prevent the communication between R7 and R8 (R. Sonneville, F. Pichaud, C. Desplan, in prep.). It is therefore likely that a repressor is expressed in the R8***p*** subtype to prevent *rh6* expression. One possible model is that a pathway allowing communication between R7***p*** and R8***p*** leads to the induction of a factor in R8***p*** that acts both as a specific activator of *rh5* and as a repressor of *rh6*. Alternatively, the activator and repressor could be distinct molecules. In these models, there is no need for a repressor of *rh5* in R8***y*** since *rh5* expression needs a paracrine input from an R7***p*** (expressing *rh3*). A global view of *rh* transcriptional regulation is presented in Fig. 2.

Acknowledgements. We wish to thank Jessica Treisman and Justin Blau for their critical comments on the manuscript.

References

Arikawa K, Inokuma K, Eguchi E (1987) Pentachromatic Visual System in a Butterfly. Naturwissenschaften 74:297–298

Armett-Kibel C, Meinertzhagen I (1985) The long visual fibers of the Dragonfly optic lobe: their cells of origin and lamina connections. J Comp Neurol 242:459–474

Asenjo A, Rim J, Oprian D (1994) Molecular determinants of human red/green color discrimination. Neuron 12:1131–1138

Berghammer AJ, Klingler M, Wimmer EA (1999) A universal marker for transgenic insects. Nature 402:370–371

Bernard GD (1979) Red-absorbing visual pigments of butterflies. Science 203:1125–1127

Braitenberg V, Strausfeld NJ (1973) Principles of the mosaic organization in the visual system's neuropil of *Musca domestica.* In: Jung AR (ed) Handbook of sensory physiology. Springer, Berlin Heidelberg New York, pp 631–659

Briscoe AD (1998) Molecular diversity of visual pigments in the butterfly *Papilio glaucus.* Naturwissenschaften 85:33–35

Campos-Ortega J, Strausfled N (1972) The columnar organization of the second synaptic region of the visual system of *Musca domestica.* Z Zellforsch 124:561–585

Chevesich J, Kreuz AJ, Montell C (1997) Requirement for the PDZ domain protein, INAD, for localization of the TRP store-operated channel to a signaling complex. Neuron 18:95–105

Chou W, Hall K, Wilson D, Wideman C, Townson S, Chadwell L, Britt S (1996) Identification of a novel opsin reveals specific patterning of the R7 and R8 photoreceptor cells. Neuron 17:1101–1115

Chou W, Huber A, Bentrop J, Schulz S, Schwab K, Chadwell L, Paulsen R, Britt S (1999) Patterning of the R7 and R8 cells of *Drosophila*: evidence for induced and default cell-fate specification. Development 126:607–616

Chow RL, Altmann CR, Lang RA, Hemmati-Brivanlou A (1999) Pax6 induces ectopic eyes in a vertebrate. Development 126:4213–4222

Clandinin TR, Zipursky SL (2000) Afferent growth cone interactions control synaptic specificity in the *Drosophila* visual system. Neuron 28:427–436

Clyne PJ, Warr CG, Freeman MR, Lessing D, Kim J, Carlson JR (1999) A novel family of divergent seven-transmembrane proteins: candidate odorant receptors in *Drosophila.* Neuron 22:327–338

Clyne PJ, Warr CG, Carlson JR (2000) Candidate taste receptors in *Drosophila.* Science 287: 1830–1834

Cronin T, Marshall N (1989) A retina with at least ten spectral types of photoreceptors in a mantis shrimp. Nature 339:137–140

Czerny T, Halder G, Kloter U, Souabni A, Gehring WJ, Busslinger M (1999) twin of eyeless, a second Pax-6 gene of *Drosophila,* acts upstream of eyeless in the control of eye development. Mol Cell 3:297–307

Desplan C (1997) Eye development: governed by a dictator or a junta? [comment]. Cell 91:861–864

Dietrich, W (1909) Die Facettenaugen der Dipteren. Z Wiss Zool 92:465–539

Eguchi E (1999) Polarized light vision and Rhabdom. In: Eguchi E, Tominaga Y (eds) Atlas of arthropods sensory receptors. Springer, Berlin Heidelberg New York, pp 33–46

Fortini ME, Rubin GM (1990) Analysis of *cis*-acting requirements of the Rh3 and Rh4 genes reveals a bipartite organization to rhodopsin promoters in *Drosophila melanogaster.* Genes Dev 4:444–463

Franceschini N (1985) Early processing of colour and motion in a mosaic visual system. Neurosci Res Suppl 2:S17–49

Franceschini N (1997) Combined optical, neuroatomical, electrophysiological and behavioural studies on signal processing in the fly compound eye. In: Taddei-Ferreti C (ed) Biocybernetics of vision. World Scientific, Singapore, pp 341–361

Franceschini N, Kirschfeld K (1971) In vivo optical study of photoreceptor elements in the compound eye of *Drosophila.* Kybernetik 8:1–13

Franceschini N, Hardie R, Ribi W, Kirschfeld K (1981a) Sexual dimorphism in a photoreceptor. Nature 291:241–244

Franceschini N, Kirschfeld K, Minke B (1981b) Fluorescence of photoreceptor cells observed in vivo. Science 213:1264–1267

Fukushi T (1994) Colour perception of single and mixed monochromatic lights in the blowfly *Lucilia cuprina*. J Comp Physiol 175:15–22
Gehring WJ, Ikeo K (1999) Pax 6: mastering eye morphogenesis and eye evolution [see comments]. Trends Genet 15:371–377
Goldsmith T (1990) Optimization, constraint, and history in the evolution of the eyes. Q Rev Biol 65:281–322
Goodman LJ (1981) Organisation and physiology of the insect dorsal ocellar system. In: Autrum H (ed) Comparative physiology and evolution of vision in invertebrates. Invertebrate visual centers and behavior II, vol VII/6 C. Springer, Berlin Heidelberg New York, pp 211–286
Hafen E, Basler K, Edstroem JE, Rubin GM (1987) Sevenless, a cell-specific homeotic gene of *Drosophila*, encodes a putative transmembrane receptor with a tyrosine kinase domain. Science 236:55–63
Halder G, Callaerts P, Gehring WJ (1995a) Induction of ectopic eyes by targeted expression of the eyeless gene in *Drosophila* [see comments]. Science 267:1788–1792
Halder G, Callaerts P, Gehring WJ (1995b) New perspectives on eye evolution. Curr Opin Genet Dev 5:602–609
Hamdorf K, Paulsen R, Schwemer J (1973) Photoregeneration and sensitivity control of photoreceptors of invertebrates. In: Langer H (ed) Biochemistry and physiology of visual pigments. Springer, Berlin Heidelberg New York
Hardie R (1979) Electrophysiological analysis of fly retina. I. Comparative properties of R1-R6 and R7-R8. J Comp Physiol 129:19–33
Hardie R (1985) Functional organization of the fly retina. In: Ottoson D (ed) Progress in sensory physiology, vol 5. Springer, Berlin Heidelberg New York
Hardie RC, Franceschini N, Ribi W, Mac Intyre P (1981) Distribution and properties of sex-specific photoreceptor in the fly *Musca domestica*. J Comp Physiol A145:139–152
Hausen K (1984) The lobula complex of the fly: Structure, function and significance in visual behavior. In: Ali MA (ed) Photoreception and vision in Invertebrates. Plenum Press, New York, pp 523–559
Hengstenberg R, Krapp H, Hengstenberg B (1997) Visual sensation of self-motions in the blowfly *Calliphora*. In: Taddei-Ferretti C (ed) Biocybernetics of vision, vol 2. World Scientific, Singapore, pp 53–70
Hiesinger PR, Scholz M, Meinertzhagen IA, Fischbach KF, Obermayer K (2001) Visualization of synaptic markers in the optic neuropils of *Drosophila* using a new constrained deconvolution method [In Process Citation]. J Comp Neurol 429:277–288
Huber A, Schulz S, Bentrop J, Groell C, Wolfrum U, Paulsen R (1997) Molecular cloning of *Drosophila* Rh6 rhodopsin: the visual pigment of a subset of R8 photoreceptor cells. FEBS Lett 406:6–10
Kirschfeld K (1967) Die Projektion der optischen Umwelt auf das Raster der rhabdomere im komplexauge von *Musca*. Exp Brain Res 3:248–270
Kirschfeld K, Franceschini N (1977) Evidence for a sensitising pigment in fly photoreceptors. Nature 269:386–390
Kitamoto J, Sakamoto K, Ozaki K, Mishina Y, Arikawa K (1998) Two visual pigments in a singe photoreceptor cell: identification and histological localization of three mRNAs encoding visual pigment opsins in the retina of the butterfly *Papilio xuthus*. J Exp Biol 201:1255–1261
Labhart T, Meyer EP (1999) Detectors for polarized skylight in insects: a survey of ommatidial specializations in the dorsal rim area of the compound eye. Microsc Res Tech 47:368–379
Lehrer M, Srinivasan MV (1994) Active vision in honeybees: task-oriented suppression of an innate behaviour. Vision Res 34:511–516
Lin S, Kochendoerfer G, Carroll K, Wang D, Mathies R, Sakmar T (1998) Mechanisms of spectral tuning in blue cone pigments. J Biol Chem 273:24583–24591
Lipson SG, Lipson H, Tannhauser DS (1995) Optical waveguides and modulated media. In: Press CU (ed) Optical physics. Cambridge University Press, Cambridge, pp 263–276
Meinertzhagen IA, O'Neil SD (1991) Synaptic organization of columnar elements in the lamina of the wild type in *Drosophila melanogaster*. J Comp Neurol 305:232–263

Menzel R (1979) Invertebrate photoreceptors. In: Autrum H (ed) Spectral sensitivity and color vision in invertebrates. Springer, Berlin Heidelberg New York, pp 503–580

Mismer D, Michael WM, Laverty TR, Rubin GM (1988) Analysis of the promoter of the Rh2 opsin gene in *Drosophila melanogaster*. Genetics 120:173–180

Mizunami M (1999) Ocelli. In: Eguchi E, Tominaga Y (eds) Atlas of arthropods sensory receptors. Springer, Berlin Heidelberg New York,, pp 71–78

Mombaerts P, Wang F, Dulac C, Vassar R, Chao SK, Nemes A, Mendelsohn M, Edmondson J, Axel R (1996) The molecular biology of olfactory perception. Cold Spring Harb Symp Quant Biol 61:135–145

Montell C, Jones K, Zuker C, Rubin G (1987) A second opsin gene expressed in the ultraviolet-sensitive R7 photoreceptor cells of *Drosophila melanogaster*. J Neurosci 7:1558–1566

Niemeyer BA, Suzuki E, Scott K, Jalink K, Zucker CS (1996) The *Drosophila* light-activated conductance is composed of the two channels TRP and TRPL. Cell 85:651–659

Okano Y, Kojima D, Fukada Y, Shichida Y, Yoshizawa T (1992) Primary structures of chicken cone visual pigments: vertebrate rhodopsins have evolved out of cone visual pigments. Proc Natl Acad Sci USA 89:5932–5936

Papatsenko D, Sheng G, Desplan C (1997) A new rhodopsin in R8 photoreceptors of *Drosophila*: evidence for coordinate expression with Rh3 in R7 cells. Development 124:1665–1673

Papatsenko D, Nazina A, Desplan C (2001) A conserved regulatory element present in all *Drosophila Rhodopsin* genes mediates Pax6 functions and participates in the fine-tuning of cell-specific expression. Mech Dev (in press)

Picaud S, Peichl L, Franceschini N (1990) Dye-induced "photo-degeneration" and "photo-permeabilization" of mammalian neurons in vivo [published errata appear in Brain Res 1991 Jan 18; 539(1):179 and 1991 Mar 1; 542(2):360]. Brain Res 531:117–126

Pichaud F, Desplan C (2001) A new visualization approach for identifying mutations that affect differentiation and organization of the *Drosophila* ommatidia. Development (in press)

Pollock J, Benzer S (1988) Transcript localization of four opsin genes in the three visual organs of *Drosophila*: RH2 is ocellus specific. Nature 333:779–782

Quiring R, Walldorf U, Kloter U, Gehring WJ (1994) Homology of the eyeless gene of *Drosophila* to the Small eye gene in mice and Aniridia in humans [see comments]. Science 265:785–789

Ready DF, Hanson TE, Benzer S (1976) Development of the *Drosophila* retina, a neurocrystalline lattice. Dev Biol 53:217–240

Ribi W (1981) The first optic ganglion of the bee IV. Synaptic fine structure and connectivity patterns of receptor cell axons and first order interneurones. Cell Tissue Res 215:443–464

Rossel S (1988) Polarization sensitivity in compound eyes. In: Stavenga DG, Hardie RC (eds) Facets of vision. Springer, Berlin Heidelberg New York, pp 298–316

Salvini-Plawen LV, Mayr E (1977) On the evolution of photoreceptors and eyes. Evol Biol 10: 207–263

Schuppe H, Hengstenberg R (1993) Optical properties of the ocelli of *Calliphora erythrocephala* and their role in the dorsal light response. J Comp Physiol 173:143–149

Sheng G, Thouvenot E, Schmucker D, Wilson DS, Desplan C (1997) Direct regulation of rhodopsin 1 by Pax-6/eyeless in *Drosophila*: evidence for a conserved function in photoreceptors. Genes Dev 11:1122–1131

Srinivasan MV (1993) How insects infer range from visual motion. Rev Oculomot Res 5:139–156

Strausfeld N (1989) Beneath the compound eye: neuroanatomical analysis and physiological correlates in the study of insect vision. In: Stavenga D, Hardie R (eds) Facets of vision. Springer, Berlin Heidelberg New York

Strausfeld N, Campos-Ortega J (1977) Vision in insects: pathways possibly underlying neural adaptation and lateral inhibition. Science 195:894–897

Toh Y, Tateda H (1991) Structure and function of the insect ocellus. Zool Sci 8:395–413

Troje N (1993) Spectral categories in the learning behaviour of blowflies. Z Naturforsch 48*c*: 96–104

Trujillo-Cenoz O (1969) Some aspect of the structural organization of the medulla in muscoid flies. J Ultrastruct Res 27:533–553

Tsunoda S, Sierralta J, Sun Y, Bodned R, Suzuki E, Bechker A, Socolich M, Zucker CS (1997) A multivalent PDZ-domain protein assembles signalling complexes in a G-protein-coupled cascade. Nature 388:243–249

Van Hateren JH (1987) Neural superposition and oscillations in the eye of the blowfly. J Comp Physiol A161:849–855

Vosshall LB, Amrein H, Morozov PS, Rzhetsky A, Axel R (1999) A spatial map of olfactory receptor expression in the *Drosophila* antenna. Cell 96:725–736

Wilcox M, Franceschini N (1984) Illumination induces dye incorporation in photoreceptor cells. Science 225:851–854

Wilson DS, Guenther B, Desplan C, Kuriyan J (1995) High resolution crystal structure of a paired (Pax) class cooperative homeodomain dimer on DNA. Cell 82:709–719

Wilson DS, Sheng G, Jun S, Desplan C (1996) Conservation and diversification in homeodomain-DNA interactions: a comparative genetic analysis. Proc Natl Acad Sci USA 93:6886–6891

Wison M (1978) The functional organisation of the locust ocelli. J Comp Physiol 124:297–316

Wolff T, Ready DF (1991) The beginning of pattern formation in the *Drosophila* compound eye: the morphogenetic furrow and the second mitotic wave. Development 113:841–850

Yang EC, Osorio D (1996) Spectral responses and chromatic processing in the dragonfly lamina. J Comp Physiol A 178:543–550

Yu X, Leconte L, Martinez JA, Barnstable CJ (1996) Ret 1, a cis-acting element of the rat opsin promoter, can direct gene expression in rod photoreceptors. J Neurochem 67:2494–2504

Zuker CS (1994) On the evolution of eyes: would you like it simple or compound? Science 265:742–743

Zuker CS, Montell C, Jones K, Laverty T, Rubin GM (1987) A rhodopsin gene expressed in photoreceptor cell R7 of the *Drosophila* eye: homologies with other signal-transducing molecules. J Neurosci 7:1550–1557

Zuker CS, Mismer D, Hardy R, Rubin GM (1988) Ectopic expression of a minor *Drosophila* opsin in the major photoreceptor cell class: distinguishing the role of primary receptor and cellular context. Cell 53:475–482

Developmental Regulation Through Protein Stability

Janice A. Fischer[1]

1 Introduction

The idea that differential gene expression is critical to the establishment of different cell identities is well-worn into the psyches of all scientists who think about the cellular dynamics of development. Equally commonplace is the idea that temporally and spatially dynamic gene expression is quite often regulated at the level of transcription initiation. More exotic forms of regulation are also well-known, including mRNA splicing, mRNA and protein localization, protein-protein interactions and protein modification. Most recently, it has become apparent that specific alteration of protein stability is a widely used mechanism for controlling the dynamics of important cellular regulators. For example, the levels of cyclins and transcription factors are controlled by specifically targeted protein degradation via the ubiquitin/proteasome pathway.

As cell cycle dynamics are important features of pattern formation in the eye, control of protein stability affects eye development through cyclin regulation. However, studies of eye development have revealed two previously unknown roles for protein degradation in cell fate determination. In the R7 cell determination pathway, Tramtrack protein (Ttk88) is degraded in response to Ras1 activation by the Sevenless receptor (Sev). In addition, a cell communication pathway that prevents ectopic photoreceptor recruitment early in eye development may involve differential stability of the endocytosis protein, Liquid facets (Lqf).

2 The Ubiquitin/Proteasome Pathway

Ubiquitin (Ub) is a ubiquitous, highly conserved (from yeast to humans) 76 amino acid polypeptide (Hershko 1998). Ub is encoded by five different *Drosophila* genes; two Ub-ribosomal protein fusion genes, one gene where Ub is fused to a non-ribosomal protein, a single monomeric Ub gene, and a poly-Ub gene (Lee et al. 1988; Flybase 1999; Rubin et al. 2000). Ub monomers are

[1] The University of Texas at Austin, Section of Molecular Cell and Developmental Biology, The Institute for Cellular and Molecular Biology, Moffett Molecular Biology Building, 2500 Speedway, Austin, Texas 78712, USA

Results and Problems in Cell Differentiation, Vol. 37
K. Moses (Ed.): Drosophila Eye Development

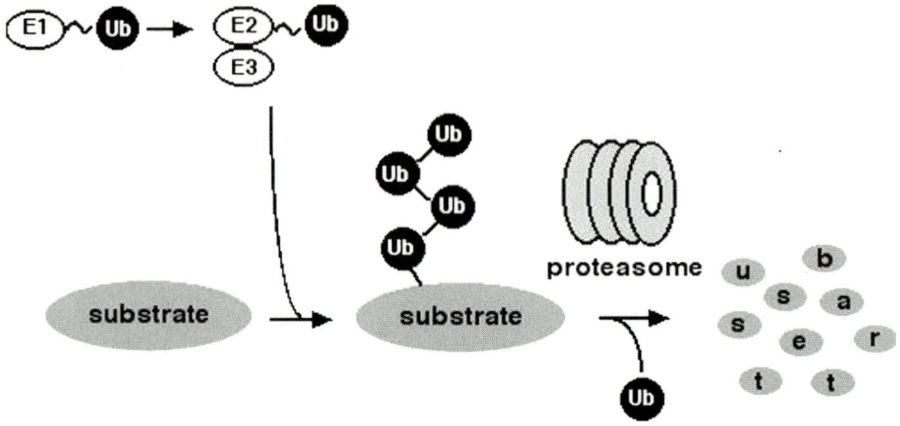

Fig. 1. The Ub pathway. Protein substrates tagged with isopeptide-linked Ub chains are recognized by the proteasome and proceed through the proteasome's catalytic core where they are degraded. Specificity of ubiquitination is provided by E2s and E3s. First, an E1 forms a high energy bond with Ub, which is then transferred to an E2. The E2 transfers Ub to a specific substrate with the help of an E3, which contacts a particular target protein The E3 can be a multi-subunit protein complex

attached covalently to other proteins, usually through isopeptide linkage with internal Lys residues of the substrate protein and the terminal Gly of Ub (Scheffner et al. 1998). Ub conjugation serves two main functions in the cell. Monoubiquitination can modulate protein function, similarly to phosphorylation (Chen et al. 1996; Hicke 1999; Strous and Govers 1999). Alternatively, additional Ub monomers can attach to the first one, usually via their internal Lys residues, to form a Ub chain, which directs the substrate protein to the proteasome, a multi-subunit proteolytic complex (Lupas and Baumeister 1998; Pickart 1998; Rechsteiner 1998).

Ubiquitination of a substrate occurs by a series of enzymatic reactions (Scheffner et al. 1998) in which Ub is transferred from an E1, or Ub-activating enzyme, to an E2, or Ub-conjugating enzyme, and then to the substrate with the help of an E3, or Ub-protein ligase (Fig. 1). E3s provide substrate specificity to ubiquitination by the E2. Some E3s are multi-subunit protein complexes. For example, in yeast, *Drosophila*, and vertebrates, regulated degradation of some proteins occurs via the Skip 1, Cullin, F-box protein (SCF) complex, which contains an E2 and a multi-subunit E3 containing an F-box/WD40 protein that recognizes a particular phosphorylated substrate (Deshaies 1999; Maniatis 1999). As in other organisms, there are large families of E2 and E3 enzymes in *Drosophila* (Flybase 1999; Rubin et al. 2000), which provide opportunities for a wide variety of specificities.

There are also two large families of deubiquitinating enzymes (DUBs) in *Drosophila* (Flybase 1999; Rubin et al. 2000), as in other organisms. All

DUBs cleave peptide and/or isopeptide Ub-protein bonds (Wilkinson and Hochstrasser 1998). The two DUB enzyme families, called Ub C-terminal hydrolases (Uchs) and Ub-specific processing proteases (Ubps), differ in the structure of their catalytic domains but may have overlapping functions (Wilkinson and Hochstrasser 1998). Many of these enzymes are likely needed to cleave newly translated Ub fusion proteins or poly-Ub into Ub monomers and a few DUBs have been shown to play other general housekeeping roles in proteolysis. As will be discussed below, the *Drosophila* Ubp, Fat facets, is thought to regulate the ubiquitination of particular substrates, by cleaving the Ub chain from the proteins, thereby preventing their proteasomal degradation (Huang et al. 1995).

3 Regulation of Ttk88 Protein Stability

The best understood example of how regulated protein degradation determines cell fate is in the R7 determination pathway (Fig. 2). The protein substrate here is Ttk88, a Broad complex, Tramtrack, Bric-a-brac/Pox virus and Zinc finger (BTB/POZ) domain family transcription factor with two Cys-His Zn fingers (Harrison and Travers 1990; Read and Manley 1992). Ttk88 is a transcriptional repressor of neural determination genes, including the R7 differentiation gene *prospero* (Xu et al. 2000). Ttk88 is normally expressed only in the undifferentiated cells surrounding the developing R-cells in the eye, and in non-neural cells of the facet (Lai et al. 1996; Lai and Li 1999). It is essential that Ttk88 remain on in the cone cells to prevent them from differentiating as R7s (Lai et al. 1996). In R7, however, Ttk88 must be eliminated or it will prevent the R7 precursor from adopting a neural fate (S. Li et al. 1997; Tang et al. 1997).

Two proteins, Sina and Phyl, function together as an E3 to ubiquitinate Ttk88, thereby eliminating it from the R7 cell (S. Li et al. 1997; Tang et al. 1997). Sina contains a RING finger domain, characteristic of E3s (Carthew and Rubin 1990; Joaziero and Weissmann 2000) and Phyl is a unique protein (Chang et al. 1995; Dickson et al. 1995). Transcription of *phyl* is repressed (although not necessarily directly) by Yan, an E twenty-six (ETS)-domain transcriptional repressor (Lai and Rubin 1992; Tei et al. 1992; O'Neill et al. 1994; Dickson et al. 1995). One response to Ras1/MAPK pathway activation by Sev is phosphorylation of Yan by MAPK, which signals Yan to leave the nucleus (O'Neill et al. 1994; Rebay and Rubin 1995). The absence of Yan allows *phyl* transcription, and in this manner, Sev signaling (and possibly also Egfr signaling) results in Ub/proteasome-mediated degradation of Ttk88.

Ttk88 stability in R7 is also controlled by an F-box/WD40 repeat-type E3 protein called Ebi (Dong et al. 1999). Ebi may function in a pathway parallel to that of Sina and Phyl, to control the stability of Ttk88 in a constitutive manner.

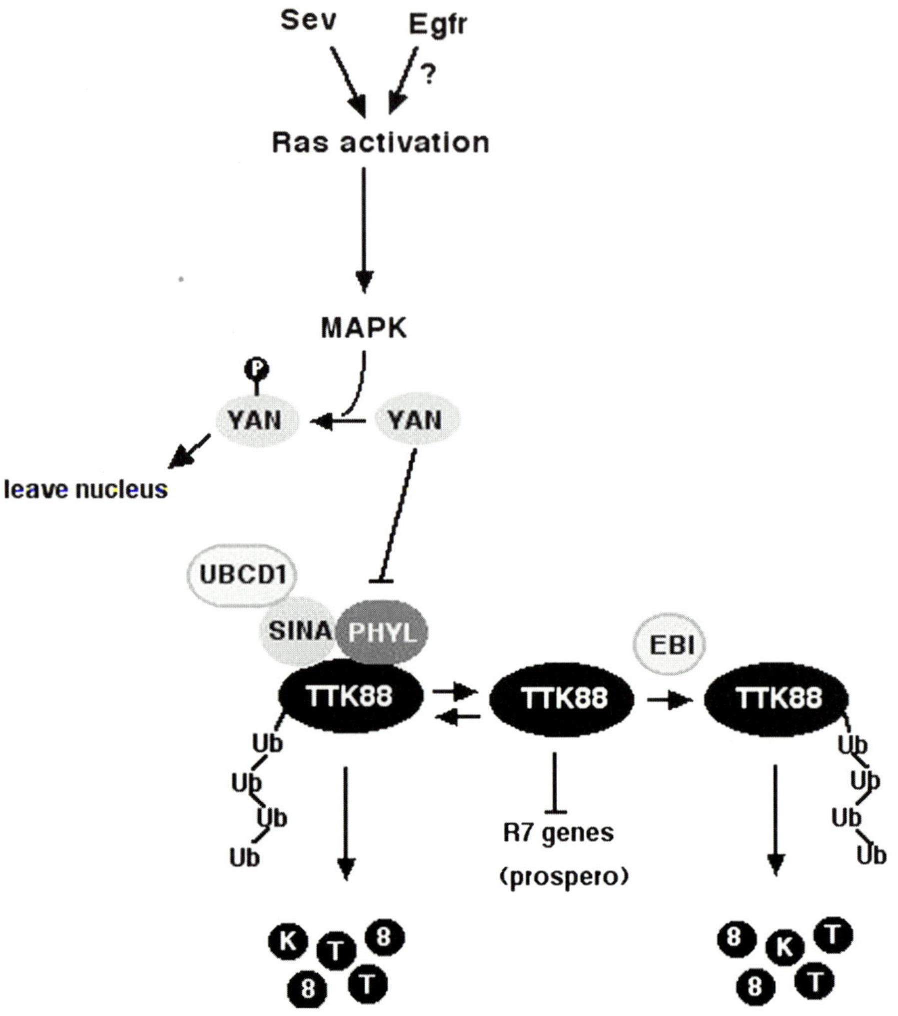

Fig. 2. Regulated Ub-mediated degradation of Ttk88. In R7, Ttk88 is a transcriptional repressor of R7 differentiation genes, such as *prospero*. Ttk88 is rapidly degraded upon Sev activation (and possibly also by Egfr activation) through the following pathway. Activation of the Ras/MAPK pathway by Sev results in the phosphorylation by MAPK of Yan, a transcriptional repressor that represses neural differentiation. Phosphorylated Yan exits the nucleus, thereby relieving (either directly or indirectly) transcriptional repression of *phyl*. Phyl and Sina constitute an E3 that with an E2, probably UbcD1, ubiquitinates Ttk88. Ebi, also an E3, may determine the constitutive level of Ttk88

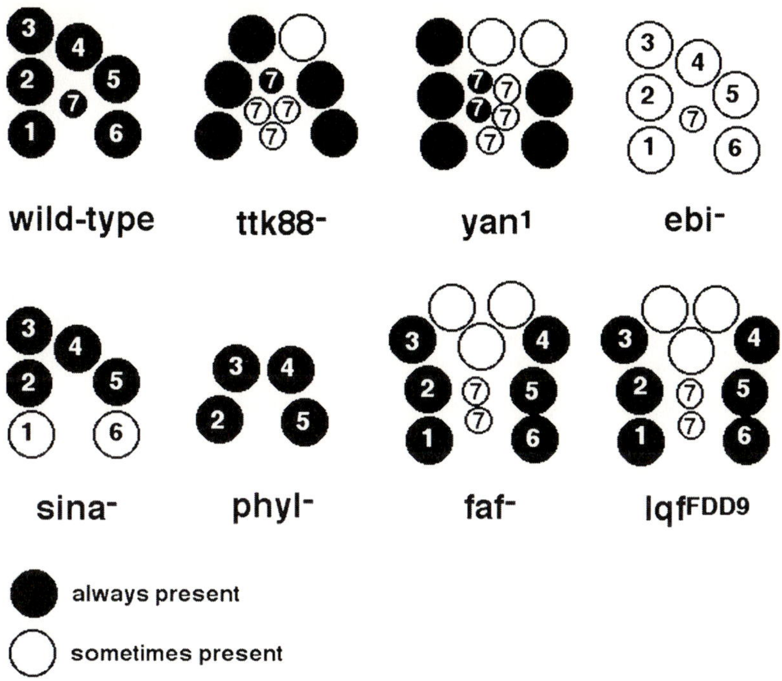

Fig. 3. Mutant ommatidial phenotypes. The R-cells present in the ommatidia of a variety of homozygous mutant adults are shown. Null phenotypes are shown except for *ttk*, *yan*, and *ebi*. The *ttk* gene encodes two proteins (Ttk88 and Ttk69) with different roles in eye development (Read and Manley 1992; Xiong and Montell 1996; Lai et al.1996; Lai and Li 1999). The phenotype of the ttk^1 allele, which specifically eliminates Ttk88 function (Xiong and Montell 1996; Lai and Li 1999), is shown. Clones of cells with null *yan* mutations have no R-cells due to an early role of *yan* in the choice between cell division and differentiation (Rogge et al. 1995). The yan^1 allele shown is hypomorphic. Cell clones with null *lqf* mutations have severely malformed ommatidia with a complex phenotype (Fischer et al. 1997). The *lqf* allele shown is hypomorphic (Cadavid et al. 2000)

3.1 Sina and Phyl Regulate Ttk88 Stability

3.1.1 Genetic Evidence

The loss-of-function mutant phenotypes of *sina*, *phyl* and *ttk88* suggest that in R7, *sina* and *phyl* antagonize the function of *ttk88* (Fig. 3). The ommatidia of *sina* and *phyl* mutants are missing R7 (Carthew and Rubin 1990; Chang et al. 1995; Dickson et al. 1995), while in *ttk88* mutants, there are one or several ectopic R7 cells (Xiong and Montell 1993; Lai and Li 1999).

Antagonism between Phyl and Ttk88 was shown more directly in experiments using transgenes that misexpress either *phyl* or *ttk88* in the eye (Tang et al. 1997). Two promoters were used: (1) the *sev* promoter, which is active in a dynamic pattern of cells that includes the M-cells, R34167 and the cone cells

(Bowtell et al. 1989), and (2) the glass multimer reporter (GMR) promoter, which is active in all cells posterior to the morphogenetic furrow (Moses and Rubin 1991; Ellis et al. 1993). Misexpression of Ttk88 in differentiating R-cells (*sev-ttk88* or *GMR-ttk88*) results in adults with externally rough eyes because the R-cells never differentiate (Tang et al. 1997). Misexpression of *phyl* in the eye with *GMR-phyl* or *sev-phyl* (*phyl* is normally expressed only in R167; Chang et al. 1995; Dickson et al. 1995) results in rough eyes also, but in this case they are due to the presence of too many R-cells (Chang et al. 1995; Tang et al. 1997). Remarkably, flies that overexpress both Phyl and Ttk88 simultaneously (flies carrying two transgenes, either *GMR-phyl* and *sev-phyl* or *GMR-ttk88* and *sev-ttk88*) have wild-type eyes, indicating that the functions of Phyl and Ttk88 cancel each other out (Tang et al. 1997).

In all of the genetic experiments described above, Phyl requires Sina to function. In a *sina* mutant background, ubiquitous expression of Phyl (*GMR-phyl*) no longer results in overneuralization and can no longer counteract the effect of Ttk88 overexpression by *GMR-ttk88* (Tang et al. 1997).

Using antibodies to Ttk88, Ttk88 protein levels were visualized in eye discs of many of the genotypes described above. The results mirrored those obtained by observation of the adult eye phenotypes, indicating that Sina and Phyl reduce Ttk88 protein levels in vivo. In larval eye discs, Ttk88 protein is not normally observed in neurons, but only in the basal nuclei of undetermined cells surrounding the developing ommatidia, and in the apical nuclei of cone cells (Lai et al. 1996; Lai and Li 1999). First, in a *phyl* or *sina* mutant background, Ttk88 protein is visible in R7 (S. Li et al. 1997). Second, in *sev-ttk88* transformants, Ttk88 is present in R3/4 and these cells do not become neurons as judged by the absence of Elav protein (Tang et al. 1997). In eye discs expressing both *sev-ttk88* and *sev-phyl*, Ttk88 in R3/4 vanishes and now these cells express Elav and undergo neural determination (Tang et al. 1997). Moreover, the effect of *sev-phyl* on Ttk88 protein was reversed in a *sina* mutant background (Tang et al. 1997). Finally, in *sev-phyl* transformant flies, the level of Tk88 protein in cone cells is much reduced and this effect is nullified in a *sina* mutant background (S. Li et al. 1997).

Several observations reveal that the effects of *phyl* and *sina* on Ttk88 protein levels are not operating at the level of *ttk88* transcription. A *ttk* enhancer trap expresses *lacZ* in all R-cells and cone cells. The level of β-galactosidase is unchanged in *sev-phyl* (S. Li et al. 1997; Tang et al.1997) and in *sev-Ras1*^{V12} (S. Li et al. 1997), although in flies with either of these transgenes, Ttk88 protein levels are altered and the eye phenotype is affected (Ttk88 is decreased in cone cells and they become R7s; S. Li et al. 1997). Similarly, using mRNA in situ, the levels of *ttk88* message in *GMR-ttk88* transformant eye discs are the same in wild-type and *sina* or *phyl* backgrounds (Tang et al. 1997).

3.1.2 Biochemical Evidence

Much biochemical data indicate that the genetic effects of Phyl and Sina on Ttk88 protein levels are direct interactions. First, all three proteins have been shown to bind to each other in a variety of in vitro assays (Kauffman et al. 1996; S. Li et al. 1997; Tang et al. 1997) and Phyl and Ttk88 expressed in S2 cells coimmunoprecipitate (Tang et al. 1997).

Second, the stability of Ttk88 expressed in S2 cells decreases in the presence of Sina and Phyl. In cells transiently co-transfected with Phyl and Ttk88, pulse chase experiments show that Phyl expression decreases the half-life of Ttk88 significantly (S. Li et al. 1997). Sina has little effect on Ttk88 stability in this assay, probably due to high endogenous Sina activity. Similarly, in cells transfected with an inducible *ttk88* expression construct and harvested 3 and 22 h post-induction, there were 15-fold fewer cells that expressed Ttk88 when Phyl and Sina were both cotransfected (Tang et al. 1997).

Finally, Sina- and Phyl-dependent Ttk88 degradation has been shown to occur via the Ub/proteasome pathway in S2 cells. In S2 cells cotransfected with Phyl and HA-Ttk88 expression constructs, among a variety of peptide aldehyde inhibitors, only proteasome inhibitors increased the half-life of Ttk88 in pulse-chase experiments (S. Li et al. 1997). Moreover, ubiquitinated Ttk88 was detected in extracts of S2 cells cotransfected with *phyl* and *HA-ttk88* expression constructs; in the presence of proteasome inhibitors, and using anti-Ub and anti-HA, a ladder of bands corresponding to Ttk88 proteins with Ub chains of various lengths was observed on Western blots (S. Li et al. 1997).

3.1.3 Involvement of UbcD1

An E2, UbcD1, appears to be involved directly in ubiquitination of Ttk88 by Sina and Phyl. Genetic evidence indicates that *UbcD1* facilitates the function of *sina*; *UbcD1* mutants were isolated as suppressors of the rough eye phenotype caused by *sina* overexpression (GMR-sina; Neufeld et al. 1998). Moreover, UbcD1 and Sina interact physically in the yeast two-hybrid assay (Tang et al. 1997).

3.2 Ebi, an E3, Also Controls Ttk88 Degradation

An E3 called Ebi, structurally distinct from Sina, is also implicated in the control of Ttk88 degradation in R7 (Dong et al. 1999). Unlike Sina, which is required in R7 and also weakly in R1 and R6 (Carthew and Rubin 1990), and Phyl, which is required absolutely in R1, 6, 7 (Chang et al. 1995; Dickson et al. 1995), Ebi functions quite broadly in the eye; in *ebi* mutants, any R-cell may be present or missing and *ebi* is not strictly required in any particular R-cell (Dong et al. 1999; Fig. 3). Also unlike Phyl and to some extent Sina, which are expressed in limited cell types in the eye, Ebi is a ubiquitous nuclear protein in the eye (Dong et al. 1999). However, a specific role for *ebi* in the R7 determination pathway is revealed in a particularly sensitive hypomorphic *sev* pathway mutant background (sev^{E4}; *Sos*^{JC2}); mutants in *ebi* behave as dominant enhancers of the R7-missing mutant phenotype while by contrast, *ttk88* is a dominant suppressor (Dong et al. 1999). Thus Ebi function works in the opposite direction in the R7 pathway to Ttk88.

Ebi function decreases the level of Ttk88 in the eye (Dong et al. 1999). In *ebi* mutant flies, the number of Ttk88-expressing nuclei increases and the number of Elav-positive nuclei decreases, indicating delayed neuronal development and persistence of Ttk88 in nucleus. In addition, *ebi* mutants act as dominant suppressors of the mutant eye phenotype caused by *GMR-phyl*. In *GMR-phyl* eyes, Ttk88 is degraded inappropriately and too many neurons form; *ebi* mutants are dominant suppressors of this phenotype and also of Ttk88 degradation in these flies. Similarly, *ebi* mutations suppress Ttk88 degradation and the overneuralization phenotype of constitutively activated *Egfr* (*sev-Tor*^{D}*Egfr*).

How does the function of Ebi relate to that of Sina/Phyl? One possibility is that Ebi forms a complex with Sina and Phyl. A RING finger protein, like Sina, and an F-box/WD40 repeat protein, like Ebi, can be subunits of a single E3, for example, in the SCF complex (Deshaies 1999; Joaziero and Weissman 2000). However, as Sina and Phyl are absolutely necessary for R7 determination and Ebi is not, Ebi probably functions in a pathway parallel to that of Sina and Phyl.

4 Control of Cell Communication by Faf, a Ubp

The Ubp encoded by *fat facets* (*faf*) regulates a cell communication pathway required to prevent extra cells present in early ommatidial preclusters – the mystery cells – from becoming ectopic photoreceptors (Fischer-Vize et al. 1992). In *faf* mutant flies, most ommatidia have more than six outer R-cells. Faf is a particularly interesting Ubp because it is the only one in any organism for which there is evidence that it removes the Ub chain from particular substrates, thereby preventing their proteasomal degradation (Huang et al. 1995). The cell communication pathway regulated by Faf is not understood, however a candidate substrate for Faf has been identified called Liquid facets (Lqf), a component of the endocytosis complex (Cadavid et al. 2000). Thus, it appears that Faf may regulate cell communication by regulating endocytosis.

4.1 Faf Prevents the Mystery Cells from Becoming R-Cells

Flies lacking all *faf* gene function are viable and nearly completely normal (*faf* mutant females are sterile) except for a rough eye phenotype due to most ommatidia having extra photoreceptors (Fig. 3; Fischer-Vize et al. 1992). The extra photoreceptors are most often outer (R1–6) cells, which originate from the mystery cells. There are also sometimes extra R7 cells and occasionally R7 is missing; the developmental origins of the R7 defects have not been determined.

4.1.1 Faf Functions Outside R-Cells

Two different kinds of experiments suggest the surprising conclusion that *faf* function is required in cells outside the facet preclusters early in eye development. Analysis of mosaic facets suggests that the focus of *faf* activity is outside of the R-cells and the mystery cells in the developing preclusters (Fischer-Vize et al. 1992). The results of transgene rescue experiments are consistent with this interpretation of the mosaic analysis (Huang and Fischer-Vize 1996). A variety of transgenes (promoter-*faf* or *UAS-faf* with different *Gal4* drivers) were tested for their ability to substitute for the endogenous *faf* gene in the eye. Only transgenes that express *faf* early in eye development, prior to the mystery cells leaving the preclusters, were able to complement the *faf* rough eye phenotype. Of particular note is the observation that *glrs-faf* (glrs stands for *gl*ass *r*esponsive *s*hort and is essentially identical to GMR (see above)) and *sev-faf* transgenes, each of which expresses *faf* in the mystery cells, do not rescue the rough eye phenotype of *faf* mutants. Of all the transgenes tested, the best rescuer is *ro-faf*, which expresses *faf* in the pattern of the *rough* gene; cells surrounding the facet preclusters in the furrow and in R2534 posterior to the furrow. The rescue by *ro-faf*, but not *glrs-faf* nor *sev-faf*, provides strong support for the idea that *faf* is required in the cells outside the R-cells. There is no known pathway by which cells surrounding the preclusters signal the mystery cells to leave the clusters.

Faf expression in the eye disc has been visualized by a *faf-lacZ* translational fusion transgene, which includes the N-terminal 15% (~400 amino acids) of the Faf protein (Fischer-Vize et al. 1992). β-galactosidase is detected most strongly in a stripe anterior to the furrow and then more weakly in all cells posterior to the furrow. Curiously, although the expression of *ro-faf* within the morphogenetic furrow is sufficient to rescue the *faf* eye phenotype, no β-galactosidase was detected in the furrow. Presumably, although Faf normally functions just anterior or posterior to the furrow, slightly later or earlier expression is sufficient.

4.1.2 Faf Indirectly Downregulates Egfr Activity in R-Cells

The results of genetic experiments suggest that in *faf* mutants, the mystery cells require Ras1 pathway activity for their determination as R-cells (Isaksson et al. 1997). Double mutants of a hypomorphic *Raf* mutation and a *faf* null display the *Raf* mutant phenotype, which is too few R-cells. In addition, mutations in three transcription factors that work downstream of Ras1 in R-cells interact with *faf* mutations genetically in a way that suggests that Faf antagonizes Ras1 activity. Mutations in *pointed* and *D-jun*, which facilitate Ras1 signal transduction, are dominant suppressors of *faf*, and mutations in the *yan* gene, which antagonizes Ras1 signaling, is a dominant enhancer of *faf*. As all of these interactions are too weak to be visible externally in the eye, none of these three genes were identified in *faf* modifier screens (see below).

The results of two additional experiments suggest that one consequence of Faf activity is the down-regulation of Egfr within the R-cells and M-cells (Isaksson et al. 1997). First, the extra R7 cells in *faf* mutants develop independently of Sevenless. As R7 cell determination requires activation of the Ras1/MAPK pathway by both Egfr and Sev, this result suggests that Egfr activity may be increased or prolonged in *faf* mutants, thus obviating the need for a second RTK signal. However, the effect of *faf* on Ras1 signaling within R-cells must be indirect; as described above, *sev-faf* and *glrs-faf*, which express *faf* within the R-cells posterior to the furrow, do not complement *faf* mutations, while *ro-faf*, which expresses *faf* in the non-precluster within the furrow does rescue both the extra outer R-cell and extra R7 phenotypes of *faf* mutants. Second, in *faf* mutants, the levels of D-Jun in R-cells and M-cells are higher and prolonged. Elevated levels of D-Jun in the M-cells cannot be responsible completely for the *faf* mutant phenotype, as D-Jun overexpression in these cells has been shown to cause no phenotype (Bohmann et al. 1994). Also, although vertebrate Jun levels are known to be regulated by ubiquitination (Treier et al. 1994), any effect of Faf on Jun levels must be indirect as Faf functions outside of the R-cells (Fischer-Vize et al. 1992; Huang and Fischer-Vize 1996; Q. Li et al. 1997).

4.2 Faf Activity Antagonizes Ubiquitination and Proteolysis

The most interesting aspect of the Faf protein is that it is appears to regulate the ubiquitination and hence the proteolysis of specific substrates. This model is based on the observation that mutations in two genes, *Pros26* (a.k.a. *l(3)73Ai*) which encodes a structural component of the proteasome (Saville and Belote 1993), and *UbcD1*, which encodes a ubiquitin-conjugating enzyme (Treier et al. 1992), each act as strong dominant suppressors of *faf* in the eye (Huang et al. 1995; Wu et al. 1999). This result indicates that Faf activity antagonizes both ubiquitination and proteolysis. Additional support for this model for Faf function comes from the identification of a candidate Faf substrate, which behaves genetically precisely as predicted by this model (see below).

4.3 The Key Substrate of Faf May Be Lqf, *Drosophila* Epsin

A candidate for the critical substrate of Faf in the eye was identified in a mutagenesis screen for enhancers of the weak eye phenotype of a hypomorphic *faf* allele (Fischer et al. 1997; Cadavid et al. 2000). If Faf deubiquitinates its substrate, thereby saving it from proteasomal degradation, halving the normal level of the substrate protein by mutagenizing one copy in a background where Faf activity is compromised (homozygotes for a hypomorphic *faf* mutation) should render the *faf* eye phenotype much worse. Thus, mutations in the gene encoding Faf's substrate should behave as strong dominant enhancers of *faf*. In addition, if the *faf* eye phenotype is the result of a decrease in the level of the substrate, then mutations in the substrate gene alone ought to have an essential role in eye development. In a screen for enhancers of *faf*, the only enhancer mutations that satisfied both conditions were those in the *lqf* gene, which encodes *Drosophila* epsin (Fischer et al. 1997; Cadavid et al. 2000). Vertebrate epsin is an essential endocytosis complex protein whose precise function is unknown (Chen et al. 1998).

There is strong genetic evidence that Lqf is the substrate of Faf. First, *lqf* mutations behave as extremely strong dominant enhancers of *faf* (Fischer et al. 1997; Cadavid et al. 2000). Second, *lqf* has an essential role in eye development. Clones homozygous for *lqf* null mutations have more severely malformed facets than in *faf* nulls, but homozygotes for a weak *lqf* allele have a similar mutant phenotype to *faf* nulls (Fig. 3), indicating that like *faf*, *lqf* plays a role in preventing ectopic neurogenesis (Fischer et al. 1997; Cadavid et al. 2000). Moreover, the similar phenotypes of weak *lqf* mutations and *faf* null mutations is precisely the result expected if Lqf is the substrate of Faf; if Faf normally increases the level of Lqf, then lowering the level of the substrate (as in homozygotes for weak *lqf* mutations) should phenocopy the *faf* null phenotype. Third, in its role in the *faf* pathway, *lqf* and *faf* function in the same cell group (Fischer-Vize et al. 1992; Huang and Fischer-Vize 1996; Cadavid et al. 2000). Similar to the results with *faf* mutants, mosaic analysis of *lqf* suggests that *lqf* works outside the R-cells. Also, while a *ro-faf* transgene rescues the *faf* eye phenotype, a *ro-lqf* transgene rescues the *lqf* mutant eye phenotype of hypomorphs. Finally, an increase in the level of Lqf completely obviates the need for Faf in the eye (Cadavid et al. 2000). One copy of a genomic *lqf* rescue fragment or one copy of a *ro-lqf* transgene completely rescues the *faf* mutant eye phenotype. This is precisely the expected result if the function of Faf is to increase the level of Lqf.

4.4 Faf Activity Facilitates Endocytosis

There are three main players in endocytosis: (1) clathrin, which forms a cage that engulfs the cell membrane, (2) AP-2, the adapter complex that brings clathrin to a particular place at the cell surface, and (3) dynamin, a GTPase required for pinching off vesicles. Many proteins have been shown to bind AP-2 and these are likely to have temporal or tissue-specific functions in endocytosis (Marsh and McMahon 1999; Mayer 1999).

Epsin is an AP-2-binding protein that is essential for endocytosis in vertebrates and yeast (Chen et al. 1998; Wendland et al. 1999), and genetic evidence suggests that *Drosphila* Lqf likewise plays a positive role in endocytosis

(Cadavid et al. 2000). Mutations in three essential *Drosophila* endocytosis genes, *a-Adaptin* (*a-Ada*, encodes an AP-2 subunit), *shibire* (*shi*, encodes dynamin), and *clathrin heavy chain* (*chc*), each act as dominant enhancers of *lqf* in the eye, with *chc* having the strongest effect (Cadavid *et al.* 2000). In addition, *shi* and *chc* are dominant enhancers of *faf*, again with *chc* having the strongest effect (Cadavid et al. 2000). Thus genetically, both *lqf* and *faf* are facilitators of endocytosis. (Neither *shi* nor *chc* could have been identified in the mutagenesis screens for enhancers of *faf* because both reside on the X chromosome, which was not screened. Also, the effects of *shi* were too weak for the gene to be identified in the screens.)

4.5 Faf also has a Redundant Function in Eye Development

In addition to the essential function of *faf* anterior to, or within the furrow, *faf* also appears to have a normally redundant role in R7 cell fate determination which is revealed only in particular mutant backgrounds (Q. Li et al. 1997). Mutations in *faf* act as enhancers of particular gain-of-function mutations in the *Ras1* and *Rap1* genes ($Ras1^{E63K}$, $Rap1^{V153M}$, $Rap1^{T53M}$), which have a sevenless mutant phenotype. This role in R7 cell fate determination is distinct from the normal role of *faf*, as the *ro-faf* transgene, which complements the *faf* null mutant eye phenotype, does not rescue the rough eye phenotype of the *faf Ras1* or *faf Rap1* double mutant combinations. By contrast, *glrs-faf*, which does not complement the *faf* null mutant eye phenotype, does rescue the sevenless phenotype of *faf Ras1* or *faf Rap1*. In this role in R7 cell fate determination, Faf also appears to function from outside R7, as *sev-faf*, which expresses *faf* in R7, does not rescue the *faf Ras1* or *faf Rap1* double mutant phenotype, but *glrs-faf*, which expresses *faf* inside *and* outside of the facet, does rescue. It is not known whether the substrate of Faf in this role is the same as in its essential earlier role.

4.6 Models for the Faf Pathway

A model for the Faf pathway, based on genetic evidence, is shown in Fig. 4. How can endocytosis in cells adjacent to the mystery cells affect the fates of the mystery cells? There are several examples of ligand/receptor interaction regulation by endocytosis. Travel of a diffusible ligand, like Wingless, through several cell distances on the way to its target receptor requires endocytosis (Bejsovec and Wieschaus 1995). Also, vertebrate Egfr is down-regulated by endocytosis of the activated receptor (Wells et al. 1990; Viera et al. 1996; Wilde et al. 1999). Finally, the Notch receptor may be up-regulated by endocytosis in two different ways. First, used Notch receptors with their intracellular domains cleaved off are known to act as dominant negatives (Lieber et al. 1993; Rebay et al. 1993) and their removal from the cell surface by endocytosis may up-

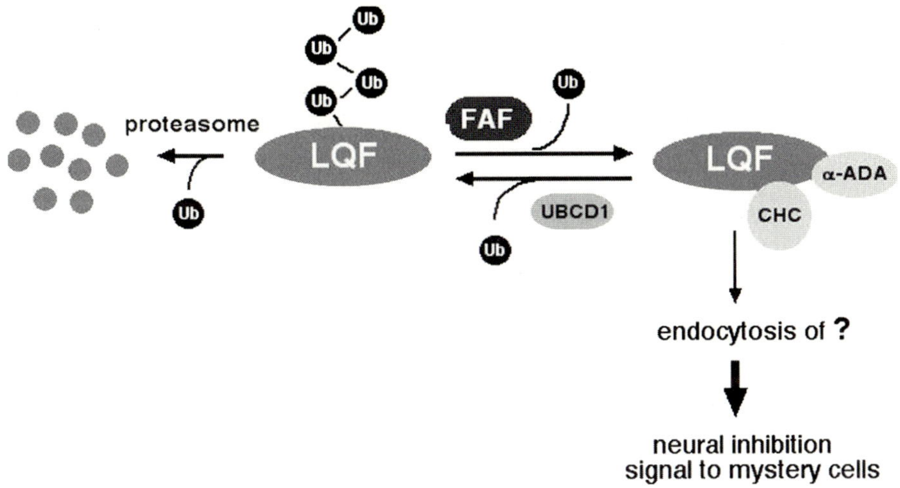

Fig. 4. A model for Faf function. The deubiquitinating enzyme Faf is required in cells surrounding the facet preclusters to send a neural inhibition signal to the mystery cells. Lqf, *Drosophila* epsin, may be the critical substrate of Faf in this pathway. It is proposed that Faf deubiquitinates Lqf, thereby increasing its concentration and enabling it to function as part of the endocytosis complex. Endocytosis of a diffusible ligand, an active or inactive cell-surface receptor, or a ligand-receptor complex would lead to signaling of neighboring cells. Lqf has motifs for binding to clathrin and toα-adaptin, a component of the core adapter complex

regulate Notch signaling (Seugnet et al. 1997). Also, a prerequisite for the cleavage of the intracellular domain of Notch and thus transduction of the activated Notch signal may be *trans*-endocytosis of the extracellular domain of Notch bound to Delta into the Delta-expressing cell (Parks et al. 2000).

Any of these mechanisms could, in theory, apply to the cell communication pathway regulated by Faf. Faf is required in the Rough-expressing cells surrounding the Atonal-expressing precluster cells in the morphogenetic furrow (Huang and Fischer-Vize 1996; Dokucu et al. 1996). Notch and Egfr are both activated in the Rough-expressing cells (Baker et al. 1996; Dominguez et al. 1998; Kumar et al. 1998). Possibly, up or down regulation of either or both of these receptors in these cells could affect the fates of the adjacent precluster cells indirectly. Similarly, it is conceivable that Faf could facilitate the ability of the Rough-expressing cells to *trans*-endocytose Delta bound to the Notch extracellular domain, thereby activating Notch in the mystery cells, which would prevent them from becoming neurons. If Lqf is indeed Faf's critical eye substrate, it appears that the endocytosis complex is a target for regulation of cell communication.

5 Future Directions

Much remains to be understood in the two Ub-regulated processes described here. In the Ttk88 pathway, the relationship between the Sina/Phyl- and Ebi-dependent protein degradation pathways needs to be clarified. It will also be interesting to know if the E3 defined by Sina and Phyl is as complex as SCF and contains additional proteins. Further experiments are needed to determine whether Lqf is indeed the critical substrate for Faf in the eye and exactly how endocytosis regulates cell communication in the morphogenetic furrow.

There are other pathways in the eye where it is likely that Ub/proteasome regulation will play an important role. For example, Ttk69, which plays a positive role in photoreceptor differentiation (Xiong and Montell 1993; Lai and Li 1999), may be regulated by the proteasome (S. Li et al. 1997). In addition, there are genes that function in eye development, like *D-Jun* (Bohmann et al. 1994; Treier et al. 1995; Kockel et al. 1997) and *armadillo* (Wehrli and Tomlinson 1998) which are known to be regulated by Ub-mediated degradation outside the eye (Treier et al. 1994; Maniatis 1999). It is not clear whether Ub regulation is important to their roles in the *Drosophila* eye. Also, the Slmb protein, which encodes an E3 similar in structure to Ebi (Jiang and Struhl 1998; Theodosiou et al. 1998) plays a role in the eye that needs to be better defined (Miletich and Limbourg-Bouchon 2000). Finally, the *non-stop* gene, required in optic ganglia to specify the correct termination site of R1–R6 neurons in the brain, encodes a Ubp whose function is relatively unknown (Martin et al. 1995).

Acknowledgements. I thank past and present members of my laboratory for their many contributions to this work and Richard Carthew for providing me with a preprint of unpublished data and discussions. Work in my laboratory on the Fat facets pathway is funded by the NIH.

References

Baker NE, Yu S, Han D (1996) Evolution of proneural atonal expression during distinct regulatory phases in the developing *Drosophila* eye. Curr Biol 6:1290–1301

Bejsovec A, Wieschaus E (1995) Signaling activities of the *Drosophila wingless* gene are separately mutable and appear to be transduced at the cell surface. Genetics 139:309–320

Bohmann D, Ellis MC, Staszewski LM, Mlodzik M (1994) *Drosophila* Jun mediates Ras-dependent photoreceptor determination. Cell 78:973–986

Bowtell DDL, Kimmel B, Simon MA, Rubin GM (1989) Regulation of the complex pattern of *sevenless* expression in the developing *Drosophila* eye. Proc Natl Acad Sci USA 86:6245–6249

Cadavid ALM, Ginzel A, Fischer JA (2000) The function of the *Drosophila* Fat facets deubiquitinating enzyme in limiting photoreceptor cell number is intimately associated with endocytosis. Development 127:1727–1736

Carthew RW, Rubin GM (1990) *seven in absentia*, a gene required for specification of R7 cell fate in the *Drosophila* eye. Cell 63:561–577

Chang HC, Solomon NM, Wassarman DA, Karim FD, Therrien, Rubin GM, Wolff T (1995) *phyllopod* functions in the fate determination of a subset of photoreceptors in *Drosophila*. Cell 80:463–472

Chen H, Fre S, Slepnev VI, Capua MR, Takei K, Butler MH, Di Fiore PP, De Camilli P (1998) Epsin is an EH-domain-binding protein implicated in clathrin-mediated endocytosis. Nature 394: 793–797

Chen ZJ, Parent L, Maniatis T (1996) Site-specific phosphorylation of IkappaBalpha by a novel ubiquitination-dependent protein kinase activity. Cell 84:853–862

Deshaies RJ (1999) SCF and Cullin/Ring H2-based ubiquitin ligases. Annu Rev Cell Dev Biol 15:435–467

Dickson BJ, Dominguez M, van der Straten A, Hafen E (1995) Control of *Drosophila* photoreceptor cell fates by Phyllopod, a novel nuclear protein acting downstream of the Raf kinase. Cell 80:453–462

Dokucu ME, Zipursky SL, Cagan RL (1996) Atonal, rough and the resolution of proneural clusters in the developing *Drosophila* retina. Development 122:4139–4147

Dominguez M, Wassarman JD, Freeman M (1998) Multiple functions of the EGF receptor in *Drosophila* eye development. Curr Biol 8:1039–1048

Dong X, Tsuda L, Zavitz KH, Lin M, Li S, Carthew RW, Zipursky SL (1999) *ebi* regulates epidermal growth factor receptor signaling pathways in *Drosophila*. Genes Dev 13:954–965

Ellis MC, O'Neill EM, Rubin GM (1993) Expression of *Drosophila* Glass protein and evidence for negative regulation of its activity in non-neuronal cells by another DNA-binding protein. Development 119:855–865

Fischer JA, Leavell SK, Li Q (1997) Mutagenesis screens for interacting genes reveal three roles for *fat facets* during *Drosophila* eye development. Dev Gen 21:167–174

Fischer-Vize JA, Rubin GM, Lehmann R (1992) The *fat facets* gene is required for *Drosophila* eye and embryo development. Development 116:985–1000

Flybase (1999) The Flybase database of the *Drosophila* genome projects and community literature. Available from http://flybase.bio.indiana.edu/. Nucleic Acids Res 27:85–88

Harrison SD, Travers AA (1990) The *tramtrack* gene encodes a *Drosophila* finger protein that interacts with *ftz* transcriptional regulatory region and shows a novel embryonic expression pattern. EMBO J 9:207–216

Hershko A (1998) The ubiquitin system. In: Peters J-M, Harris JR, Finley D (eds) Ubiquitin and the biology of the cell. Plenum Press, New York, pp 1–17

Hicke L (1999) Gettin' down with ubiquitin: turning off cell-surface receptors, transporters and channels. Trends Cell Biol 9:107–111

Huang Y, Fischer-Vize JA (1996) Undifferentiated cells in the developing *Drosophila* eye influence facet assembly and require the Fat facets deubiquitinating enzyme. Development 122: 3207–3216

Huang Y, Baker RT, Fischer-Vize JA (1995) Control of cell fate by a deubiquitinating enzyme encoded by the *fat facets* gene. Science 270:1828–1831

Isaksson A, Peverali FA, Kockel L, Mlodzik M, Bohmann D (1997) The deubiquitination enzyme Fat facets negatively regulates RTK/Ras/MAPK signalling during *Drosophila* eye development. Mech Dev 68:59–67

Jiang J, Struhl G (1998) Regulation of the Hedgehog and Wingless signalling pathways by the F-box/WD40-repeat protein Slimb. Nature 391:493–496

Joaziero CAP, Weissman AM (2000) Ring finger proteins: mediators of ubiquitin ligase activity. Cell 102:549–552

Kauffmann RC, Li S, Gallagher PA, Zhang J, Carthew RW (1996) Ras1 signaling and transcriptional competence in the R7 cell of *Drosophila*. Genes Dev 10:2167–2178

Kockel L, Zeitlinger J, Staszewski LM, Mlodzik M, Bohmann D (1997) Jun in *Drosophila* development: redundant and nonredundant functions and regulation by two MAPK signal transduction pathways. Genes Dev 11:1748–1758

Kumar JP, Tio M, Hsiung F, Akopyan S, Gabay L, Seger R, Shilo B-Z, Moses K (1998) Dissecting the roles of the *Drosophila* EGF receptor in eye development and MAP kinase activation. Development 125:3875–3885

Lai Z-C, Rubin GM (1992) Negative control of photoreceptor development in *Drosophila* by the product of the *yan* gene, an ETS domain protein. Cell 70:609–620

Lai Z-C, Harrison SD, Karim F, Li Y, Rubin GM (1996) Loss of *tramtrack* gene activity results in ectopic R7 cell formation, even in a *sina* background. Proc Natl Acad Sci USA 93:5025–5030

Lai Z-C, Li Y (1999) Tramtrack69 is positively and autonomously required for *Drosophila* photoreceptor development. Genetics 152:299–305

Lee HS, Simon JA, Lis JT (1988) Structure and expression of ubiquitin genes of *Drosophila melanogaster.* Mol Cell Biol 11:4727–2735

Li Q, Hariharan IK, Chen F, Huang Y, Fischer JA (1997) Genetic interactions with *Rap1* and *Ras1* reveal a second function for the *fat facets* gene in *Drosophila* eye development. Proc Natl Acad Sci USA 94:12515–12520

Li S, Li Y, Carthew RW, Lai Z-C (1997) Photoreceptor cell differentiation requires regulated proteolysis of the transcriptional repressor Tramtrack. Cell 90:469–478

Lieber T, Kidd S, Alcamo E, Corbin V, Young MW (1993) Antineurogenic phenotypes induced by truncated Notch proteins indicate a role in signal transduction and may point to a novel function for Notch in nuclei. Genes Dev 7:1949–1965

Lupas A, Baumeister W (1998) The ubiquitin system. In: Peters J-M, Harris JR, Finley D (eds) Ubiquitin and the biology of the cell. Plenum Press, New York, pp 127–146

Maniatis T (1999) A ubiquitin ligase complex essential for the NF-kB, Wnt/Wingless, and Hedgehog signaling pathways. Genes Dev 13:505–510

Marsh M, McMahon HT (1999) The structural era of endocytosis. Science 285:215–220

Martin KA, Poeck B, Roth H, Ebens AJ, Ballard LC, Zipursky LS (1995) Mutations disrupting neuronal connectivity in the *Drosophila* nervous system. Neuron 14:229–240

Mayer BJ (1999) Endocytosis: EH domains lend a hand. Curr Biol 9:R70-R73

Miletich I, Limbourg-Bouchon B (2000) *Drosophila* null *slimb* clones transiently deregulate Hedgehog-independent transcription of *wingless* in all limb discs, and induce *decapentaplegic* transcription linked to imaginal disc regeneration. Mech Dev 93:15–26

Moses K, Rubin GM (1991) *glass* encodes a site-specific DNA-binding protein that is regulated in response to positional signals in the developing *Drosophila* eye. Genes Dev 5:583–593

Neufeld TP, Tang AH, Rubin GM (1998) A genetic screen to identify components of the *sina* signaling pathway in *Drosophila* eye development. Genetics 148:277–286

O'Neill EM, Rebay I, Tjian R, Rubin GM (1994) The activities of two ETS-related transcription factors required for *Drosophila* eye development are modulated by the Ras/MAPK pathway. Cell 78:137–147

Parks AL, Klueg KM, Stout JR, Muskavitch MAT (2000) Ligand endocytosis drives receptor dissociation and activation in the Notch pathway. Development 127:1373–1385

Pickart CM (1998) The ubiquitin system. In: Peters J-M, Harris JR, Finley D (eds) Ubiquitin and the biology of the cell. Plenum Press, New York, pp 19–63

Rebay I, Rubin GM (1995) Yan functions as a general inhibitor of differentiation and is negatively regulated by activation of the Ras/MAPK pathway. Cell 81:857–866

Read D, Manley JL (1992) Alternatively spliced transcripts of the *Drosophila tramtrack* gene encode zinc finger proteins with distinct DNA binding specificities. EMBO J 11:1035–1044

Rebay I, Fehon RG, Artavanis-Tsakonas S (1993) Specific truncation of *Drosophila* Notch define dominant activated and dominant negative forms of the receptor. Cell 74:319–329

Rechsteiner M (1998) The ubiquitin system. In: Peters J-M, Harris JR, Finley D (eds) Ubiquitin and the biology of the cell. Plenum Press, New York, pp 147–189

Rogge R, Green PJ, Urano J, Horn-Saban S, Mlodzik M, Shil B-Z, Hartenstein V, Banerjee U (1995) The role of *yan* in mediating the choice between cell division and differentiation. Development 121:3947–3958

Rubin GM, Yandell MD, Wortman JR, Miklos GLG, Nelson CR et al. (2000) Comparative genomics of the eukaryotes. Science 287:2204–2215

Saville KJ, Belote JM (1993) Identification of an essential gene, *l(3)73Ai*, with a dominant temperature-sensitive lethal allele, encoding a *Drosophila* proteasome subunit. Proc Natl Acad Sci USA 90:8842–8846

Scheffner M, Smith S, Jentsch S (1998) The ubiquitin system. In: Peters J-M, Harris JR, Finley D (eds) Ubiquitin and the biology of the cell. Plenum Press, New York, pp 65–91

Seugnet L, Simpson P, Haenlin M (1997) Requirement for dynamin during Notch signaling in *Drosophila* neurogenesis. Dev Biol 192:585–598

Strous GJ, Govers R (1999) The ubiquitin-proteasome system and endocytosis. J Cell Sci 112:1417–1423

Tang AH, Neufeld TP, Kwan E, Rubin GM (1997) Phyl acts to down-regulate Ttk88, a transcriptional repressor of neuronal cell fates, by a Sina-dependent mechanism. Cell 90:459–467

Tei H, Nihonmatsu I, Yokokura T, Ueda R, Sano Y, Okuda T, Sato K, Hirata K, Fujita SC, Yamamoto D (1992) *pokkuri*, a *Drosophila* gene encoding an E-26-specific (ETS) domain protein, prevents overproduction of the R7 photoreceptor. Proc Natl Acad Sci USA 89:6856–6860

Theodosiou NA, Zhang S, Wang W-Y, Xu T (1998) *slimb* coordinates *wg* and *dpp* expression in the dorsal-ventral and anterio-posterior axes during limb development. Development 125: 3411–3416

Treier M, Seufert W, Jentsch S (1992) *Drosophila UbcD1* encodes a highly conserved ubiquitin-conjugating enzyme involved in selective protein degradation. EMBO J 11:367–372

Treier M, Staszewski LM, Bohmann D (1994) Ubiquitin-dependent c-Jun degradation in vivo is mediated by the δ domain. Cell 78:787–798

Treier M, Bohmann D, Mlodzik M (1995) JUN cooperates with ETS domain protein Pointed to induce photoreceptor R7 fate in the *Drosophila* eye. Cell 83:753–760

Viera AV, Lamaze A, Schmid SL (1996) Control of EGF receptor signaling by clathrin-mediated endocytosis. Science 274:2086–2089

Wehrli M, Tomlinson A (1998) Independent regulation of anterior/posterior and equatorial/polar polarity in the *Drosophila* eye; evidence for the involvement of Wnt signaling in the equatorial/polar axis. Development 125:1421–1432

Wells A, Welsh JB, Lazar CS, Wiley HS, Gill GN, Rosenfeld MG (1990) Ligand-induced transformation by a noninternalizing epidermal growth factor receptor. Science 247:962–964

Wendland B, Steece KE, Emr SD (1999) Yeast epsins contain an essential N-terminal ENTH domain, bind clathrin and are required for endocytosis. EMBO J 18:4383–4393

Wilde A, Beattie EC, Lem L, Riethof DA, Liu SH, Mobley WB, Soriano P, Brodsky FM (1999) EGF receptor signaling stimulates SRC kinase phosphorylation of clathrin, influencing clathrin redistribution and EGF uptake. Cell 96:677–687

Wilkinson KD, Hochstrasser M (1998) The deubiquitinating enzymes. The ubiquitin system. In: Peters J-M, Harris JR, Finley D (eds) Ubiquitin and the biology of the cell. Plenum Press, New York, pp 99–125

Wu Z, Li Q. Fortini M, Fischer JA (1999) Genetic analysis of the role of the *Drosophila fat facets* gene in the ubiquitin pathway. Dev Gen 25:312–320

Xiong W-C, Montell C (1993) *tramtrack* is a transcriptional repressor required for cell fate determination in the *Drosophila* eye. Genes Dev 7:1085–1096

Xu C, Kauffman RC, Zhang J, Kaladny S, Carthew RC (2000) Overlapping activators and repressors delimit transcriptional response to receptor tyrosine kinase signals in the *Drosophila* eye. Cell 103:87–97

Programmed Death in Eye Development

Rebecca Hays, Caroline Craig, and Ross Cagan[1]

1 Introduction

Apoptosis, a morphologically distinct form of programmed cell death, is an essential component of both vertebrate and invertebrate development. The functions of apoptosis vary widely and include the removal of obsolete cells and tissues, the elimination of aberrant, potentially harmful cells, and the sculpting and maintenance of tissue architecture (reviewed in Vaux and Korsmeyer 1999). Many examples of apoptosis in patterning have been described including the separation of digits during limb development (Mori et al. 1995), the ordering of axonal projections to the mammalian visual cortex (So et al. 1990), and interommatidial lattice refinement in the *Drosophila* retina (Cagan and Ready 1989a; Wolff and Ready 1991), the focus of this review.

The fly eye is an excellent system for the study of apoptosis for a number of reasons. The ease of scoring adult phenotypes makes it an ideal subject for genetic screens, and because of the extreme regularity of the retina it is exceptionally sensitive to perturbation. Even minute disruptions of pattern generate an adult rough eye phenotype. As such, a host of mutations have been identified that result in the dysregulation of cell death in the eye, many of them in genes not known to function overtly in apoptosis. Mutations in genes required for specification of the eye field, morphogenetic furrow progression, cell division and differentiation, and neural maintenance all result in increased apoptosis in the retina (reviewed in Bonini and Fortini 1999).

The availability of tissue-specific promoters has made the fly eye a convenient and effective system for the identification and characterization of death molecules. Indeed, much of the study of *Drosophila* apoptosis takes the form of overexpression studies in the retina. However, these studies do not address the regulation of apoptosis in **patterning** the retina.

This review addresses the morphology and regulation of spatially restricted apoptosis in the developing *Drosophila* retina, and visual system mutants that display apoptotic neurodegeneration. It begins with a brief discussion of the molecules that regulate and execute apoptosis. For comprehensive reviews of apoptotic pathways in *Drosophila*, see reference (Abrams 1999; Bangs and White 2000; Meier et al. 2000a; Rusconi et al. 2000).

[1] Department of Molecular Biology and Pharmacology, Washington University School of Medicine, St. Louis, Missouri 63110, USA

Results and Problems in Cell Differentiation, Vol. 37
K. Moses (Ed.): Drosophila Eye Development

2 Downstream Components: Molecules of Death

The genetic basis of apoptosis was first recognized in the nematode C. elegans, made possible due to the invariant number of cells present in the adult worm. Genetic screens for genes which alter the number of cells present in the adult led to the identification of four genes required for the regulation and execution of apoptosis in the worm: *egl-1* (egl, egg laying defective; Conradt and Horvitz 1998), *ced-3* (Yuan and Horvitz 1990), *ced-4* (Yuan and Horvitz 1992), and *ced-9* (ced, cell death abnormal; Hengartner et al. 1992). Loss-of-function mutations in three of these genes, *egl-1*, *ced-3* and *ced-4*, result in a complete loss of developmentally regulated cell death, suggesting that these factors function in the induction of apoptosis (reviewed in Metzstein et al. 1998). In contrast, loss-of-function mutations in *ced-9* cause widespread ectopic apoptosis and early lethality, implicating *ced-9* in the suppression of cell death. All of these genes are highly conserved throughout metazoan development and are critical components of the pathways that regulate and execute apoptosis (Fig. 1). A schematic diagram of apoptotic pathways in *Drosophila* is shown in Fig. 2.

The core mediators of apoptosis are a family of cysteine proteases typified by C. elegans Ced-3. Caspases are present in all cells as inactive zymogens and are central to all known examples of apoptotic cell death (reviewed in Thornberry 1998; Thornberry and Lazebnik 1998). Death stimuli lead to oligomerization of caspases and activation of the enzymes through intrinsic proteolytic activity that removes the N-terminal prodomain and separates large and small caspase subunits (Salvesen and Dixit 1999). Caspase-initiated proteolytic cascades result in many of the biochemical and morphological changes characteristic of apoptosis including cell shrinkage, membrane blebbing, and chromosome fragmentation.

Two classes of caspases have been characterized based on functional and structural differences. Initiator, or apical, caspases act upstream in the transduction of death stimuli. They bear long prodomains containing caspase recruitment domains (CARDs) that serve to mediate procaspase assembly and activation (reviewed in Budihardjo et al. 1999; Kumar 1999). Effector, or executioner, caspases are downstream substrates of initiator caspases and are dependent on them for activation. These enzymes typically bear shorter prodomains than enzymes of the initiator class and act downstream in the proteolytic cascade.

Caspases are by far the largest family of death-associated molecules identified to date. For example, 14 mammalian caspases have been identified, 10 of which have initiator class pro-domains (reviewed in Budihardjo et al. 1999). Two initiator caspases have been characterized in *Drosophila*: Dcp-2/Dredd and Dronc (Inohara et al. 1997; Chen et al. 1998; Dorstyn et al. 1999a). A third, Dream, has been predicted by genome sequence analysis (Vernooy et al. 2000). Three effector class caspases have also been described: Dcp-1, DrICE, and Decay (Fraser and Evan 1997; Song et al. 1997; Dorstyn et al. 1999b). A fourth,

C. elegans	*Drosophila*	Mammals
Ced-3	Dcp-2/Dredd Dronc (Dream) DrICE Dcp-1 Decay (Daydream)	Caspase-8, -10 Caspase-2, -9
Ced-4	Dark/Dapf-1/Hac-1	Apaf-1
Ced-9	Drob-1/Debcl/Dborg-1 (pro-apoptotic) Dborg-2/Buffy (action unknown)	Bcl-2 family members pro-apoptotic anti-apoptotic
Bir-1 Bir-2	Diap1/Thread Diap2 Deterin	cIAP-1 cIAP-2 NIAP XIAP Survivin
? ? ?	Reaper Grim Hid	Diablo/Smac?

Fig. 1. Components of the apoptotic machinery conserved in *C. elegans*, *Drosophila*, and mammals

Daydream, has been predicted based on genome sequence analysis (Vernooy et al. 2000).

Recent work has detailed the role of Apaf-1, the homologue of *C. elegans* Ced-4, in caspase activation. Apaf-1/Ced-4 is a cytosolic adapter protein that promotes the activation of mammalian caspase-9 and *C. elegans* Ced-3. In both cases, Apaf-1 and initiator caspases interact directly through N-terminal CARD domains. Together with cytochrome c released from mitochondria, Apaf-1-caspase forms the apoptosome, an activational complex that initiates activation of the caspase proteolytic cascade and the final transition to cell death (Saleh et al. 1999; Zou et al. 1999; Cain et al. 2000).

One Apaf-1 homologue has thus far been identified in *Drosophila*, variously referred to as Dark (Rodriguez et al. 1999), Dapaf-1 (Kanuka et al. 1999), and Hac-1 (Zhou et al. 1999). Dark binds to the *Drosophila* initiator caspases Dredd and Dronc (Kanuka et al. 1999; Rodriguez et al. 1999), suggesting that, similar to Ced-4 and Apaf-1, Dark mediates the activation of initiator caspases in response to death stimuli. Similar to mammalian Apaf-1, Dark contains WD repeats that are positively regulated by cytochrome c.

Mutations in Dark greatly reduce killing by ectopically expressed *reaper*, *grim*, and *hid* (head involution defective; Bergmann et al. 1998; Kanuka et al. 1999; Rodriguez et al. 1999; Zhou et al. 1999), key regulators of apoptosis in *Drosophila* (White et al. 1994; Grether et al. 1995; Chen et al. 1996b). All of

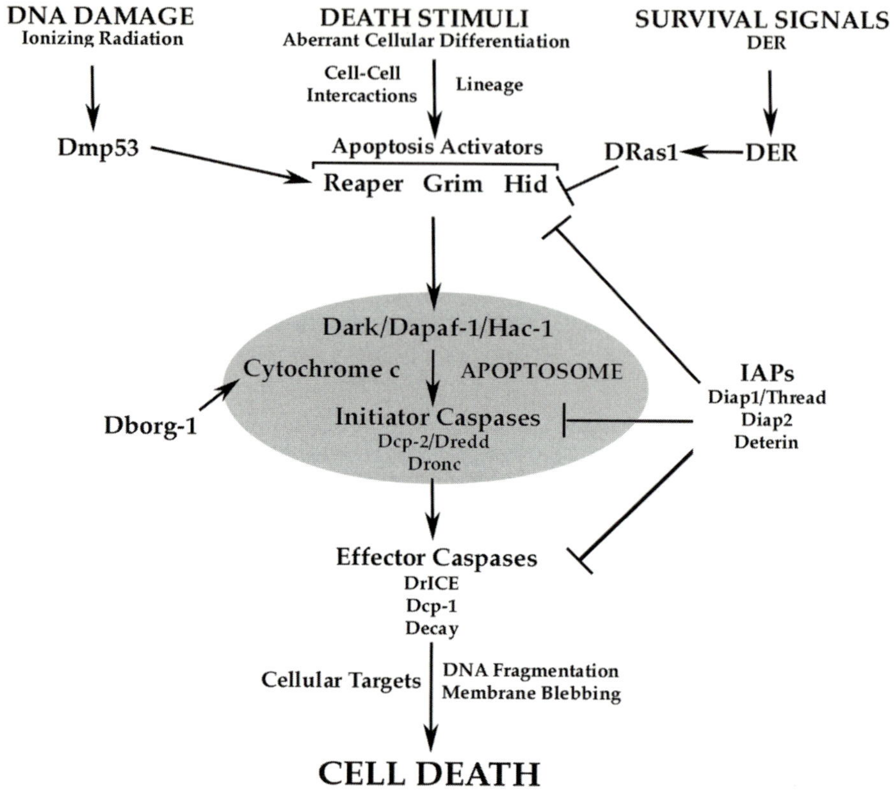

Fig. 2. Schematic diagram of apoptotic pathways in *Drosophila*

these genes map to the 75C region of the *Drosophila* genome, and lie within the H99 chromosomal deficiency. H99 mutant animals lack apoptosis and die late in embryogenesis (White et al. 1994). Reaper, Grim, and Hid share a similar 14 amino acid sequence at their N-termini, referred to as the RHG motif, but no similarity outside of this region (Wing et al. 1998), and their functions in vivo do not completely overlap. Expression of *reaper* and *grim* is restricted to cells fated to die, whereas *hid* is also expressed in many cells that will live (White et al. 1994; Grether et al. 1995; Chen et al. 1996a; Jiang et al. 1997; Robinow et al. 1997) and is tightly regulated by the Ras signal transduction pathway. Ras signaling both reduces the level of its transcription and inactivates Hid protein through direct phosphorylation by the *Drosophila* MAP kinase Rolled (Bergmann et al. 1998; Kurada and White 1998).

Reaper, Grim, and Hid are each sufficient to induce caspase-mediated cell death in transgenic models and tissue culture (Grether et al. 1995; Hay et al. 1995; Chen et al. 1996b; Nordstrom et al. 1996; Pronk et al. 1996; White et al. 1996; Vucic et al. 1997b, 1998; Zhou et al. 1997; Wing et al. 1998). This occurs, at least in part, through disruption of caspase interactions with inhibitor of apoptosis proteins (IAPs; Wang et al. 1999; Goyal et al. 2000; Meier et al. 2000b; Song et al. 2000).

No vertebrate homologues of *reaper*, *grim*, or *hid* have been identified thus far, though Diablo/Smac is thought to act as a functional homologue. Diablo/Smac bears no sequence similarity to the *Drosophila* proteins, but shares a similar function in promoting cytochrome c-dependent caspase activation by antagonizing IAP function (Chai et al. 2000; Du et al. 2000; Srinivasula et al. 2000; Verhagen et al. 2000).

First identified in baculovirus (Crook et al. 1994), IAPs bind to caspases and prevent their assembly into activational complexes (reviewed in Deveraux and Reed 1999). Three IAPs have been identified in *Drosophila*: Diap1/Thread (Hay et al. 1995), Diap2 (Hay et al. 1995; Duckett et al. 1996; Uren et al. 1996), and Deterin (Jones et al. 2000). Each of these proteins suppresses caspase-mediated cell death when expressed at high levels (Hay et al. 1995; Vucic et al. 1997a, 1998; Bergmann et al. 1998; Jones et al. 2000; Wing et al. 1998). Diap1 and Diap2 can promote cell survival by binding to and inhibiting the action of caspases (Kaiser et al. 1998; Wang et al. 1999; Goyal et al. 2000; Hawkins et al. 2000; Meier et al. 2000b). They also interact physically with and block the pro-apoptotic activity of Reaper, Grim, and Hid (Hay et al. 1995; Vucic et al. 1997a, 1998). Expression of Deterin in tissue culture cells inhibits Reaper-induced death (Jones et al. 2000).

The vertebrate Bcl-2 family consists of both pro- and anti-apoptotic upstream regulators that control cell death decisions through regulation of mitochondrial cytochrome c release, Apaf-1-mediated caspase activation, and neutralization of opposing Bcl-2 family members (Adams and Cory 1998; Gross et al. 1999). After considerable speculation about the existence of Bcl-2/Ced-9 homologues in *Drosophila*, two have been identified: Drob-1/Debcl/Dborg-1 (Brachmann et al. 2000; Colussi et al. 2000; Igaki et al. 2000; Zhang et al. 2000) and Dborg-2/Buffy (Brachmann et al. 2000; Colussi et al. 2000). RNA interference and overexpression studies have demonstrated that Drob-1 promotes caspase-mediated cell death (Brachmann et al. 2000; Colussi et al. 2000; Igaki et al. 2000) and suggest that, similar to Bcl-2, it may function in an anti-apoptotic capacity under certain conditions (Brachmann et al. 2000). Dborg-2 has been identified only by sequence homology; no genetic or biochemical data is available regarding the role of this molecule in apoptosis.

3 Adaptive Apoptosis: DNA Damage

In addition to its functions in development, multicellular organisms routinely employ apoptosis in an adaptive response to cellular assaults such as DNA damage. Many of the components of developmental and adaptive apoptosis overlap with respect to radiation-induced cell death. For example, mice with targeted knockouts of cytochrome c, Apaf-1, caspase-9, and caspase-3 all display defects in UV-induced apoptosis (reviewed in Davis 2000), and *Drosophila* embryos deficient for specific components of the death machinery show resistance to radiation-induced death (Abrams et al. 1993; White et al. 1994).

Similar to its vertebrate counterpart, the *Drosophila* p53 homologue Dmp53 is a key regulator of adaptive apoptosis in response to DNA damage. Overexpression of wild-type Dmp53 induces apoptosis in naive tissue (Jin et al. 2000; Ollmann et al. 2000), and expression of a dominant-negative form of the protein blocks radiation-induced apoptosis (Brodsky et al. 2000; Ollmann et al. 2000). Dmp53 mediates DNA damage-induced death in part through transcriptional regulation of the *reaper* locus, where it binds a p53 consensus site in a radiation-inducible enhancer element (Brodsky et al. 2000). Two additional observations suggest that *reaper* is involved in mediating cell death in response to DNA damage: chromosomal deletions removing *reaper*, *grim*, and *hid* block radiation-induced apoptosis (White et al. 1994), and *reaper* transcription is induced by ionizing radiation (Nordstrom et al. 1996). Dark and Drob-1 have also been implicated in radiation-induced apoptosis. *Dark* transcription is induced in response to both X-ray and UV-irradiation (Zhou et al. 1999), and killing by exogenous Drob-1 in the *Drosophila* retina is potentiated by exposure to UV irradiation (Brachmann et al. 2000).

Sensitivity to irradiation varies during the course of development at the levels of cellular response and tissue patterning. In the wing imaginal disc, for example, mitotically active cells preferentially undergo apoptosis in response to X-ray irradiation both early and late in development (Milan et al. 1997). Tissue patterning is largely unaffected, however. Compensatory increases in cell proliferation following widespread death restore cell numbers and rescue potential patterning errors (Milan et al. 1997).

The eye imaginal disc is most sensitive to pattern disruption when irradiated during the third larval instar (Fryxell and Kumar 1993), as cells undergo their terminal rounds of division. Patterning is disrupted when too few cells remain to differentiate all cell types and assemble ommatidial repeats. By contrast, when irradiated earlier in larval development, patterning errors do not occur in the eye (Fryxell and Kumar 1993). This may be due to the restoration of cell numbers through increased proliferation, though such events have not been described. Sensitivity to irradiation is again elevated in the pupal eye at the time of lattice cell death (Brachmann et al. 2000), and may reflect a general increase in apoptotic potential or competence during this period.

4 Upstream Signals: Death Decisions in the Fly Eye

The *Drosophila* compound eye is composed of approximately 750 individual unit eyes, or ommatidia, precisely positioned within a highly ordered crystalline lattice (Fig. 3). Each ommatidium is an assembly of 14 cells: 8 photoreceptors, 4 cone cells, and 2 primary pigment cells. Elaboration of the

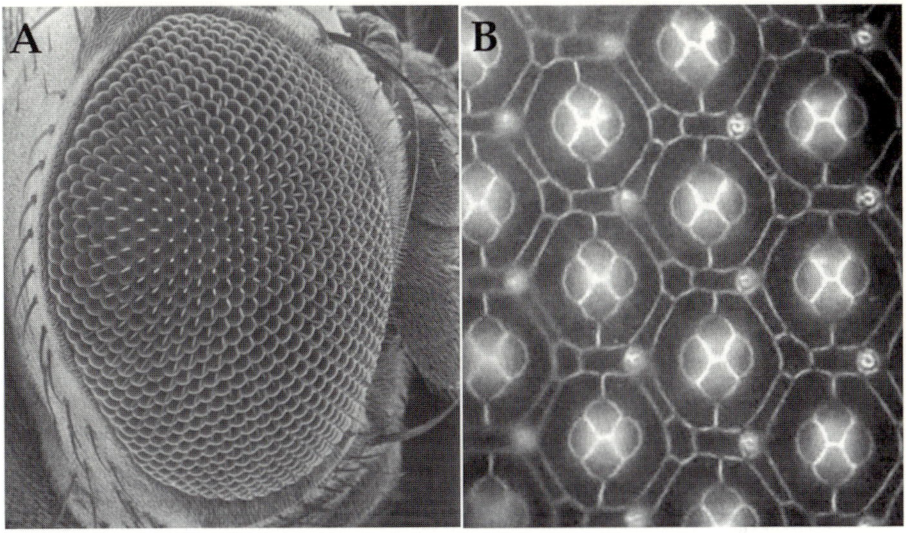

Fig. 3. The *Drosophila* compound eye is a highly ordered array of ommatidial repeats. **A** Scanning electron micrograph of an adult eye showing the ommatidial rows. Note the precision with which each row is aligned. **B** Apical surface of the retina 42 h after puparium formation. In the mature retina, individual ommatidia are positioned within a hexagonal array of shared lattice cells. Cells are stained for Armadillo, which localizes to apical adherens junctions

ommatidial lattice begins late in larval development as ommatidial precursor clusters emerge in a wave of patterned differentiation referred to as the morphogenetic furrow (Ready et al. 1976). The furrow is initiated at the posterior margin of the eye disc and sweeps across the eye field in the anterior direction. As the furrow passes, cells enter G1 arrest and the ommatidial lattice begins to form with the differentiation of the first photoreceptor neuron, R8. Additional cells are sequentially recruited to the growing ommatidial cluster through signals emanating from previously differentiated cells, culminating in the 14-cell group by early pupal development. Ultimately, individual ommatidia are positioned within a hexagonal lattice of shared pigment cells and mechanosensory bristles (Fig. 3).

Proper assembly of the pigment cell lattice is dependent on spatially restricted programmed cell death. By 24 h of pupal development (25°C) all 14 ommatidial cells have differentiated, leaving a large number of undifferentiated cells in the interommatidial space (Fig. 4A,B). Approximately one-third of these cells are eliminated through selective apoptosis between 24 and 40 h of pupal development in order to refine a hexagonal pigment cell lattice (Fig. 4C). Remaining cells adopt secondary and tertiary pigment cell fates and bring individual ommatidia into register within the lattice.

Morphological studies suggest that, in general, interommatidial cells making direct contact with primary pigment cells survive the period of apop-

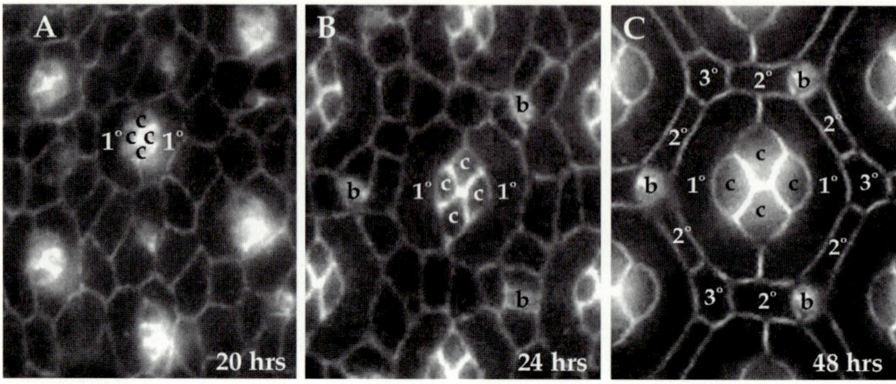

Fig. 4. Apical surface of the retina at **A** 20 h, **B** 24 h, and **C** 48 h after puparium formation (APF). Cells are stained for Armadillo, which localizes to apical adherens junctions. Posterior is to the *left*. **A** At 20 h APF the cone (*c*) and primary pigment cells (*1°*) have formed; lattice precursor cells and mechanosensory bristles are distributed randomly between ommatidia. **B** Prior to the onset of apoptosis, lattice cells reorganize into single-cell chains surrounding each ommatidia. **C** In the mature retina, individual ommatidia are brought into register through a hexagonal array of shared secondary and tertiary pigment cells (*2°,3°*) and bristles (*b*) Approximately one-third of the initial population of interommatidial cells have been removed through selective cell death

tosis, while those removed from the primaries are fated to die (Cagan and Ready 1989a; Reiter et al. 1996). Laser ablation studies support this view. Direct ablation of the cone or primary pigment cells results in the death of neighboring lattice cells (Miller and Cagan 1998), suggesting that cells within the ommatidia provide a survival signal to neighboring 2°/3° pigment cell precursors.

Genetic studies are consistent with this finding and reveal that life and death decisions in the lattice involve the interplay of opposing signals by two cell surface receptors: Notch and DER (*Drosophila* EGF Receptor). Notch signaling within the lattice promotes cell death (Cagan and Ready 1989b; Muskavitch 1994), while DER signaling promotes cell survival (Baker and Rubin 1989; Freeman 1994, 1996; Miller and Cagan 1998; Sawamoto et al. 1998) through up-regulation of Ras signal transduction (Bergmann et al. 1998; Kurada and White 1998). In the laser ablation studies described above, for example, lattice cell death in the absence of cone and primary cells can be rescued by reducing Notch function or by up-regulating DER signaling (Miller and Cagan 1998).

Together, these data support a general model in which apoptosis is promoted by Notch signaling within the interommatidial lattice cells and opposed by DER/Ras signaling initiated by the cone or primary pigment cells (Fig. 5). Consistent with this, the expression of Notch and DER is restricted to interommatidial lattice cells during the period of retinal apoptosis (Kooh et al. 1993; Miller and Cagan 1998). Spitz, a diffusible ligand of DER, is expressed in the cone and primary pigment cells, and may represent the ommatidial life signal (Miller and Cagan 1998).

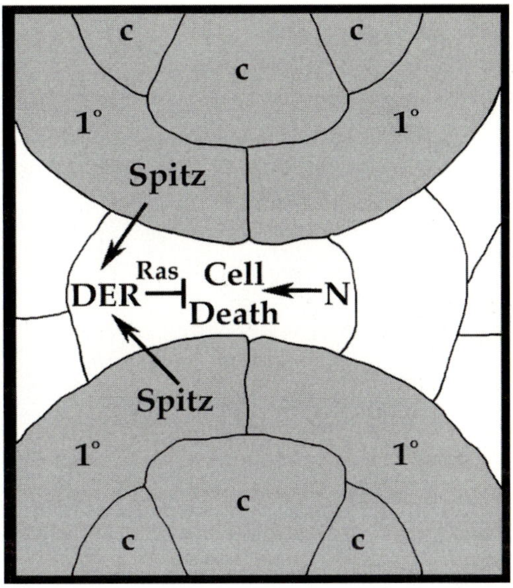

Fig. 5. A model for the spatial regulation of apoptosis in the developing *Drosophila* retina. Notch activity in the interommatidial lattice promotes cell death and is opposed by DER signaling. Spitz secreted by primary pigment (*1°*)and cone cells (*c*) activates DER in a subset of lattice cells and promotes their survival through up-regulation of Ras signal transduction

The involvement of Notch signaling in lattice refinement suggests that cell-cell contacts are important in the selection of cells to undergo apoptosis. Additional support for this view comes from mutational analysis of *Irregular chiasm C-roughest* (*IrreC-rst*). *IrreC-rst* encodes a transmembrane protein with extracellular immunoglobulin-like repeats similar to those of known cell adhesion molecules (Ramos et al. 1993), and is known to be involved in mediating cell movement, axonal pathfinding, and cell death (Wolff and Ready 1991; Ramos et al. 1993; Schneider et al. 1995; Reiter et al. 1996). Prior to the onset of lattice cell death, interommatidial lattice cells undergo substantial reorganization into single cell rows surrounding each ommatidium (Figs. 4B, 7D–F; Reiter et al. 1996). Concomitantly, IrreC-rst protein accumulates at the borders between primary pigment cells and interommatidial lattice cells (Reiter et al. 1996) in a Notch-dependent manner (Gorski et al. 2000). Mutations in *IrreC-rst* impair lattice cell reorganization and subsequent apoptosis, suggesting that proper cell sorting is required for death to proceed normally (Reiter et al. 1996).

Loss of cell death in *IrreC-rst* mutants is not specific to the pigment cell lattice or the eye, however, suggesting that cell-cell contacts are generally important in the induction of apoptosis. During the wave of morphogenesis in the third instar eye disc, two significant zones of apoptotic cell death are visible: one at the posterior margin of the eye, and one just anterior to the mor-

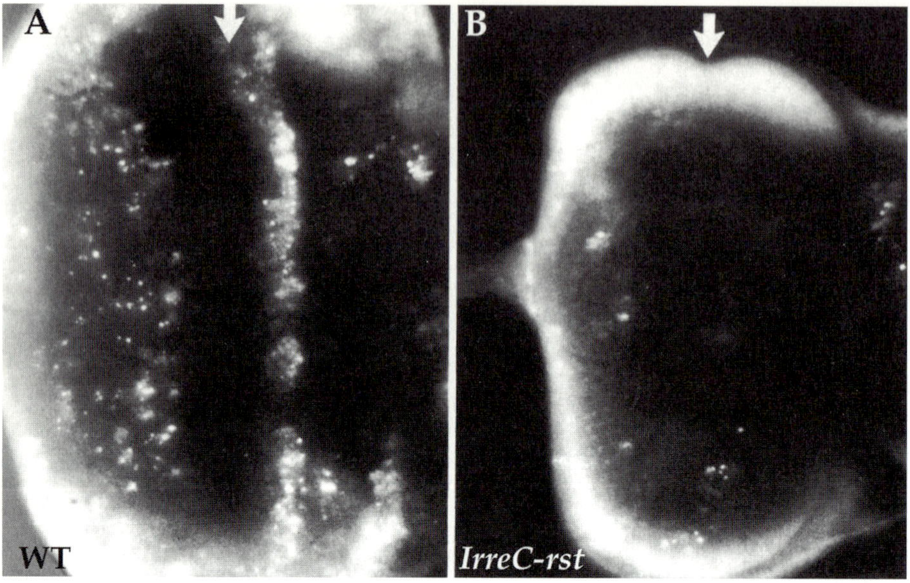

Fig. 6. Acridine orange staining in third larval instar eye imaginal discs of **A** wild-type and **B** *IrreC-rst* mutant animals. Posterior is to the *left*. The position of the morphogenetic furrow is indicated by an *arrow* in each panel. **A** During this period, two zones of apoptosis are present in wild-type discs: one at the posterior margin, and one just anterior to the morphogenetic furrow. **B** Cell death in the third instar eye is absent in an *IrreC-rst* mutant background. (Taken from Wolff and Ready 1991)

phogenetic furrow (Fig. 6A; Bonini et al. 1990; Wolff and Ready 1991). The significance of this cell death is not known; it is possibly involved in regulating the number and quality of ommatidial precursor cells. *IrreC-rst* mutants are devoid of this cell death (Fig. 6B), as well as normally occurring death in the third instar antennal disc (Wolff and Ready 1991). Recently, several mutations that specifically affect cell death in the eye were identified in a genetic screen for enhancers and suppressers of the *IrreC-rst* loss-of-function phenotype (Tanenbaum et al. 2000). Characterization of these loci will allow for a greater understanding of this aspect of retinal patterning and apoptosis in general.

5 Morphogenesis of Lattice Patterning: Making a Hexagon

One of the advantages of studying apoptosis in the *Drosophila* retina is that the morphogenesis of this tissue has been extremely well characterized, allowing for observation of this process in the context of a developing neuroepithelium. Careful consideration of morphological aspects of cell death such as spatial restrictions and cell–cell contacts may yield important clues as to the mechanisms of this remarkably selective cell death. This section describes some of the more striking morphological features of lattice patterning.

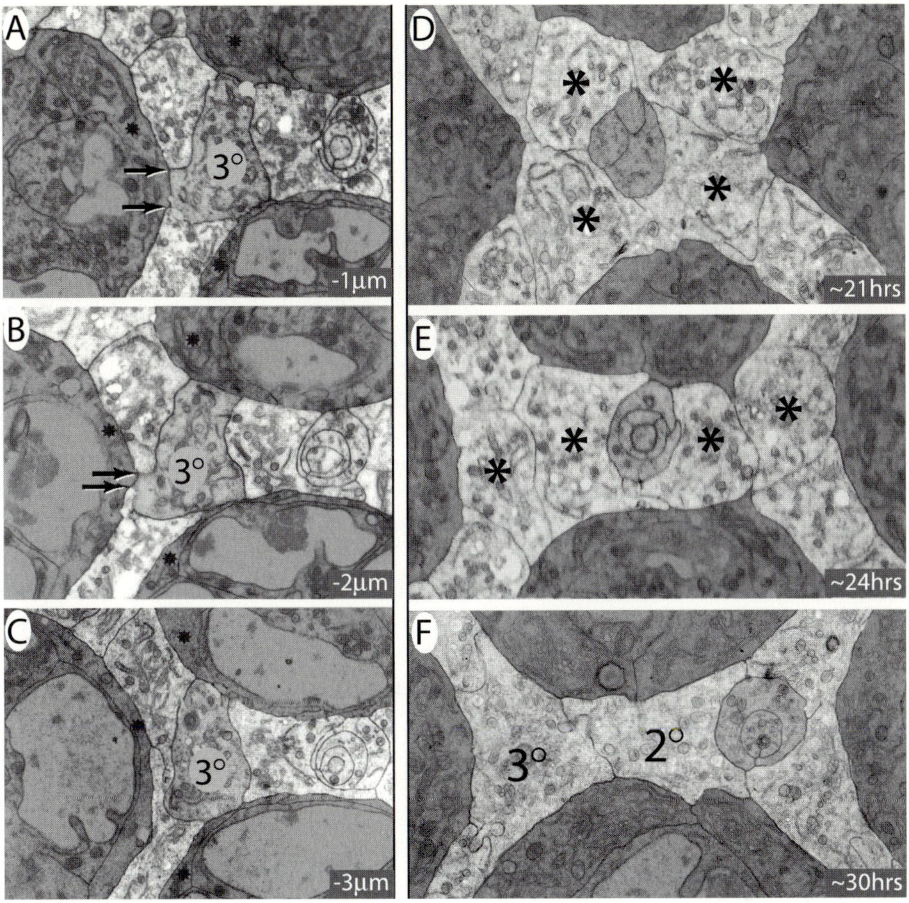

Fig. 7. Cell morphogenesis and cell death work together to pattern the ommatidial array. Electron micrographs at **A**, **D–F** or near **B**,**C** the apical surface of a maturing pupal retina. The cone cell/primary pigment cell ommatidial cores are *darkly shaded* to distinguish the interommatidial lattice (*unshaded*). **A** New cell contacts are made first apically. An example of a tertiary pigment cell (*3°*, *light shading*) that has just reached across to contact (*arrows*) a third ommatidium. Descending 1 (**B**) and 2 μm (**C**) reveals the tertiary's "earlier" position away from the ommatidium. **D** Re-organizing the lattice. The bristle group (*light shading*) emerges at the center of a group of interommatidial lattice cells (*asterisks*). **E** The interommatidial lattice cells re-arrange to lie end-to-end between the ommatidia. In this case, two cells anterior to the bristle group will need to be removed. **F** Removal of the two anterior lattice cells re-locates the bristle group to the anterior side of the horizontal face of the hexagon. The posterior two cells become the secondary (*2°*) and tertiary (*3°*) pigment cells

Interommatidial lattice cell death occurs by apoptosis. Membrane blebbing and nuclear condensation are easily recognizable, and dying cells within the pigment cell lattice can be labeled with acridine orange and TUNEL (Cagan and Ready 1989a; Wolff and Ready 1991). As lattice cells die, they usually lose contact with the basal surface of the retinal epithelium, and drift upward to the apical surface as the cell body fragments. Remnants of dead cells are rapidly engulfed by their neighbors.

During the reorganization of lattice cells prior to the onset of cell death (Figs. 4B, 7D–F), interommatidial cells extend membranous projections across one or more cell diameters to make contacts with cells in the new niche (Fig. 7A–C; Cagan and Ready 1989a). These apical contacts are then zippered downward to the basal surface of the retina and the cell assumes its new position. As interommatidial cells extend their projections, they often occlude their immediate neighbors from making the same transition. On a microscopic level, cells are rapidly altering their apical contacts and appear to compete with each other for specific niches. Successful cells assume positions within the hexagonal lattice, while occluded cells typically die. It is likely, though not demonstrated, that this competition occurs in response to determinative factors supplied by cells of the ommatidium, such as Spitz (see above).

Studies of lattice cell patterning by fixed tissue microscopy (Cagan and Ready 1989a; Wolff and Ready 1991; Reiter et al. 1996) and live visualization (Brachmann and Cagan, unpubl. observ.) reveal that cell death does not occur randomly throughout the interommatidial space. Rather, death proceeds in a spatially restricted fashion, occurring at high levels on the equatorial and polar positions of each ommatidium (six and twelve o'clock positions), and lower levels on the lateral ommatidial faces. Removal of a preponderance of cells on the equatorial and polar faces draws ommatidia together on this axis and reorganizes the square array of the larval eye into the hexagonal array of the mature retina. Following the initial burst of death, low level death over the next several hours refines the pigment cell lattice to the wild-type compliment of cells.

How cell death occurs with such spatial precision remains a key question. It has been suggested that the hexagonal nature of the lattice merely represents the most efficient packing of ommatidia, similar to the packing of soap bubbles. Laser ablation studies, however, do not support this model. Even in a field of undifferentiated cells, a single, isolated ommatidia is surrounded by a hexagonal array of lattice cells (Brachmann and Cagan, unpubl. observ.), suggesting that each ommatidium autonomously provides signals required for proper patterning.

The relative importance of specific lattice cell types in the spatial restriction of apoptosis is unknown. Most death occurs in areas immediately adjacent to sensory bristles, suggesting that perhaps bristles provide positional information to neighboring cells (Wolff and Ready 1991). Apoptosis occurs fairly normally, however, in the absence of this cell type. Misexpression of *wingless* under control of the *sevenless* promoter inhibits the formation of lattice bristles (Cadigan and Nusse 1996), yet the hexagonal array is preserved and a nearly wild-type number of lattice cells are present. In contrast, ablation of cone or primary pigment cells results in ectopic death of interommatidial cells and, necessarily, disruption of the hexagonal lattice (Miller and Cagan 1998). Ectopic death under these conditions can be rescued by up-regulation of DER/Ras signaling or down-regulation of Notch signaling, though lattice patterning is not rescued (Miller and Cagan 1998). These data suggest that cells of the ommatidia supply both life signals and critical positional information to cells within the lattice, and raise interesting questions about the relationship between these two processes in patterning the hexagonal pigment cell lattice.

Apoptosis also occurs at mid-pupal development (60–70 h) to eliminate stunted ommatidia at the periphery of the eye (Wolff and Ready 1991). These ommatidia, referred to as the perimeter clusters, are present at the end of nearly every ommatidial row. They frequently contain the full compliment of photoreceptors, cone cells, and primary pigment cells, but they are reduced in size compared to internal ommatidia. Secondary and tertiary pigment cells surrounding perimeter clusters survive and serve to extend the interommatidial lattice to the edge of the retina (Wolff and Ready 1991).

6 Retinal Degeneration

6.1 Loss of Trophic Support

Cell-cell signaling is a common mechanism for regulating cell survival in the vertebrate nervous system. Many neural and glial cell types are produced in excess, with surplus cells eliminated by apoptosis. Often, neurons undergo apoptosis after failure to innervate appropriate target cells or as a result of trophic deprivation (reviewed in Mahalik and Owens 1997). This is also true of axonal projections from the *Drosophila* retina, as illustrated by several mutations that result in non-productive axonal pathfinding or impaired optic lobe development.

One example of this is the degeneration of supernumerary eyes in flies carrying mutations in *extra eyes* (*ee*), which send axonal projections that never reach the optic ganglia (Marcey and Stark 1985). Morphologically, the eyes develop normally and differentiate a full complement of photoreceptors (Marcey and Stark 1985; Campos et al. 1992). Failure to establish the proper synaptic connections, however, leads to degeneration of the photoreceptors shortly after eclosion (Campos et al. 1992).

Retinal degeneration in *disconnected* (*disco*) and *reversed polarity* (*repo*) mutants results from inadequate optic lobe development. Photoreceptor projections in *disco* mutants fail to reach their target cells during larval development, disrupting development of the optic lobe from which they ultimately require trophic support (Steller et al. 1987). They typically innervate a mass muscle that often replaces the optic lobe and progressively degenerate following eclosion (Campos et al. 1992). *repo* mutants exhibit an initial degeneration of the laminar portion of the optic lobes that leads to subsequent death of retinal cells. *repo* encodes a homeodomain protein expressed in the laminar and medullar portions of the optic lobes that is required for terminal differentiation of glial cells (Campbell et al. 1994; Xiong et al. 1994). In the absence

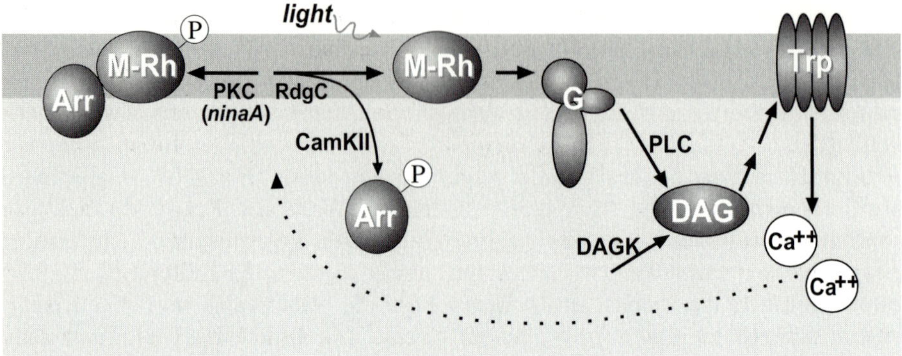

Fig. 8. Schematic diagram of the phototransduction pathway. Light induces a conformational change in the NinaE rhodopsin (Rh1), allowing Rh1 to interact with a G-protein (Dgq), to activate the NorpA phospholipase C (PLC). PLC catalyzes the formation of diaclylglycerol (DAG) and the subsequent opening of the transient receptor potential (TRP) and transient receptor potential-like (TRPL) channels. Deactivation is initiated following phosphorylation of activated rhodopsin by the InaC eye-specific protein kinase C (eye-PKC). This allows Arrestins (Arr1 and Arr2) to bind directly to Rh1 Arrestin itself is then phosphorylated by calcium-calmodulin dependent protein kinase (CamKII) and released from Rh1, whereupon Rh1 is dephosphorylated by the rdgC Ca^{2+}-dependent serine/threonine phosphatase

of trophic support from associated glial cells, laminar neurons degenerate and fail to support retinal axonal projections (Xiong and Montell 1995).

6.2 Phototransduction Mutants

A second class of degeneration mutants involves components of the phototransduction machinery. Impairment of the pathway in which Rhodopsin (Rh1/NinaE) undergoes cyclic activation and inactivation in response to light (Fig. 8; for comprehensive reviews see Zuker 1996; Montell 1999; Tsunoda and Zuker 1999) has been shown to result in retinal degeneration through both necrosis and apoptosis.

Mutations that result in constitutive activation of the phototransduction pathway result in pathologically elevated intracellular calcium levels and lytic, necrotic death. These include mutations in Rhodopsin that render it refractory to inactivation (reviewed in Bentrop 1998), and hypomorphic alleles of the arrestin *arr2* that abate the inactivation of Rh1 (Dolph et al. 1993). Similarly, mutations that result in constitutive activity of the TRP and TRPL calcium channels result in the accumulation of intracellular calcium to toxic levels (Raghu et al. 2000; Yoon et al. 2000). The relative toxicity of these disruptions is dependent on the function of other components of the pathway.

Apoptotic cell death is typically the result of improper processing or localization of Rh1. For example, certain dominant mutations in Rh1 result in its accumulation within the endoplasmic reticulum (ER) and apoptotic cell death

(Colley et al. 1995; Kurada and O'Tousa 1995). Remarkably, Colley et al. (1995) isolated four Rh1 alleles with mutations corresponding to the same amino acid changes implicated in *retinitis pigmentosa*, a leading cause of human degenerative vision loss (reviewed in van Soest et al. 1999). Mutations in the *ninaA* cyclophilin, which is required to transport Rh1 from the ER to the rhabdomere, similarly result in apoptotic death (Baker et al. 1994).

Mutations that promote the formation of stable Arr2-Rh1 complexes also lead to apoptotic neurodegeneration. These include mutations in the phospholipase C (PLC) ortholog *norpA* (Meyertholen et al. 1987; Stark et al. 1989) and the rhodopsin-specific phosphatase *rdgC* (Steele and O'Tousa 1990; Kiselev et al. 2000), loss of which prevents the dissociation of Arr2 and Rh1. The nature of Arr-2-Rh1 cytotoxicity is unclear, though it has been suggested to result from impaired Ras/MAPK signaling that normally provides a life signal to cells (Kiselev et al. 2000). Mutations that prevent inactivation of Arr2 also induce apoptotic degeneration, including missense mutations in Arr2 itself and loss of RdgB, which results in lowered intracellular calcium levels (Alloway et al. 2000). Finally, the findings by Alloway et al. (2000) and Kiselev et al. (2000) that reduction of dynamin activity partially rescues the Arr-Rh1 degeneration phenotype suggests that endocytosis of Arr-Rh1 complexes is a necessary step in the induction of apoptosis.

Many of the *Drosophila* retinal degeneration phenotypes have been found to parallel human disease states, including those caused by mutations in human PLCβ4 (Rao et al. 1995), Arrestin (Fuchs et al. 1995), and rhodopsin (described above). Given the remarkable similarity between the genetic loci responsible for these conditions in flies and humans, *Drosophila* is an excellent system for modeling a variety of human retinal degenerative disorders. Therefore, it is particularly exciting to find that the retinal degeneration and loss of visual function caused by mutations in *Drosophila ninaE, rdgC, arr2,* and *norpA* is rescued by overexpression of the baculovirus caspase inhibitor P35 (Davidson and Steller 1998; Alloway et al. 2000). These studies suggest important avenues of investigation for the development of effective treatments.

7 Concluding Remarks

The importance of understanding the mechanisms of apoptosis extends far beyond general scientific interest. Dysfunctional programmed cell death results in a number of human abnormalities and disease states including autoimmune disorders, cancer, and neurodegenerative conditions such as Parkinson's and Alzheimer's diseases. In many cases, induction or suppression of apoptosis may restore tissue functionality and delay or halt disease progression.

The *Drosophila* retina is a superb system for the study of apoptosis and the characterization of new molecules involved in this process. Several human degenerative diseases are now being modeled in the fly eye, with the potential for very rapid advancement in this powerful genetic system. Eventually, this approach is certain to lead to the development of new and more effective therapies.

For additional discussion of disease modeling in the *Drosophila* eye, see Bonini and Fortini (this Vol.).

References

Abrams JM (1999) An emerging blueprint for apoptosis in *Drosophila*. Trends Cell Biol 9:435–440

Abrams JM, White K, Fessler LI, Steller H (1993) Programmed cell death during *Drosophila* embryogenesis. Development 117:29–43

Adams JM, Cory S (1998) The Bcl-2 protein family: arbiters of cell survival. Science 281:1322–1326

Alloway PG, Howard L, Dolph PJ (2000) The formation of stable rhodopsin-arrestin complexes induces apoptosis and photoreceptor cell degeneration. Neuron 28:129–138

Baker EK, Colley NJ, Zuker CS (1994) The cyclophilin homolog NinaA functions as a chaperone, forming a stable complex in vivo with its protein target rhodopsin. EMBO J 13:4886–4895

Baker NE, Rubin GM (1989) Effect on eye development of dominant mutations in the *Drosophila* homologue of the EGF receptor. Nature 340:150–153

Bangs P, White K (2000) Regulation and execution of apoptosis during *Drosophila* development. Dev Dynam 218:68–79

Bentrop J (1998) Rhodopsin mutations as the cause of retinal degeneration. Classification of degeneration phenotypes in the model system *Drosophila melanogaster*. Acta Anat 162:85–94

Bergmann A, Agapite J, McCall K, Steller H (1998) The *Drosophila* gene *hid* is a direct molecular target of Ras-dependent survival signaling [see comments]. Cell 95:331–341

Bonini N, Leiserson W, Benzer S (1990) A mutation in compound eye development in *Drosophila* that results in cell death rather than differentiation. J Cell Biochem 14E:502

Bonini NM, Fortini ME (1999) Surviving *Drosophila* eye development: integrating cell death with differentiation during the formation of a neural structure. BioEssays 21:991–1003

Brachmann CB, Jassim OW, Wachsmuth BD, Cagan RL (2000) The *Drosophila* bcl-2 family member dBorg-1 functions in the apoptotic response to UV-irradiation. Curr Biol 10:547–550

Brodsky MH, Nordstrom W, Tsang G, Kwan E, Rubin GM, Abrams JM (2000) *Drosophila* p53 binds a damage response element at the *reaper* locus. Cell 101:103–113

Budihardjo I, Olive RH, Lutter M, Luo X, Wang X (1999) Biochemical pathways of caspase activation during apoptosis. Annu Rev Cell Dev Biol 15:269–290

Cadigan KM, Nusse R (1996) wingless signaling in the *Drosophila* eye and embryonic epidermis. Development 122:2801–2812

Cagan RL, Ready DF (1989a) The emergence of order in the *Drosophila* pupal retina. Dev Biol 136:346–362

Cagan RL, Ready DF (1989b) Notch is required for successive cell decisions in the developing *Drosophila* retina. Genes Dev 3:1099–1112

Cain K, Bratton SB, Langlais C, Walker G, Brown DG, Sun XM, Cohen GM (2000) Apaf-1 oligomerizes into biologically active approximately 700-kDa and inactive approximately 14-MDa apoptosome complexes. J Biol Chem 275:6067–6070

Campbell G, Goring H, Lin T, Spana E, Andersson S, Doe CQ, Tomlinson A (1994) RK2, a glial-specific homeodomain protein required for embryonic nerve cord condensation and viability in *Drosophila*. Development 120:2957–2966

Campos AR, Fischbach KF, Steller H (1992) Survival of photoreceptor neurons in the compound eye of *Drosophila* depends on connections with the optic ganglia. Development 114:355–366

Chai J, Du C, Wu JW, Kyin S, Wang X, Shi Y (2000) Structural and biochemical basis of apoptotic activation by Smac/DIABLO. Nature 406:855–862

Chen P, Lee P, Otto L, Abrams J (1996a) Apoptotic activity of REAPER is distinct from signaling by the tumor necrosis factor receptor 1 death domain. J Biol Chem 271:25735–25737

Chen P, Nordstrum W, Gish B, Abrams J (1996b) *grim*, a novel cell death gene in *Drosophila*. Genes Dev 10:1773–1782

Chen P, Rodriguez A, Erskine R, Thach T, Abrams JM (1998) Dredd, a novel effector of the apoptosis activators *reaper*, *grim*, and *hid* in *Drosophila*. Dev Biol 201:202–216

Colley NJ, Cassill JA, Baker EK, Zuker CS (1995) Defective intracellular transport is the molecular basis of rhodopsin-dependent dominant retinal degeneration. Proc Natl Acad Sci USA 92:3070–3074

Colussi PA, Quinn LM, Huang DC, Coombe M, Read SH, Richardson H, Kumar S (2000) Debcl, a proapoptotic Bcl-2 homologue, is a component of the *Drosophila melanogaster* cell death machinery [see comments]. J Cell Biol 148:703–714

Conradt B, Horvitz HR (1998) The C. elegans protein *EGL*-1 is required for programmed cell death and interacts with the Bcl-2-like protein CED-9 Cell 93:519–529

Crook NE, Clem RJ, Miller LK (1994) An apoptosis inhibiting baculovirus gene with a zinc finger-like motif. J Virol 67:2168–2174

Davidson FF, Steller H (1998) Blocking apoptosis prevents blindness in *Drosophila* retinal degeneration mutants. Nature 391:587–591

Davis RJ (2000) Signal transduction by the JNK group of MAP kinases. Cell 103:239–252

Deveraux QL, Reed JL (1999) IAP family proteins – suppressors of apoptosis. Genes Dev 13:239–252

Dolph PJ, Ranganathan R, Colley NJ, Hardy RW, Socolich M, Zuker CS (1993) Arrestin function in inactivation of G protein-coupled receptor rhodopsin in vivo. Science 260:1910–1916

Dorstyn L, Colussi PA, Quinn LM, Richardson H, Kumar S (1999a) DRONC, an ecdysone-inducible *Drosophila* caspase. Proc Natl Acad Sci USA 96:4307–4312

Dorstyn L, Read SH, Quinn LM, Richardson H, Kumar S (1999b) DECAY, a novel *Drosophila* caspase related to mammalian caspase-3 and caspase-7 [published erratum appears in J Biol Chem 2000 May 19;275(20):15600]. J Biol Chem 274:30778–30783

Du C, Fang M, Li Y, Li L, Wang X (2000) Smac, a mitochondrial protein that promotes cytochrome c-dependent caspase activation by eliminating IAP inhibition. Cell 102:33–42

Duckett CS, Nava VE, Gedrich RW, Clem RJ, Van Dongen JL, Gilfillan MC, Shiels H, Hardwick JM, Thompson CB (1996) A conserved family of cellular genes related to the *baculovirus* iap gene and encoding apoptosis inhibitors. EMBO J 15:2685–2694

Fraser AG, Evan GI (1997) Identification of a *Drosophila melanogaster* ICE/CED-3-related protease, drICE. EMBO J 16:2805–2813

Freeman M (1994) Misexpression of the *Drosophila argos* gene, a secreted regulator of cell determination. Development 120:2297–2304

Freeman M (1996) Reiterative use of the EGF receptor triggers differentiation of all cell types in the *Drosophila* eye. Cell 87:651–660

Fryxell KJ, Kumar JP (1993) Characterization of the radiation-sensitive stage in the development of the compound eye of *Drosophila*. Mutat Res 285:181–189

Fuchs S, Nakazawa M, Maw M, Tamai M, Oguchi Y, Gal A (1995) A homozygous 1-base pair deletion in the arrestin gene is a frequent cause of Oguchi disease in Japanese. Nat Genet 10: 360–362

Gorski S, Brachmann CB, Tanenbaum S, Cagan RL (2000) Delta and Notch promote correct localization of IrreC-rst. Cell Death Differ 7:1011–1013

Goyal L, McCall K, Agapite J, Hartwieg E, Steller H (2000) Induction of apoptosis by *Drosophila reaper*, *hid* and *grim* through inhibition of IAP function. EMBO J 19:589–597

Grether M, Abrams J, Agapite J, White K, Steller H (1995) The head involution defective gene of *Drosophila melanogaster* functions in programmed cell death. Genes Dev 9:1674–1708

Gross A, McDonnell JM, Korsmeyer SJ (1999) BCL-2 family members and the mitochondria in apoptosis. Genes Dev 13:1899–1911

Hawkins CJ, Yoo SJ, Peterson EP, Wang SL, Vernooy SY, Hay BA (2000) The *Drosophila* caspase DRONC cleaves following glutamate or aspartate and is regulated by DIAP1, HID, and GRIM. J Biol Chem 275:27084–27093

Hay BA, Wassarman DA, Rubin GM (1995) *Drosophila* homologs of *baculovirus* inhibitor of apoptosis proteins function to block cell death. Cell 83:1253–1262

Hengartner MO, Ellis RE, Horvitz HR (1992) *Caenorhabditis elegans* gene *ced*-9 protects cells from programmed cell death. Nature 356:494–499

Igaki T, Kanuka H, Inohara N, Sawamoto K, Nunez G, Okano H, Miura M (2000) Drob-1, a *Drosophila* member of the Bcl-2/CED-9 family that promotes cell death. Proc Natl Acad Sci USA 97:662–667

Inohara N, Koseki T, Hu Y, Chen S, Nunez G (1997) CLARP, a death effector domain-containing protein interacts with caspase-8 and regulates apoptosis. Proc Natl Acad Sci USA 94:10717–10722

Jiang C, Baehrecke EH, Thummel CS (1997) Steroid regulated programmed cell death during *Drosophila* metamorphosis. Development 124:4673–4683

Jin S, Martinek S, Joo WS, Wortman JR, Mirkovic N, Sali A, Yandell MD, Pavletich NP, Young MW, Levine AJ (2000) Identification and characterization of a p53 homologue in *Drosophila melanogaster*. Proc Natl Acad Sci USA 97:7301–7306

Jones G, Jones D, Zhou L, Steller H, Chu Y (2000) Deterin, a new inhibitor of apoptosis from *Drosophila melanogaster*. J Biol Chem 275:22157–22165

Kaiser WJ, Vucic D, Miller LK (1998) The *Drosophila* inhibitor of apoptosis D-IAP1 suppresses cell death induced by the caspase drICE. FEBS Lett 440:243–248

Kanuka H, Sawamoto K, Inohara N, Matsuno K, Okano H, Miura M (1999) Control of the cell death pathway by Dapaf-1, a *Drosophila* Apaf-1/CED-4- related caspase activator. Mol Cell 4:757–769

Kiselev A, Socolich M, Vinos J, Hardy RW, Zuker CS, Ranganathan R (2000) A molecular pathway for light-dependent photoreceptor apoptosis in *Drosophila* [In Process Citation]. Neuron 28:139–152

Kooh PJ, Fehon RG, Muskavitch MA (1993) Implications of synamic patterns of Delta and Notch expression for cellular interactions during *Drosophila* development. Development 117: 493–507

Kumar S (1999) Regulation of caspase activation in apoptosis: implications in pathogenesis and treatment of disease. Clin Exp Pharmacol Physiol 26:295–303

Kurada P, O'Tousa JE (1995) Retinal degeneration caused by dominant rhodopsin mutations in *Drosophila*. Neuron 14:571–579

Kurada P, White K (1998) Ras promotes cell survival in *Drosophila* by downregulating *hid* expression [see comments]. Cell 95:319–329

Mahalik TJ, Owens GP (1997) Cell death in the nervous system. J Invest Dermatol Symp Proc 2:14–18

Marcey DJ, Stark WS (1985) The morphology, physiology, and neural projections of supernumerary compound eyes in *Drosophila melanogaster*. Dev Biol 107:180–197

Meier P, Finch A, Evan G (2000a) Apoptosis in development. Nature 407:796–801

Meier P, Silke J, Leevers SJ, Evan GI (2000b) The *Drosophila* caspase DRONC is regulated by DIAP1 EMBO J 19:598–611

Metzstein MM, Stanfield GM, Horvitz HR (1998) Genetics of programmed cell death in C. elegans: past, present and future. Trends Genet 14:410–416

Meyertholen EP, Stein PJ, Williams MA, Ostroy SE (1987) Studies of the *Drosophila* norpA phototransduction mutant. II. Photoreceptor degeneration and rhodopsin maintenance. J Comp Physiol [A] 161:793–798

Milan M, Campuzano S, Garcia-Bellido A (1997) Developmental parameters of cell death in the wing disc of *Drosophila*. Proc Natl Acad Sci USA 94:5691–5696

Miller DT, Cagan RL (1998) Local induction of patterning and programmed cell death in the developing *Drosophila* retina. Development 125:2327–2335

Montell C (1999) Visual transduction in *Drosophila*. Annu Rev Cell Dev Biol 15:231–268

Mori C, Nakamura N, Kimura S, Irie H, Takigawa T, Shiota K (1995) Programmed cell death in the interdigital tissue of the fetal mouse limb is apoptosis with DNA fragmentation. Anat Rec 242:103–110

Muskavitch MA (1994) Delta-notch signaling and *Drosophila* cell fate choice. Dev Biol 166:415–430

Nordstrom W, Chen P, Steller H, Abrams JM (1996) Activation of the *reaper* gene during ectopic cell killing in *Drosophila*. Dev Biol 180:213–226

Ollmann M, Young LM, Di Como CJ, Karim F, Belvin M, Robertson S, Whittaker K, Demsky M, Fisher WW, Buchman A et al. (2000) *Drosophila* p53 is a structural and functional homolog of the tumor suppressor p53. Cell 101:91–101

Pronk GJ, Ramer K, Amiri P, Williams LT (1996) Requirement of an ICE-like protease for induction of apoptosis and ceramide generation by REAPER. Science 271:808–810

Raghu P, Usher K, Jonas S, Chyb S, Polyanovsky A, Hardie RC (2000) Constitutive activity of the light-sensitive channels TRP and TRPL in the *Drosophila* diacylglycerol kinase mutant, rdgA. Neuron 26:169–179

Ramos RG, Igloi GL, Lichte B, Baumann U, Maier D, Schneider T, Brandstatter JH, Frohlich A, Fischbach KF (1993) The irregular chiasm C-roughest locus of *Drosophila*, which affects axonal projections and programmed cell death, encodes a novel immunoglobulin-like protein. Genes Dev 7:2533–2547

Rao PN, Pettenati MJ, Butler MG, Shaffer LG, Baehr W, Alvarez RA, Anderson RE (1995) Submicroscopic deletion of the human retinal phosphoinositide-specific phospholipase C-beta-4 (plc-beta-4) in Alagille syndrome patients. (Abstract) Am J Hum Genet 57 (4 Suppl):A35

Ready DF, Hanson TE, Benzer S (1976) Development of the *Drosophila* retina, a neurocrystalline lattice. Dev Biol 53:217–240

Reiter C, Schimansky T, Nie Z, Fischbach KF (1996) Reorganization of membrane contacts prior to apoptosis in the *Drosophila* retina: the role of the IrreC-rst protein. Development 122: 1931–1940

Robinow S, Draizen TA, Truman JW (1997) Genes that induce apoptosis: transcriptional regulation in identified, doomed neurons of the *Drosophila* CNS. Dev Biol 190:206–213

Rodriguez A, Oliver H, Zou H, Chen P, Wang X, Abrams JM (1999) Dark is a *Drosophila* homologue of Apaf-1/CED-4 and functions in an evolutionarily conserved death pathway [see comments]. Nat Cell Biol 1:272–279

Rusconi JC, Hays R, Cagan RL (2000) Programmed cell death and patterning in *Drosophila*. Cell Death Differ 7:1063–1070

Saleh A, Srinivasula SM, Acharya S, Fishel R, Alnemri ES (1999) Cytochrome c and dATP-mediated oligomerization of Apaf-1 is a prerequisite for procaspase-9 activation. J Biol Chem 274:17941–17945

Salvesen GS, Dixit VM (1999) Caspase activation: the induced-proximity model. Proc Natl Acad Sci USA 96:10964–10967

Sawamoto K, Taguchi A, Hirota Y, Yumada C, Jin M, Okano H (1998) Argos induces programmed cell death in the developing *Drosophila* eye by inhibition of the Ras pathway. Cell Death Differ 5:262–270

Schneider T, Reiter C, Eule E, Bader B, Lichte B, Nie Z, Schimansky T, Ramos RG, Fischbach KF (1995) Restricted expression of the irreC-rst protein is required for normal axonal projections of columnar visual neurons. Neuron 15:259–271

So KF, Campbell G, Lieberman AR (1990) Development of the mammalian retinogeniculate pathway: target finding, transient synapses and binocular segregation. J Exp Biol 153:85–104

Song Z, McCall K, Steller H (1997) DCP-1, a *Drosophila* cell death protease essential for development [published erratum appears in Science 1997 Jul 11;277(5323):167]. Science 275:536–540

Song Z, Guan B, Bergman A, Nicholson DW, Thornberry NA, Peterson EP, Steller H (2000) Biochemical and genetic interactions between *Drosophila* caspases and the proapoptotic genes rpr, *hid*, and *grim*. Mol Cell Biol 20:2907–2914

Srinivasula SM, Datta P, Fan XJ, Fernandes-Alnemri T, Huang Z, Alnemri ES (2000) Molecular determinants of the caspase-promoting activity of Smac/DIABLO and its role in the death receptor pathway. J Biol Chem 275:36152–36157

Stark WS, Sapp R, Carlson SD (1989) Photoreceptor maintenance and degeneration in the norpA (no receptor potential-A) mutant of *Drosophila melanogaster*. J Neurogenet 5:49–59

Steele F, O'Tousa JE (1990) Rhodopsin activation causes retinal degeneration in *Drosophila* rdgC mutant. Neuron 4:883–890

Steller H, Fischbach KF, Rubin GM (1987) Disconnected: a locus required for neuronal pathway formation in the visual system of *Drosophila*. Cell 50:1139–1153

Tanenbaum SB, Gorski SM, Rusconi JC, Cagan RL (2000) A screen for dominant modifiers of the irreC-rst cell death phenotype in the developing *Drosophila* retina. Genetics 156:205–217

Thornberry NA (1998) Caspases: key mediators of apoptosis. Chem Biol 5:R97–103

Thornberry NA, Lazebnik Y (1998) Caspases: enemies within. Science 281:1312–1316

Tsunoda S, Zuker CS (1999) The organization of INAD-signaling complexes by a multivalent PDZ domain protein in *Drosophila* photoreceptor cells ensures sensitivity and speed of signaling. Cell Calcium 26:165–171

Uren AG, Pakusch M, Hawkins CJ, Puls KL, Vaux DL (1996) Cloning and expression of apoptosis inhibitory protein homologs that function to inhibit apoptosis and/or bind tumor necrosis factor receptor-associated factors. Proc Natl Acad Sci USA 93:4974–4978

Van Soest S, Westerveld A, de Jong PT, Bleeker-Wagemakers EM, Bergen AA (1999) *Retinitis pigmentosa*: defined from a molecular point of view. Surv Ophthalmol 43:321–334

Vaux DL, Korsmeyer SJ (1999) Cell death in development. Cell 96:245–254

Verhagen AM, Ekert PG, Pakusch M, Silke J, Connolly LM, Reid GE, Moritz RL, Simpson RJ, Vaux DL (2000) Identification of DIABLO, a mammalian protein that promotes apoptosis by binding to and antagonizing IAP proteins. Cell 102:43–53

Vernooy SY, Copeland J, Ghaboosi N, Griffin EE, Yoo SJ, Hay BA (2000) Cell death regulation in *Drosophila*: conservation of mechanism and unique insights. J Cell Biol 150:F69–76

Vucic D, Kaiser WJ, Harvey AJ, Miller LK (1997a) Inhibition of *reaper*-induced apoptosis by interaction with inhibitor of apoptosis proteins (IAPs). Proc Natl Acad Sci USA 94:10183–10188

Vucic D, Seshagiri S, Miller LK (1997b) Characterization of *reaper*- and FADD-induced apoptosis in a lepidopteran cell line. Mol Cell Biol 17:667–676

Vucic D, Kaiser WJ, Miller LK (1998) Inhibitor of apoptosis proteins physically interact with and block apoptosis induced by *Drosophila* proteins HID and GRIM. Mol Cell Biol 18:3300–3309

Wang SL, Hawkins CJ, Yoo SJ, Muller HA, Hay BA (1999) The *Drosophila* caspase inhibitor DIAP1 is essential for cell survival and is negatively regulated by HID. Cell 98:453–463

White K, Grether M, Abrams J, Young L, Farrell K, Steller H (1994) Genetic control of programmed cell death in *Drosophila*. Science 264:677–683

White K, Tahaoglu E, Steller H (1996) Cell killing by the *Drosophila* gene *reaper*. Science 271:805–807

Wing JP, Zhou L, Schwartz LM, Nambu JR (1998) Distinct cell killing properties of the *Drosophila reaper*, head involution defective, and *grim* genes [published erratum appears in Cell Death Differ 1999 Feb; 6(2):212–3]. Cell Death Differ 5:930–939

Wolff T, Ready DF (1991) Cell death in normal and rough eye mutants of *Drosophila*. Development 113:825–839

Xiong WC, Montell C (1995) Defective glia induce neuronal apoptosis in the repo visual system of *Drosophila*. Neuron 14:581–590

Xiong WC, Okano H, Patel NH, Blendy JA, Montell C (1994) repo encodes a glial-specific homeo domain protein required in the *Drosophila* nervous system. Genes Dev 8:981–994

Yoon J, Ben-Ami HC, Hong YS, Park S, Strong LL, Bowman J, Geng C, Baek K, Minke B, Pak WL (2000) Novel mechanism of massive photoreceptor degeneration caused by mutations in the trp gene of *Drosophila*. J Neurosci 20:649–659

Yuan J, Horvitz HR (1992) The *Caenorhabditis elegans* cell death gene *ced*-4 encodes a novel protein and is expressed during the period of extensive programmed cell death. Development 116:309–320

Yuan JY, Horvitz HR (1990) The *Caenorhabditis elegans* genes *ced*-3 and *ced*-4 act cell autonomously to cause programmed cell death. Dev Biol 138:33–41

Zhang H, Huang Q, Ke N, Matsuyama S, Hammock B, Godzik A, Reed JC (2000) *Drosophila* pro-apoptotic Bcl-2/Bax homologue reveals evolutionary conservation of cell death mechanisms. J Biol Chem 275:27303–27306

Zhou L, Schnitzler A, Agapite J, Schwartz LM, Steller H, Nambu JR (1997) Cooperative functions of the *reaper* and head involution defective genes in the programmed cell death of *Drosophila* central nervous system midline cells. Proc Natl Acad Sci USA 94:5131–5136

Zhou L, Song Z, Tittel J, Steller H (1999) HAC-1, a *Drosophila* homolog of APAF-1 and CED-4 functions in developmental and radiation-induced apoptosis. Mol Cell 4:745–755

Zou H, Li Y, Liu X, Wang X (1999) An APAF-1cytochrome c multimeric complex is a functional apoptosome that activates procaspase-9 J Biol Chem 274:11549–11556

Zuker CS (1996) The biology of vision of *Drosophila*. Proc Natl Acad Sci USA 93:571–576

Drosophila Compound Eye Morphogenesis: Blind Mechanical Engineers?

Donald F. Ready[1]

1 Introduction

D'Arcy Thompson's 1917 treatise, *On Growth and Form*, articulated what might reasonably be considered the First Law of Morphogenesis: "... the form of an object is a 'diagram of forces' ... which have been impressed upon it ..." (Thompson 1917). This chapter is an informal, speculative essay on the emergence of form during *Drosophila* eye development and the forces that impress this form. It reviews selected events of retinal morphogenesis and considers mechanisms potentially responsible. In most instances, the molecular basis of morphogenesis in the eye is poorly understood; in many cases, correlation and causality remain to be distinguished. An integrated, realistic force diagram of the fly eye lies in the future.

This chapter draws primarily from Ready et al. (1976), Tomlinson and Ready (1987), Cagan and Ready (1989) Wolff and Ready (1991), Longley and Ready (1995), Kumar and Ready (1995), Fan and Ready (1997), Chang and Ready (2000) and refers to unpublished work of S. Karigiosis, T.-K. Sang, and A.K. Satoh. I am indebted to these collaborators for their many discussions of fly eye morphogenesis. Recent discussions with U. Tepass, Univ. Toronto, contribute to an understanding of the photoreceptor stalk. Morphogenesis of eye mechanosensory bristles is not considered. Waddington's 1962 treatise, *New Patterns in Genetics and Development*, contains a discussion of fly eye morphogenesis by a pioneer in this field; it includes numerous informative electron micrographs of developing eyes taken by Margaret Perry.

Readers are encouraged to consider Wolff and Ready (1993), a summary of pattern formation in the *Drosophila* eye; its discussions and illustrations are a useful background and accompaniment to the following chapter.

2 Morphogenic Forces are Grounded in Epithelial Tension

Tension across the retinal epithelium is the foundation of retinal morphogenesis. Although pattern formation in the third instar eye disc takes place within an epithelium that appears relatively "relaxed", eversion of the disc at the

[1] Department of Biological Sciences, Purdue University, Lilly Hall, West Lafayette Indiana 47907, USA

Results and Problems in Cell Differentiation, Vol. 37
K. Moses (Ed.): Drosophila Eye Development

beginning of pupation and its subsequent stretching to cover half of the head results in pronounced thinning of the epithelium; an eye disc approximately 38 μm deep thins to approximately 18 μm. The tension stretching the epithelium appears to arise largely in the increasing hydrostatic pressure which "inflates" the pupa. Epithelial tension is distributed principally across the apical *zonula adherens* (*z.a.*) junctions and the basal anchorage of cells to the underlying extracellular matrix (ECM). Between these specialized cell-cell and cell-ECM connections, retinal cell surfaces follow free curves that appear to accommodate cell volume in a fluid manner, resembling the interfaces between bubbles in a froth. The shapes of retinal cell types are thus importantly determined by their anchorage to the apical and basal tensile surfaces of the epithelium.

3 Apical Cell–Cell Contacts

Pattern formation establishes the web of cell–cell contacts that distributes apical tension across the disc. As cells are recruited into developing ommatidia, their adherens junctions with already fated cells become fixed; with certain important exceptions, these stabilized junctions persist throughout morphogenesis, likely contributing an actomyosin-based circumferential tension around the cell apex. The mechanisms which render an adherens junction stable or unstable in the developing eye are unknown but appear to be controlled locally along individual cell–cell boundaries; a cell can maintain fixed contacts with one partner while releasing contact with another. For example, transformation of the symmetrical eight cell cluster into the asymmetrical cluster of the two cone cell stage requires R4 to selectively release its *z.a.* with R8 while maintaining *z.a.* contact with R3 and R5. Ommatidial rotation within the third instar eye disc likewise appears to require differential control of individual cell-cell contacts, allowing slippage between stably connected cells of the R-cell cluster and surrounding undifferentiated cells.

Alignment of rhabdomeres to the ommatidial optical axis, a fundamental step of compound eye morphogenesis, is achieved in a program of conserved and changing *z.a.* contacts. Following cone cell recruitment, photoreceptor apical surfaces are involuted into the retinal epithelium as cone cells close "above" them (Fig. 1A,B). As the cone cells are recruited, their apical endfeet encircle photoreceptor apices in a four cell ring (Fig. 2A). Shortly thereafter, while retaining their contacts to the more centrally placed photoreceptors that have recruited them, cone cells increase their mutual contacts, zipping shut above the photoreceptor apices and inpocketing them into a trapped cavity, the future inter-rhabdomeral space (IRS). It is not known what closes cone cells over the photoreceptors, but later in development, contacts between cone cells are marked by massive septate junctions; it is possible that septate junction assembly mediates their closure.

The trapezoid of adult rhabdomeres has its origin early in pupal life. Within the trapped apical cavity, photoreceptors, while retaining the *z.a.* junctions established during pattern formation, form a new set of face-to-face contacts that prefigure the adult trapezoid. Contact between photoreceptors R2, R4, and R7 at the center of the cavity occludes R3, displacing its apical membrane to the future "point" of the trapezoid. Adhesions between future rhabdomeres are highly stereotyped, suggesting a simple "trapezoid code" of apical adhesivities. Distally, R7 contacts R2, R4 and R5; R2 contacts R4; R1, R3 and R6 do not engage in rhabdomere–rhabdomere contacts. Proximally, R4, R5 and R6 contact R8. By 55% pd, when photoreceptor apical surfaces are anchored to the cone cell feet, face-to-face contacts between photoreceptors are relinquished, opening the IRS.

Until approximately 37% pd (pd = pupal development; 0% pd = white prepupupa, 100% pd = eclosion), photoreceptor apices remain relatively small, lining the shallow pocket below the cone (Fig. 2B). At this stage, photoreceptor apical surfaces do not show obvious differentiation; they show irregular minor infoldings, but the future rhabdomere is not morphologically apparent (Fig. 3A). Beginning about 37% pd, photoreceptor apical surfaces expand, far outstripping growth of the baso-lateral cell surface. This differential growth within the conserved mesh of cell contacts is plausibly the proximal cause of the future rhabdomere's extension towards the retinal floor.

As the pocket deepens to the retinal floor, it encounters a basal nexus of the cone cell feet (Fig. 2C). Earlier in pupal life, the basal projections of the cone cell, while retaining their ECM contacts, intrude between the photoreceptors: the equatorial cone cell foot moves between R7 and R1, the polar between R3 and R4, the posterior between R5 and R6, and the anterior between R2 and R8. The result is a basket that partitions the R-cell cluster in four. (The topology of this basket resembles the stitching of a baseball, with two additional stitches across the waist of each flap.)

Where the deepening R-cell pocket encounters the cone cell feet, photoreceptors exchange *z.a.* junctions with each other for junctions with the cone cell feet (Fig. 1C). R8, which had contributed to the pocket's floor, loses contact with R-cells 3, 5 and 6 while retaining contact with R1 and R2. As R8's cell body moves off the central axis anteriorly between R1 and R2, its apical surface becomes aligned between the proximal R1–R2 junction and the cone cell foot complex which now defines the floor of the apical chamber. The cone cell end feet are anchored to the basal ECM and attachment of R-cell *z.a.* junctions to the cone cell feet has the consequence that R-cell apical surfaces are sprung between the tensile surfaces of the distal and proximal epithelium.

Almost nothing is known of the forces that orchestrate the movements described above. Interesting questions concerning cell polarity signaling are raised by the emergence in cone cell feet of a second, apparently apical domain that forms the floor of the IRS. An armadillo-positive junctional ring ultrastructurally resembling a *z.a.* connects the feet to each other and to R-cells. In the EM, unusual structures, resembling ordered vesicle clusters, are

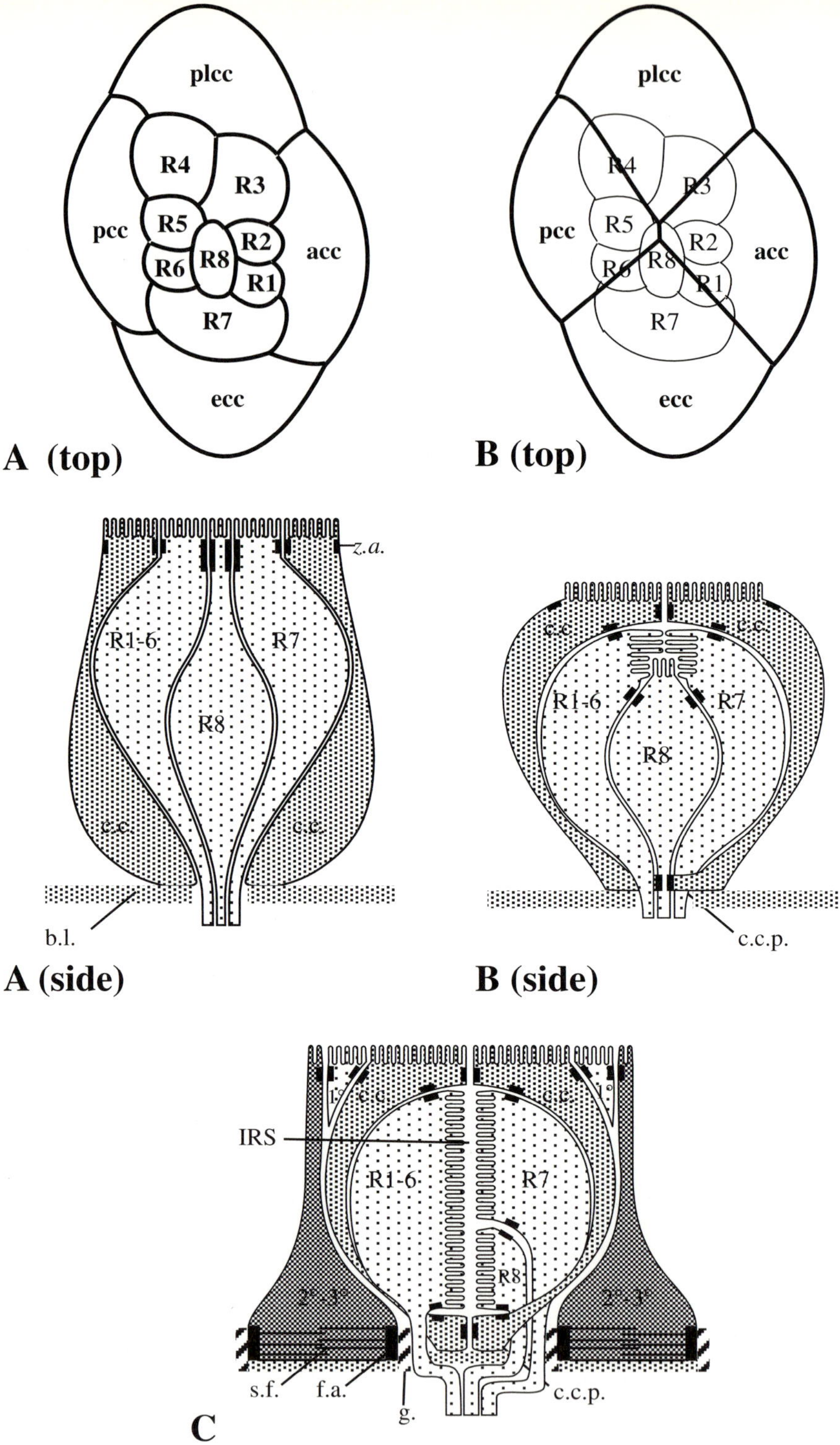

plcc
R4
R3
pcc
R5
R2
acc
R8
R6
R1
R7
ecc
A (top)
plcc
R4
R3
pcc
R5
R2
acc
R6
R8
R1
R7
ecc
B (top)
z.a.
R1-6
R7
R8
c.c.
c.c.
b.l.
A (side)
c.c.
c.c.
R1-6
R7
R8
c.c.p.
B (side)
1°
c.c.
c.c.
1°
IRS
R1-6
R7
R8
2°-3°
2°-3°
s.f.
f.a.
g.
c.c.p.
C

observed in cone cells at button-like contacts with the distal ends of the photoreceptors.

4 Apical-ECM Contacts

Cell-ECM contacts on the apical surface are modest until the overt elaboration of the corneal cuticle beginning at about 65% pd. Prior to this time, an amorphous and uncharacterized ECM overlies the epithelium. It is unclear what mechanical role, if any, is played by this layer. One interesting possibility is that cells may spread their apical end feet by traction against this surface. In early pupal life, the apices of cone cells and primary (1°) pigment cell apical surfaces expand coincident with the development of prominent actin cables resembling stress fibers facing the apical ECM. Alternatively, it is possible that these actin cables balance the stretching forces of eversion across the distal epithelial surface. Once the corneal cuticle is established, it is probable that distal tensions are grounded by local contacts with this rigid carapace.

Apical ECM secretion by the cone cells and primary pigment cells establishes the familiar faceted surface of the compound eye. The secondary and tertiary pigment cells retain a strong anchorage to the early distal ECM, but do not secrete extensive additional cuticle. In contrast, the cone cells and primary pigment cells engage in two phases of massive ECM deposition to stack two lenses above the photoreceptors. The cornea, a cuticular specialization, is deposited first, followed by secretion of the cone. The sequential secretion of

Fig. 1. Photoreceptor apical surfaces are involuted during development. **A** (*top*) Viewed from above, *z.a.* of an ommatidial cluster nine rows behind the furrow lie in a stereotyped, planar configuration (*dark cell outlines*). Note that R4 is not connected to R8. **A** (*side*) Viewed from the side, the photoreceptors of a cluster at this stage are exposed on the apical surface of the epithelium, facing the lumen of the disc. Cells are joined distally by *z.a.* junctions. Bundled axons of an ommatidium penetrate the basal lamina (*b.l.*) ECM to which the cone cell feet are anchored. **B** (*top*) Closure of the cone cells "above" the apical tips of the photoreceptors involutes them into the epithelium in a trapped cavity. Cone cell *z.a.* junctions are shown as *dark outlines*; photoreceptor *z.a.* junctions, lying below, are shown in *lighter outline*. Contacts between photoreceptors and cone cells are not shown in this schematic. The details of R-cell/cone cell contacts and their behavior during development are not understood. **B** (*side*) In a pupal eye at 37% pd, the apical tips of the photoreceptors face the trapped apical cavity, the future IRS. Cone cell processes have met mid-center at the floor of the retina in a nexus, the cone cell plate (*c.c.p*), "below" the *z.a.* junctions of the photoreceptors. **C** In a topologically mature ommatidium, at 55% pd, expansion of photoreceptor apical surfaces has brought them into contact with the cell feet; via exchange of junctions between R-cells and cone cells (see text), cone cell feet have become the floor of the IRS. Photoreceptor apical surfaces are linked to the retinal floor via the cone cell plate, aligning them to the future optical axis. For clarity of the diagram, R7 has been swung to the anterior*b.l.* Basal lamina; *z.a.* zonula adherens junctions; c.c. cone cells; *IRS* inter-rhabdomeral space; *1°, 2°, 3°* primary secondary and tertiary pigment cells; *s.f.* stress fibers; *f.a.* focal adhesions; *c.c.p.* cone cell plate; *g.* grommet

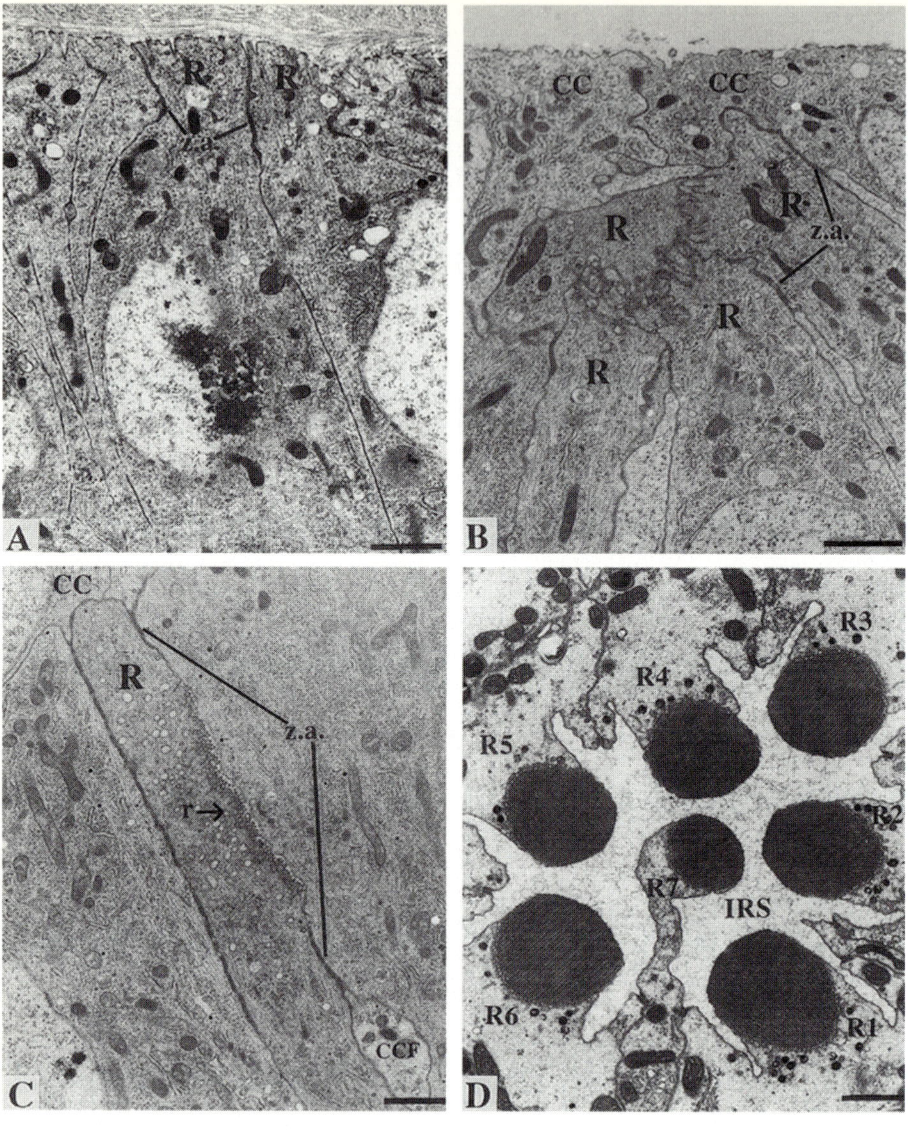

Fig. 2. Developing rhabdomeres are bounded by *z.a.* junctions. **A** In third instar eye discs, photoreceptors (*R*) expose a microvillar coxcomb at the apical surface of the epithelium. Adherens junctions (*z.a.*) form an apicolateral belt which joins adjacent cells and separates apical and basolateral domains. **B** At 37% pd, photoreceptor apical surfaces face the trapped apical cavity, the IRS, formed by the closure of the overlying cone cells (*CC*). Adherens junctions surround the microvillus photoreceptor apical membrane. **C** By 55% pd, photoreceptor apical surfaces have elongated and anchored to the cone cell feet (*CCF*); the outer, distal, ends of the photoreceptors are anchored to the overlying cone cells (*CC*). Abundant vesicles feeding the growing rhabdomere are evident, as are "vertically" oriented microtubules. The microtubule array appears to be an essential component of the vesicle delivery machinery. Whether it also plays a more direct mechanical role in morphogenesis is not known. The forming rhabdomere (*r*) is bracketed by *z.a.* junctions. **D** Cross section of an adult wild-type ommatidium displaying the rhabdomeres of photoreceptors R1–R7 in an open, trapezoidal pattern. The rhabdomere of R8 lies below that of R7. Rhabdomeres and the stalks which bear them face the inter-rhabdomeral space (IRS). Adherens junctions connect photoreceptors to their neighbors and close the IRS. *Scale bars* = 1 μm. (Longley and Ready 1995)

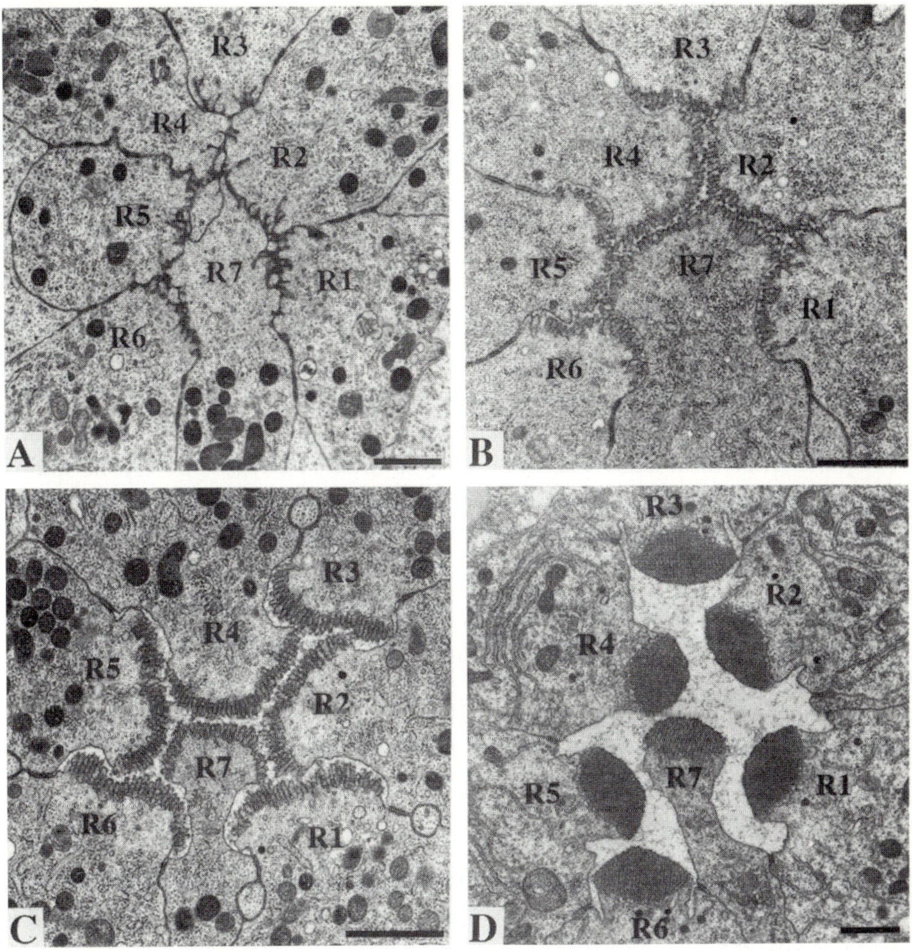

Fig. 3. Rhabdomere morphogenesis. **A** A cross section from a 37% pd ommatidium showing the rhabdomeres of photoreceptors R1–R7. **B** Rhabdomeres at 55% pd. The apical surfaces now possess distinct microvillar and stalk domains. Note the plane of separation that lies between neighboring microvilli. **C** Rhabdomeres at 67% pd. The microvilli have elongated and are more tightly packaged. **D** Rhabdomeres at 78% of pupal life. Microvilli at the sides of the rhabdomere are shorter relative to those in the center and the rhabdomere has taken on a more oval profile. *Scale bars* = 1 μm. (Longley and Ready 1995)

the cornea and cone raises the eye's outer surface, the epicuticle, like an inflatable dome, anchored to the hexagonal perimeter of secondary (2°) and tertiary (3°) pigment cells.

Increased pressure below the epicuticle is evident as the primary pigment and cone cell apices are depressed into the retina. The hexagonal plate formed by the actin-reinforced apices of these cells is deformed into a teacup shape in a manner resembling the pressing of a metal disc into a cup by a die. Primary pigment cells form the walls of the cup and the cone cells its floor. It is likely that the "blueberry" phenotype of erupted facets points to inability of the epicuticle to contain the pressure of lens formation. In Moire mutant eyes, the corneal surface detaches from its normal attachment to the lattice of secondary and tertiary pigment cell apices, allowing light to scatter laterally below the lenslets, resulting in the "watered-silk" luster of the mutant eye.

5 Basal-ECM Contacts

Tension across the retinal floor plays a significant role in morphogenesis. A delamination of the epithelium behind the furrow, the retinal floor, or fenestrated membrane, is a specialized plane of pigment cell basal end feet apposing an underlying ECM. Until the close of pattern formation at approximately 55% pd, the apical "headprints" of cells tiling the distal epithelium, principally the primary pigment and cone cells, occupy an area equivalent to the basal footprints of the secondary and tertiary pigment cells that tile the floor.

Beginning about 55% pd, tension developed by the basal end feet of the secondary and tertiary pigment cells results in an approximately fourfold contraction of the retinal floor. Following the establishment of the grommet (g; Fig. 1C), a distinctive, laminin-containing ring of ECM through which each ommatidial bundle of axons exits the retina, pigment cell end feet organize planar arrays of actomyosin stress fibers which anchor to the grommet in integrin-dependent contacts that resemble the focal adhesions (f.a.) of fibroblasts.

Grommets appear to serve as a restraint, holding photoreceptor cell bodies above the fenestrated membrane. A common phenotype, "ommatidia falling through the floor", is associated with floor defects. Whether forces arising from ommatidial elongation push photoreceptors through the floor, or if tension along R-cell axons pulls photoreceptors through the floor, or both, is undetermined.

Between the close of pattern formation and eclosion, ommatidia elongate approximately threefold, from 31 to 100 μm. Whether this elongation is a simple by-product of cell growth (i.e., cell volume increase) or relies on a dedicated extensory force-generating mechanism is not known. Ommatidia centrally placed in the eye are longer than those at the periphery, suggesting a currently unknown mechanism that regulates ommatidial elongation. Since the apical "headprint" of an ommatidium on the corneal ECM remains constant while pigment cell basal end feet contract, mature ommatidia taper towards the retinal floor. Packing proximally tapered ommatidia contributes to the interommatidial angle, an essential determinant of the optics of the eye.

Primary pigment cells are unique in releasing their attachment to basal ECM. Coincident with their domination on the apical surface, their basal end feet detach from the ECM (Fig. 1C). It is not known how this release is effected or the fate of the anchoring complex.

6 Rhabdomere Morphogenesis

Rhabdomeres, the "thread-like bodies" of classical anatomists, are columnar stacks of approximately 60,000 closely packed photosensitive membrane microvilli aligned to the optical axis of the eye. The lenses of the cornea and cone focus the fly's visual world onto rhabdomere distal tips. Much of rhabdomere morphogenesis occurs during the latter half of pupal life as abundant vesicular traffic enormously expands the photosensitive membrane. Directed membrane traffic, a fundamental contributor to rhabdomere morphogenesis, is not considered here.

Rhabdomeres are apical plasma membrane specializations. Prior to approximately 35% pd, R-cell apices are thrown into irregular folds without evident morphological differentiation (Fig. 3A). Between 35 and 50% pd, R-cell apices expand within the cone cell cage, elongating to the retinal floor and connecting to the cone cell feet as described above (Fig. 2C). Although the forces driving apical elongation are not known, simple expansion by targeted membrane delivery seems a parsimonious explanation. Given the prevailing R-cell contacts, differential apical expansion can be accommodated by floorward elongation. Consistent with such a mechanism are unpublished observations by Fan and Ready that this stage of rhabdomere morphogenesis is particularly sensitive to perturbation of normal membrane traffic by expression of dominant negative Glued, the fly dynactin homologue. Instead of elongating in a well-bounded apical domain, *z.a.* integrity is compromised and apical membrane is delivered to inappropriate basolateral locations resulting in isolated, ectopic rhabdomeres. The "main" rhabdomere expands in irregular, shallow accumulations. It is notable that the growing apical surface is not accommodated by infolding into the cell cytoplasm, possibly an early manifestation of a barrier-like specialization of the subapical cortical cytoskeleton.

A center-surround organization, marking the definitive rhabdomere and surrounding supporting membrane of the stalk, emerges within the R-cell apical membrane coincident with its anchorage to the cone cell complex (Fig. 3B). The future rhabdomere is evident as a fringe of short microvilli, the surrounding stalk membrane takes on a distinctly "stiffened" appearance. Apical membrane proteins are redistributed at this stage, with Crumbs and beta-Heavy Spectrin becoming restricted to the stalk. The establishment of a cytocortical membrane scaffold likely gives the cytoplasmic surface of the stalk its increased density in electron micrographs. We speculate that the stalk is a key constraint during rhabdomere morphogenesis, imparting the elongated loop of the R-cell *z.a.* junction to the future rhabdomere and constraining its lateral expansion like a mandarin collar.

Beyond its mechanical role, it is likely that the stalk supports distinctive regional physiology. The Na^+–K^+ATPase, initially distributed uniformly over the apical surface, becomes localized to the stalk as it is established (Baumann et al. 1994). It is unknown if regional physiological specialization contributes to morphogenesis.

Following establishment of the definitive rhabdomere primordium as a fringe of short microvilli by 55% pd, microvilli progressively elongate and pack more regularly, reaching adult morphology shortly before eclosion (Fig. 3C,D). The forces that drive microvillar elongation are not known. Rhabdomere microvilli do not project as free-standing individuals, but associate with neighboring microvilli over their entire length. Nor do they elongate "away" from the cell, pushing the plasma membrane closer to the ommatidial center. Rather, rhabdomere microvilli grow back into the R-cell cytoplasm along a unified front. Initially, this front is concave, running below the curving lawn of short microvilli. As microvilli elongate, this front first becomes flat and then convex as central microvilli grow longer than more lateral neighbors. The progress of microvillar elongation approximates a sieve or colander pressed into a deformable sheet possessing the important property that, where not contacted by the sieve, the sheet retains its position as though supported from within. Membrane adhesion and microfilament extension are likely to contribute to microvillar elongation.

Microvilli elongate "between" two specialized, apparently tensile surfaces. At their distal tips, microvilli terminate along a thin, dense ECM of unknown composition; proximal microvillar ends terminate along a distinctive rhabdomere base that marks the boundary between the rhabdomere and underlying photoreceptor cytoplasm. The degree to which expansion of self-adhesive membrane trapped between such surfaces contributes to microvillar elongation is not known. Chaoptin, a membrane-associated homophilic adhesion molecule whose loss results in microvilli splaying apart, appears a principal mediator of inter-microvillar adhesion (Van Vactor et al. 1988).

Axial microfilaments extend the length of growing microvilli with plus ends distal (Arikawa et al. 1990). It is appealing to consider that, analogous to membrane extension in other systems, microfilament polymerization contributes to microvillar elongation; to date there is no evidence addressing this possibility. The striking regularity of the rhabdomeral column over supramolecular dimensions, evident in its round cross-sectional profile, suggests an optimal design in which forces are distributed through a single mechanical system. Tension-dependent plus end growth, a "Brownian ratchet" (Mogilner and Oster 1996), at distal microvillar tips may take up the "slack" in the surface that occurs as each bolus of membrane adds to the growing rhabdomere, reestablishing uniform tension across the surface. We speculate that the catenary-like curve of the rhabdomere base (and tip) describes a balance between the forces of microvillar extension and tensile sheets bounding the microvilli.

7 The Rhabdomere Terminal Web

The rhabdomere base, the curving boundary between the microvilli and underlying photoreceptor cytoplasm, is defined by a striking specialization of the cortical actin cytoskeleton, the rhabdomere terminal web (RTW), a profusion of microfilaments emanating from the base of the rhabdomere and gathering into cables as it projects deep into the cytoplasm. The RTW resembles the splayed fibers of an artist's paintbrush pressed against a window. Intriguingly, the deepest ends of the RTW, its "dorsal spines" often turn into the long axis of the cell and appear to join into a "vertical" bundle. When developing eyes expressing a chimeric GFP-actin binding protein are fixed and counterstained using rhodamine-phalloidin, dorsal spines appear red only in the confocal microscope, suggesting binding by the chimeric protein may be diminished by heavy decoration by other proteins. Microvillar axial microfilaments likewise show little chimeric protein staining.

The extent and manner in which microvillar axial microfilaments engage with and contribute to the RTW are not known. RTW actin cables terminate in distinct bundles on the rhabdomere base and do not appear to originate in the uniform manner that might be expected if all or most axial microfilaments extended into it. S1 decoration experiments by Arikawa and Williams indicate RTW microfilaments terminate with "+" ends at the rhabdomere base. Moesin, a membrane/cytoskeletal linking protein, is concentrated at the rhabdomere base and we postulate that it anchors the RTW to unknown membrane proteins. Rhabdomere morphogenesis is disrupted by misexpression of mutant moesins in developing R-cells. Non-muscle myosin II and alpha-actinin are associated with the rhabdomere base; it is reported that alpha-actinin null photoreceptors have normal (Fyrberg 1998).

The branching actin mesh of the RTW resembles the actin networks driving membrane protrusion in migrating cells such as fish keratinocytes (Pollard et al. 2000); we speculate that the RTW provides a similar force-generating network which, instead of pushing the membrane outward, resists intrusion of the expanding rhabdomere into the photoreceptor cytoplasm. It will be interesting to see if the actin-associated proteins that regulate the dynamic microfilament networks of migratory cells operate during photoreceptor morphogenesis.

Rhodopsin is essential for rhabdomere morphogenesis. Consistent with the observation that substantial rhabdomere development, including the establishment of stalk and rhabdomere domains and considerable microvillar development occurs before the onset of rhodopsin synthesis at 70% pd, early stages of R1–6 rhabdomere morphogenesis are executed normally in Rh1 null mutants. A crisis of morphogenesis occurs in nulls at approximately 90% pd. At this stage, normal microvilli have elongated to the degree that the rhabdomere is now beginning to curve strongly into the cytoplasm; the distinctive catacomb-like membrane architecture of the base develops at this time. In rhodopsin null mutants, the rhabdomere base does not develop normally and rhabdomere membrane intrudes deep into photoreceptor cytoplasm in convoluted sheets of apposed membrane. The phenotype suggests a failure to consolidate the barrier that normally provides a subapical constraint. It is possible that prior to 90% pd, the intrusive "pressure" is insufficient to overwhelm the compromised barrier of the null.

It is notable that a rhodopsin replacement can rescue rhabdomere morphogenesis in genetically Rh1 null animals, but only when expressed at a critical window of eye development coinciding with the time of normal Rh1 expression, suggesting rhodopsin may contribute an activity necessary to organize the barrier of the RTW(Kumar et al. 1997). A similar rescue of rhodopsin null rhabdomere morphogenesis can be effected by timed expression of dominant active Drac1, suggesting rhodopsin may exert its effect via pathways including Drac1.

8 Constrained Expansion

We propose a simple model for rhabdomere morphogenesis, constrained expansion (Fig. 4). The model postulates that targeted delivery of rhodopsin-containing membrane vesicles differentially expands the rhabdomere within constraints imposed "laterally", in the plane of the membrane, by the reinforced membrane of the stalk and subapically by the RTW. Self adhesion of rhabdomere membrane, perhaps organized in conjunction with the axial microfilaments, folds the rhabdomere membrane into closely packed microvilli. Trapped between tensile surfaces at their proximal and distal ends, membrane addition "inflates" the rhabdomere cylinder. We further suggest that rhodopsin acts via the small GTPase, Drac1 to organize the RTW. This has the attractive property that the distinctive cytoskeletal organization that supports rhabdomere morphogenesis is locally controlled by the photosensitive membrane.

9 Conclusion

The present survey suggests that several forces shape the geometrical regularity of the *Drosophila* compound eye. Like a game of cat's cradle, sequential manipulation of cell-cell and cell-ECM contacts, establishes a "knot" of coupled mechanical systems within the retinal epithelium. Differential growth of these distinct environments, fed by directed membrane traffic, constrained by membrane scaffolds and cytoskeletal barriers and shaped by adhesive forces generates form. Gene expression is known to be conditioned by a cell's mechanical state and it is interesting to wonder if the evolving shapes of morphogenesis in some way signal the progress of morphogenesis to the genome. Few signals would seem more useful to a genetic program of eye development than those which could communicate a current "goal" has been achieved.

Structural modeling of *Drosophila* compound eye morphogenesis, informed and tested by molecular genetic methods available in the fly, seems an attainable and informative goal. The ability of *Drosophila* studies to investigate and integrate an ever-widening realm of development, makes fly eye morphogenesis an unparalleled opportunity to look over the shoulder of the blind mechanical engineers of development. Given that the Blind Watchmaker is a

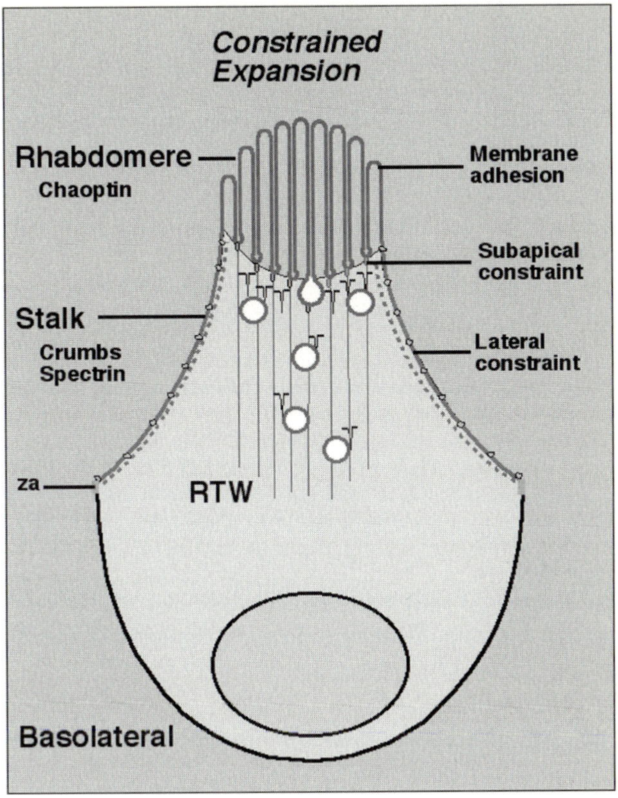

Fig. 4. A constrained expansion model for rhabdomere morphogenesis. Three important forces are considered: (1) membrane adhesion between microvilli; (2) a constraint on lateral expansion provided by the stalk, and (3) a subapical constraint provided by the RTW: It is proposed that massive delivery of self-adhesive membrane to the developing rhabdomere is ordered into closely packed microvilli in a manner resembling a raft of bubbles blown into a Langmuir trough. Lateral expansion of the surface is contained by the stalk, a collar of stiffened membrane which is grounded in the encircling zonula adherens junctions; the stalk and junctions constitute the walls of the trough. Intrusion of rhabdomere membrane into photoreceptor cytoplasm is prevented by the RTW, a specialization of the apical cortical cytoskeleton; the RTW is the water surface which imposes planar order upon the raft. Some key proteins mediating this organization are indicated (see text)

fundamentalist conservative, lessons learned in *Drosophila* are likely to be useful elsewhere.

References

Arikawa K, Hicks JL, Williams DS (1990) Identification of actin-filaments in the rhabdomeral microvilli of *Drosophila* photoreceptors. J Cell Biol 110:1993

Baumann O, Lautenschlager B, Takeyasu K (1994) Immunolocalization of Na,K-Atpase in blowfly photoreceptor cells. Cell Tissue Res 275:225

Cagan RL, Ready DF (1989) The emergence of order in the *Drosophila* pupal retina. Dev Biol 136:346–362

Chang HY, Ready DF (2000) Rescue of photoreceptor degeneration in rhodopsin-null *Drosophila* mutants by activated Rac1. Science 290:1978–1980

Fan SS, Ready DF (1997) Glued participates in distinct microtubule-based activities in *Drosophila* eye development. Development 124:1497

Fyrberg C, Ketchum A, Ball E, Fyrberg E (1998) Characterization of lethal *Drosophila melanogaster* alpha-actinin mutants. Biochem Genet 36:299

Kumar JP, Ready DF (1995) Rhodopsin plays an essential structural role in *Drosophila* photoreceptor development. Development 121:4359

Kumar JP, Bowman J, Otousa JE, Ready DF (1997) Rhodopsin replacement rescues photoreceptor structure during a critical developmental window. Dev Biol 188:43–47

Longley RL, Ready DF (1995) Integrins and the development of 3-dimensional structure in the *Drosophila* compound eye. Dev Biol 171:415

Mogilner A, Oster G (1996) Cell motility driven by actin polymerization. Biophys J 71:3030–3045

Pollard TD, Blanchoin L, Mullins RD (2000) Molecular mechanisms controlling actin filament dynamics in nonmuscle cells. Annu Rev Biophys Biomol Struct 29:545–576

Ready DF, Hanson TE, Benzer S (1976) Development of the *Drosophila* retina, a neurocrystalline lattice. Dev Biol 53:217–240

Thompson (1917) On growth and form. Cambridge University Press, Cambridge

Tomlinson A, Ready DF (1987) Neuronal differentiation in the *Drosophila* ommatidium. Dev Biol 120:366–376

Van Vactor D, Krantz DE, Reinke R, Zipursky SL (1988) Analysis of mutants in chaoptin, a photoreceptor cell-specific glycoprotein in *Drosophila*, reveals its role in cellular morphogenesis. Cell 52:281–290

Waddington CH (1962) New patterns in genetics and development. Columbia University Press, New York

Wolff T, Ready DF (1991) The beginning of pattern-formation in the *Drosophila* compound eye – the morphogenetic furrow and the 2nd mitotic wave. Development 113:841–846

Wolff T, Ready DF (1993) Pattern formation in the *Drosophila* retina. In: Bate M, Martinez-Arias A (eds) The development of *Drosophila melanogaster*. Cold Spring Harbor Press, Cold Spring Harbor

The Establishment of Retinal Connectivity

Ulrike Gaul[1]

1 Introduction

This chapter describes the developmental, cellular and molecular mechanisms involved in establishing neural connectivity in the adult *Drosophila* visual system. It concerns itself primarily with the projections of photoreceptor axons into the optic lobe, with an emphasis on the more recent literature.

2 The Structure of the Adult Visual System

The visual system of the adult fly consists of the compound eye, with 800 ommatidia containing eight photoreceptors each (Ready et al. 1976), and four visual processing centers in the optic lobe, the lamina, medulla, lobula and lobula plate (for a detailed review, see Meinertzhagen and Hanson 1993). The optic lobe processing centers are organized in a retinotopic fashion, which is essential for their function in interpreting visual input. Unlike in vertebrates, the photoreceptors of the fly project directly onto the optic lobe. Based on their different spectral sensitivity, the photoreceptor (R) cells select different areas of the optic lobe as their synaptic targets: R1–R6, which express Rhodopsin 1 (Rh1), synapse in the first optic ganglion, the lamina; R7 cells, which express Rh3 or Rh4, and R8 cells, expressing Rh5 or Rh6, terminate in different layers of the medulla (see also Pichaud and Desplan, this Vol.). Both lamina and medulla contain 800 columnar units, precisely matching the number of ommatidial units in the eye. The mapping of visual space to the lamina is direct, but the map is inverted in the medulla along the anterior-posterior axis as a result of a chiasm. An intriguing feature of the fly visual system is the neural superposition governing the projection of R1–R6 onto the lamina (Braitenberg 1967): During the pupal stage, the global retinotopic map is refined such that the axons of the photoreceptors with the same visual axis connect to a single unit of interneurons in the lamina. Since the photoreceptors R1–R6 in each ommatidium have slightly different visual axes, this means that photoreceptors from the same ommatidium connect to separate cartridges in the lamina;

[1] Laboratory of Developmental Neurogenetics, Rockefeller University, 1230 York Avenue, New York, New York 10021, USA

Results and Problems in Cell Differentiation, Vol. 37
K. Moses (Ed.): Drosophila Eye Development

conversely, each lamina cartridge receives input from six different neighboring ommatidia.

3 The Development of the Photoreceptor Projection

The differentiation of the adult visual system begins during the third instar larval stage with the onset of photoreceptor differentiation in the eye disc. As described in detail in Lee and Treisman (this Vol.), R cell differentiation sweeps across the eye disc in a posterior to anterior progression that is tightly regulated both temporally and spatially. The newly generated photoreceptors begin to send out axons, forming fascicles that join axons from the same ommatidium. The axons first grow towards the base of the eye disc and then turn posteriorly. Retaining their relative positions, the axon fascicles converge and funnel into the optic stalk, a transient tube-like structure connecting the eye disc with the developing optic lobe. After exiting the stalk, the fascicles separate again, assuming a fan-like shape with strict sorting according to their positions in the retina, and attach to the surface of the lamina. Having thus secured their proper retinotopic position, they turn medially into the lamina, where they trigger the differentiation of lamina target cells and their maturation into cartridges. This innervation process proceeds in a posterior to anterior progression that mirrors the progression of photoreceptor differentiation in the eye disc, with fibers arriving later assuming more anterior positions in the optic lobe. The R1–R6 axons terminate within the lamina, between two layers of glial cells, while the R7 and R8 axons project deeper and terminate in the medulla (for review see Meinertzhagen and Hanson 1993; see Fig. 1).

The complexity and precision of the neural connectivity in the adult visual system has fascinated *Drosophila* researchers for a long time. The first wave of investigations led to a detailed description of the development of the system and established some of its governing principles (Kunes and Steller 1993; Meinertzhagen and Hanson 1993; Cutforth and Gaul 1997; Wolff et al. 1997). More recently, the study of the development of the visual system has enjoyed a renaissance owing to technical advances that permit the generation of mosaics through site-specific mitotic recombination (FLP/FRT technology; see Golic 1991; Xu and Rubin 1993). The identification of relevant genes, through both candidate gene approaches and mutagenesis, has greatly spurred the understanding of the cellular and molecular mechanisms underlying the guidance and targeting of photoreceptor axons.

The first mutagenesis screens were F2 screens, in which homozygous mutant third instar larvae were examined for defects in R axon projections using R cell-specific lacZ transgenes as markers; one of the drawbacks of this approach is its bias against early lethal genes (Martin et al. 1995). This problem has been overcome by an elegant mosaic approach in which mitotic clones are induced efficiently in the eye by expressing the FLP recombinase under an early acting and eye-specific promoter element of *eyeless* (Stowers and Schwarz 1999;

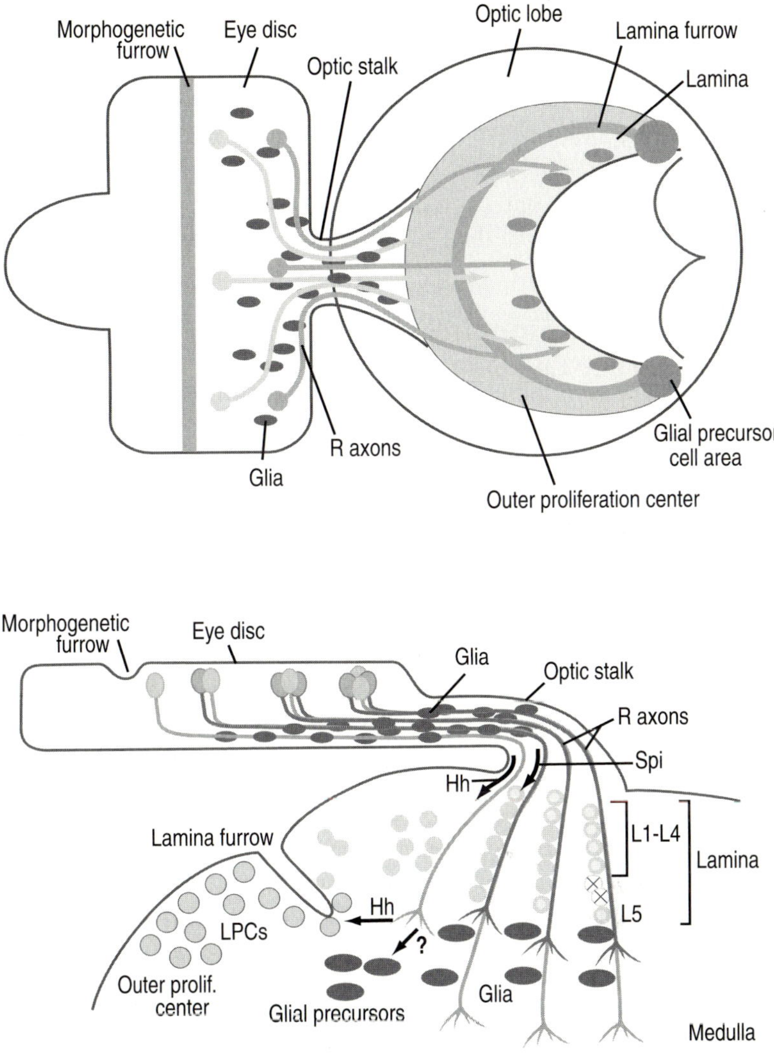

Fig. 1. Schematic representing the development of the adult *Drosophila* visual system: Lateral view (*top*) and horizontal view (*bottom*). See text for explanation

Newsome et al. 2000a). To increase the size of the mutant clones, Minute or cell lethal mutations are placed on the opposite chromosome, leading to eyes that consist largely of homozygous mutant cells and are readily analyzed for phenotypic defects. One drawback of this approach is that it uncovers only genes that act autonomously in the photoreceptors.

Two major themes are beginning to emerge from these recent studies. (1) Besides the neuronal cells themselves, glia play a crucial role in establishing retinal connectivity. (2) The establishment of connectivity relies on a series of complex interactions between R axons and their environment: Anterograde signals from growing R axons trigger the proliferation and differentiation of lamina precursor cells into lamina neurons, as well as the migration and differentiation of different glial cell populations; retrograde signals from glial cells and from the lamina precursors provide guidance cues to the R axons. We will first discuss the anterograde influence of R axons on their (intermediate) targets and then turn to the retrograde influence on the R axons.

4 Anterograde Signals: Incoming Retinal Fibers Trigger Target Cell Development

The primary target of the R axons in the optic lobe is the lamina. The lamina neurons derive from neuroblasts in the outer proliferation center (OPC), a curved band on the surface of the developing brain (White and Kankel 1978; Hofbauer and Campos-Ortega 1990; see Fig. 1). Much like the differentiation of the eye disc, the differentiation of the lamina takes place in a posterior to anterior progression that sweeps across the lamina anlage and whose front is marked by a fold in the brain, called the lamina furrow (Selleck and Steller 1991; Selleck et al. 1992). Spawned by the OPC, the lamina precursor cells (LPCs) undergo two cell divisions, the first anterior to the lamina furrow, the second posterior to it. After the second division, the LPCs start differentiating into lamina neurons and become incorporated into columns which then mature into lamina cartridges. Newly incoming retinal fibers grow into the area between the posterior edge of the lamina furrow and the already assembled lamina columns and thus come in close contact with the LPCs.

It has long been recognized that the differentiation of the optic lobe depends on innervation from the compound eye (Power 1943; Meyerowitz and Kankel 1978; Macagno 1979; Fischbach and Technau 1984). In adult mutants that lack eyes, the lamina is completely absent and the two other ganglia, medulla and lobula complex, are considerably reduced in size. Mosaic analysis has confirmed that the maturation of the optic lobes requires proper differentiation of R cells in the eye and normal innervation by their axons (Meyerowitz and Kankel 1978). In the case of the lamina, there is a direct coupling of neurogenesis to the arrival of retinal axons: The newly arriving retinal fibers are in close proximity to LPCs posterior to the lamina furrow that have divided once and are now in G1. In mutants lacking axonal ingrowth, LPCs arrest in G1 and thus fail to undergo the second cell division and to differentiate into lamina neurons. Thus, the R axons trigger the final cell division and terminal differentiation of the LPCs (Selleck et al. 1992).

In a series of beautiful studies, Kunes and colleagues (Huang and Kunes 1996, 1998; Huang et al. 1998) have uncovered the molecular basis of this induction. They were able to show that Hedgehog (Hh) protein, which is expressed in young R cells in the eye disc and transported down the axons as they grow into the lamina target, is the critical factor inducing both the final cell division and the onset of terminal differentiation in the LPCs (Huang and Kunes 1996). Given the role of Hh in R cell development in the eye, experiments aimed at assessing its function in optic lobe development required separating the two processes as much as possible. Relying on temperature-sensitive and tissue-specific alleles of *hh*, Kunes and coworkers were able to generate retinal axons that arrive at the brain without Hh activity. In such animals, the LPCs arrest in G1 and fail to express both the early neuronal differentiation marker Dac and the late differentiation marker Elav – as is observed in mutants that lack axonal ingrowth altogether. Mosaic experiments using FLP-out constructs to induce Hh expression show that Hh is also sufficient to trigger lamina proliferation and the onset of differentiation. Expression of Hh in small patches in the eyes of animals bearing hh^{ts} under restrictive conditions results in a subset of retinal axons carrying Hh activity into the brain and locally inducing a patch of lamina differentiation. Similarly, ectopic expression of Hh in the optic lobes of animals lacking photoreceptors and R axons restores LPC cell division and the expression of early lamina differentiation markers. The phenotypic effects of lack of Hh in the R axons can be mimicked/rescued by expressing mutant versions of components of the Hh signal transduction pathway in the LPCs: LPCs mutant for the positive Hh effector *smoothened* (*smo*) (and thus lacking Hh signal transduction) fail to proliferate and differentiate, despite retinal innervation and the presence of Hh (Huang and Kunes 1998). Conversely, LPCs mutant for *patched* (*ptc*), a negative regulator of Hh signaling, undergo differentiation even in the absence of retinal innervation. In all cases, the effects are strictly cell autonomous, indicating that Hh is acting directly on the LPCs.

The powers of Hh are, however, limited at least two ways: Hh is not able to induce differentiation in LPCs anterior to the lamina furrow, suggesting that LPCs have to undergo their final cell division before gaining the competence to differentiate into lamina neurons. Secondly, Hh is able to induce early lamina differentiation markers, such as Dac and FasII, but not the late differentiation marker Elav, indicating that at least one other factor has to be released by retinal fibers to complete the maturation of the lamina neurons (Huang and Kunes 1996).

Huang et al. (1998) show that this factor is the EGF family member Spitz (Spi). They demonstrate that Hh induces high levels of EGFr protein expression in the LPC progeny, which is followed by activation of *argos*, a known transcriptional target of EGFr-mediated signaling. EGFr activity in these cells is both necessary and sufficient for their maturation into lamina neurons: expression of a dominant negative form of EGFr blocks Elav expression, while expression of an activated form of EGFr leads to ectopic Elav expression. The EGF ligand Spi, like Hh, is expressed in R cells and transported along the axons into the lamina. Partial loss of *spi* gene function leads to a loss of Elav expression in the lamina, which is restored by photoreceptor-specific expression of Spi. Finally, ectopic expression of Spi leads to the generation of an excess number of Elav-positive cells in the lamina. Thus, the differentiation of lamina precursor cells into mature neurons is achieved in two steps, which are both triggered by anterograde signals released by the incoming photoreceptor axons: Hh triggers the terminal cell division and onset of differentiation of the LPCs, Spi effects their maturation into lamina neurons (see Fig. 1).

The second cell type that receives anterograde signals from the photoreceptors are the visual system glia. The first glial population the R axons encounter on their journey to the optic lobe are the subretinal glia, which originate from the optic stalk and migrate into the eye disc with the onset of photoreceptor differentiation (Choi and Benzer 1994; Rangarajan et al. 1999). These glia require the presence of differentiating photoreceptors in the retina for their proliferation and migration into the eye disc, but not for their expression of the differentiation marker Repo. It seems likely that the signal emanating from the differentiating photoreceptors is diffusible, since even photoreceptors whose axons remain stuck in the eye disc and thus do not make direct contact with the glia in the optic stalk are able to induce proliferation and migration of the glia into the eye disc (Rangarajan et al. 1999). In the lamina, the R axons encounter several layers of glia, called the satellite, marginal and epithelial glia, which derive from the glial precursor cell area at the lateral edges of the lamina (Winberg et al. 1992; Huang and Kunes 1996; Perez and Steller 1996). Some of these glia populate the lamina prior to innervation, but the continued migration of glia into the target field appears to be dependent on the continuous ingrowth of R axons. For both glial populations, the nature of the anterograde signal(s) is currently not known. Although the Hh receptor Ptc is transcriptionally upregulated in the subretinal and lamina glia, indicating that the cells respond to the Hh signal, Hh does not seem to be required for their proliferation, migration, or differentiation as judged by expression of the glial differentiation marker Repo (Huang and Kunes 1998; Rangarajan et al. 2001).

5 Retrograde Signals

5.1 Retinotopic Map Formation

One of the striking features of the R cell projections is the precision with which they form topographic connections within the optic lobe. This feature, also called retinotopy, is common to both the vertebrate and the insect visual systems and is essential for their function. Various models have been proposed to account for the formation of retinotopic maps, including morphogenetic assembly and chemoaffinity (Cowan and Hunt 1985). The morphogenetic assembly model holds that, due to a defined spatiotemporal order of outgrowth and passive fasciculation with neighbors, ommatidial fibers maintain their topographic relationship while growing toward and innervating the target. In this view, fiber-fiber interactions are crucial for establishing the map; the target is merely passively filled. The chemoaffinity model proposes that position-dependent chemical labels on both retinal axons and target cells mediate target recognition. In this view, fiber-target interactions are crucial for establishing the retinotopic map. In vertebrates, chemoaffinity plays a crucial role in establishing the topography of the retino-tectal projections. Ephrin receptors and their ligands, Ephrins, are expressed in complementary gradients in the retina and the tectum, and are thought to serve as matching positional labels that guide axons to their correct termination point based on their point of origin (Drescher et al. 1997; Flanagan and Vanderhaeghen 1998).

How about *Drosophila*? The few studies that have been carried out do not provide conclusive evidence, but suggest that morphogenetic assembly and chemoaffinity both contribute to the establishment of retinotopy in the fly visual system, perhaps with differing importance for the two relevant axes.

Along the anteroposterior (a-p) axis, the differentiation and growth of R cell axons follows a strict temporal progression, such that differences in arrival time might suffice to ensure the proper order along this axis (see Fig. 1). Ablation experiments in lower arthropods (Anderson 1978; Macagno 1978) provide support for this idea. In *Daphnia*, later arriving (= anterior) R axons will connect with target cells normally reserved for earlier (= posterior) axons if those axons have been ablated earlier in development. This result suggests that, along the a-p axis, positions in the lamina anlage are filled according to time of arrival of axons and not according to the position of their cell bodies in the retina and that therefore the lamina precursor cells do not bear guidance labels specifying the a-p position.

Along the dorsoventral (d-v) axis, the differentiation and growth of R axons occurs more or less simultaneously, but axons retain their spatial order as they grow toward and through the optic stalk into the lamina (Fig. 1). Theoretically, fiber-fiber interactions might again suffice to ensure order along the d-v axis; however, several genetic experiments in *Drosophila* argue against it (Kunes et al. 1993). *In sine oculis* or *Ellipse* mutants, which have only a few, dispersed ommatidia, the axons emerging from the isolated ommatidia are still able to project to their proper dorsoventral positions in the lamina despite the absence of neighboring retinal fibers. Further, wild-type retinal fibers are able to project to their proper dorsoventral position even when surrounded by mis-projecting *glass* mutant fibers. Thus, at least along the d-v axis, retinal fibers do not depend on the presence of neighboring fibers to find their retinotopic position in the lamina, suggesting that lamina precursors bear d-v position-specific labels that guide R axons.

Further evidence that morphogenetic assembly is insufficient to establish the retinotopic map comes from the analysis of Frazzled, a *Drosophila* Netrin receptor (Gong et al. 1999). Fra is found on both retinal fibers and lamina target cells, but is required only in the latter. Wild-type retinal fibers are unable to innervate *fra* mutant lamina target cells, but reroute into *fra*$^+$ areas along both axes. This shows that innervation depends on the presence of attractive cues on the lamina target cells, and thus on interactions between retinal fibers and their targets, contrary to a strictly morphogenetic model. In avoiding *fra* mutant regions, retinal fibers do not scramble randomly, but rather reroute in an orderly fashion to reach *fra*$^+$ areas, retaining their relative order. This suggests that the process of retinotopic map formation relies on two functionally separable mechanisms, one mediating generic attraction to the target, the other providing specific positional information.

At present, we have little insight into the molecular machinery that provides positional information in the fly. *Drosophila* has only one Ephrin receptor (Scully et al. 1999), but to date there is no functional evidence implicating the molecule in retinal connectivity. In time, the comprehensive genetic screens for R axon guidance mutants underway in several laboratories will provide the relevant molecular entry points.

5.2 Axon Guidance and Target Layer Selection

To reach their proper targets in the optic lobe, photoreceptor axons have to make multiple pathfinding decisions along the way. Once they have reached the basal surface of the eye disc, they have to grow posteriorly, converge funnel-like to enter the optic stalk, then fan out again to find their retinotopically appropriate position on the surface of the lamina, and finally grow medially and terminate in their appropriate target layer (Fig. 1). The axons from the same ommatidium make most of this journey together. As they are born, axons from the same ommatidial cluster fasciculate with one another and form an ommatidial bundle. In the eye disc, the formation of the bundle reflects the order in which the photoreceptor cells differentiate: the R8 axon grows out first, it is then joined by the R1–R6 axons, and finally by the R7 axon. However, as the axons grow medially into the optic lobe, they exit the bundle and terminate in a different order: R1–R6 terminate first, in the lamina, then R8 in a superficial layer of the medulla (M3) and R7 in a deeper layer of the medulla (M7; Meinertzhagen and Hanson 1993).

We are far from a comprehensive picture of how this complex pathfinding behavior is regulated. However, it has become clear that glial cells play an important role in the guidance of retinal axons (as they do, e.g., along the midline in the central nerve cord of the embryo). The first glia that the R axons encounter are the subretinal and optic stalk glia (Fig. 1). When the migration of subretinal glia from the stalk into the eye disc is suppressed, R axons grow out and navigate posteriorly as in wild type, but are unable to exit the eye disc and enter the optic stalk (Rangarajan et al. 1999). On the other hand, a small number of glia in the eye disc suffices to ensure the normal exit of R axons. The entrance to the optic stalk thus represents a critical point of choice; the presence of glia at this point is required for R axons to make the passage from the eye disc to the stalk properly. It is not clear whether the glia simply form a physical bridge at the juncture between eye disc and optic stalk or whether they in fact provide specific molecular cues.

In the lamina, the R1–R6 axons terminate between the epithelial and marginal glial layers, which has long suggested that it is the glia that provide the stop signal for these axons (Winberg et al. 1992; see Fig. 1). This idea has been confirmed through the phenotypic analysis of the *nonstop* gene (Poeck et al. 2001). In *nonstop* mutants, many R1–R6 axons fail to terminate in the lamina and project into the medulla instead. *nonstop* encodes a ubiquitin-specific protease and is not required for the development of either R axons or lamina neurons, but is required for the normal migration of the epithelial and marginal glia into the lamina target area. These glia originate from glial precursor cell areas which are located at the dorsal and ventral edges of the developing lamina. In wild type, a small number of migrating glial cells express the differentiation marker Repo as they enter the lamina. In *nonstop* mutants, the number of Repo-positive cells is greatly increased; in mosaics, *nonstop* mutant glial cells fail to migrate into the lamina proper. In contrast, target layer selection is unaffected when axons encounter *nonstop* mutant lamina neurons or a lamina lacking differentiated neurons entirely. The nature of the stop signal provided by the glial cells is so far unknown.

Two additional genes, *brakeless* (*bks*) and the *receptor protein tyrosine phosphatase 69D* (*Ptp 69D*), have been shown to control the selection of the target layer; both are required in the R axons themselves (Garrity et al. 1999; Newsome et al. 2000a; Rao et al. 2000; Senti et al. 2000). In *bks* mutants, cell fate determination of the R1–R6 cells is normal, but most of their axons fail to stop in the lamina; the projections of the R7 and R8 axons are unaffected (Rao et al. 2000; Senti et al. 2000). *bks* encodes two isoforms of a novel nuclear protein, whose function is exclusively required in the photoreceptors. Overexpression of either isoform in all photoreceptors does not lead to a retargeting of R7 and R8 axons, indicating that the Bks proteins are necessary but not sufficient for lamina targeting. Most likely, Bks is needed to generate the components that receive the targeting signal presented by the glial cells.

In *Ptp69D* mutants, the R axon projections make multiple targeting errors (Garrity et al. 1999; Newsome et al. 2000a). 5–20% of the R1–R6 axons misproject into the medulla and terminate at the level of either the R7 or the R8 axon termini. In addition, about 50% of the R7 axon fail to reach their proper layer in the medulla and instead terminate at the level of the R8 termini. Only the R8 axons appear to terminate correctly. These phenotypic defects suggest that Ptp69D has a permissive role in the targeting of retinal fibers by effecting de-adhesion, allowing retinal axons to peel off the ommatidial fascicle and successfully terminate in their appropriate layer. Such a role has been ascribed to Ptp69D in the context of motor-axon guidance in the embryo. Thus, the proper targeting of retinal fibers seems to require not only the presence of target cues and their correct interpretation by the fibers, but also the appropriate regulation of adhesive and deadhesive forces in the ommatidial bundle to switch from fasciculative to independent navigation.

6 Signal Transduction in Photoreceptor Axon Growth Cones

The analysis of mutants isolated in R axon guidance screens has also led to important insights into the general signal transduction machinery involved in photoreceptor growth cone guidance. A combination of genetic and biochemical approaches led to the identification of a complex of proteins that is thought to regulate the actin cytoskeleton in the growth cones of R axons in response to guidance cues in their environment.

The story began with the isolation of the *dreadlocks* (*dock*) mutation (Garrity et al. 1996). In *dock* mutants, R cell projections are affected in multiple ways. R axons show gross alterations in retinotopy, including regions of hyper- and hypoinnervation in both lamina and medulla. Target layer selection is affected, as well: R1–R6 axons terminate at different levels in the lamina, leading to an uneven neuropil; some of them misproject into the medulla; and some R7 and R8 axons misproject beyond the medulla. Mosaic analysis and rescue experiments place the requirement for *dock* function into the R cell axons; consistent with these results, the protein is found predominantly in the growth cones of R cell axons. The *dock* gene encodes an adapter molecule, consisting of three SH3 domains and one SH2 domain, and is homologous to the human Nck protein. Nck had been shown to interact through its SH2 domain with a number of receptor and non-receptor tyrosine kinases, and through its SH3 domains with mPAK3, a serine/threonine kinase activated by Cdc42 and Rac, Rho family GTPases involved in regulating the actin cytoskeleton (reviewed in Hall 1998; Dickson 2001). Thus, the homology with Nck raised the intriguing possibility that Dock regulates growth cone motility through physical interaction with PAK and Rho family GTPases.

This idea was confirmed when Dock was shown to physically interact with *Drosophila* Pak (Hing et al. 1999). Through its second SH3 domain, Dock binds to an N-terminal PXXP site in Pak, and the two proteins colocalize in the axons and growth cones of R cells. Loss-of-function mutants in *Pak* cause R axon projection defects that are indistinguishable from those of *dock*; like *dock*, *Pak* gene function is required in the R cells. The phenotypic similarity between *dock* and *Pak* mutants strongly supports the idea that the direct interaction between Dock and Pak is essential for R axon guidance. Moreover, the domains through which the two proteins interact, namely the second SH3 domain in Dock and the PXXP motif of Pak, are critical for in vivo function of their proteins (Rao and Zipursky 1998; Hing et al. 1999). The finding that constitutively membrane-tethered Pak is able to rescue the phenotype of *dock* loss-of-function mutations confirms the notion that Pak acts downstream of Dock and suggests that Dock functions to uniformly recruit Pak to the plasma membrane (Hing et al. 1999). How does Pak effect cytoskeletal changes? The N-terminal region of Pak contains, in addition to its Dock binding motif, a CRIB (Cdc42/Rac interactive binding) motif, which binds to GTP-bound forms of Cdc42 and Rac, followed by the C-terminal kinase domain. Binding of Rho family GTPases to the CRIB domain disables an intramolecular association between the kinase domain and an autoinhibitory region that overlaps the CRIB domain, thereby activating the kinase domain (Frost et al. 1998; Zhao et al. 1998). Both the CRIB and the kinase domains are essential for Pak function in R axons, suggesting that Pak has to be activated by Cdc42 or Rac in order to signal (Hing et al. 1999).

The importance of Rac in regulating Pak activity was demonstrated by the analysis of Trio, a multidomain guanine nucleotide exchange factor containing two tandem Dbl homology (DH) domains flanked by pleckstrin homology (PH) domains (Newsome et al. 2000b). Trio, like Pak and Dock, is present in photoreceptor growth cones and required autonomously for growth cone guidance. Mutations in *trio* display phenotypic defects similar to those of *dock* and *Pak*. Further, *trio* displays strong dosage-sensitive interactions with both *Pak* and *dock*, strongly suggesting that the three genes act in a common pathway in vivo. Trio function in axon guidance requires the activity of its GEF1 domain. This domain activates the Rac GTPases (Rac1 and Rac2) in vitro, and increasing Rac levels potentiate GEF1 signaling in vivo. Finally, the Rac GTPases bind directly to the CRIB domain of Pak and strongly stimulate Pak catalytic activity. Thus, Trio activates Rac GTPases, which in turn activate Pak.

While the epistatic relationships between the three components appear well defined, it is less clear how the differential localization of Pak activity within the growth cone is achieved. Dock appears to function primarily to recruit Pak to the plasma membrane in a uniform fashion (Hing et al. 1999), but there is evidence that Trio provides local activation. Constitutive membrane-tethering of the Trio GEF1 leads to very severe R axon growth and guidance defects, indicating that Trio activity must be spatially regulated, possibly through its PH domains (Newsome et al. 2000b). In other systems, membrane localization of PH domain proteins has been shown to be regulated by PIP_3, whose local concentration is controlled by a number of membrane receptors acting through PI3K (Meili et al. 1999; Servant et al. 2000).

Which guidance receptors act upstream of the Dock-Trio-Pak pathway? In a biochemical approach to this question, Schmucker et al. (2000) have identified a novel guidance receptor that is highly related to a human protein called Down syndrome cell adhesion molecule (Dscam). Dscam is a 270-kDa transmembrane protein whose extracellular portion contains ten immunoglobulin domains and six fibronectin type II domains and whose novel cytoplasmic domain contains multiple potential tyrosine phosphorylation sites and several putative SH3 binding sites. Dscam was isolated from Schneider cell extracts as a phosphoprotein that binds to the SH2 domain of Dock; however, the subsequent analysis showed that it also interacts with the SH3 domains through two of its proline-rich repeats. The role of Dscam in axon guidance has so far not been evaluated in the developing adult visual system. However, in the larval visual system, which in many ways represents a simple version of the adult system, *Dscam* loss-of-function mutants display phenotypic defects similar to those of *dock* and *Pak*, namely a mistargeting of the larval optic nerve at its second intermediate target. *Dscam* shows strong dominant genetic interactions with both *dock* and *Pak*, strongly suggesting that the three genes act in a common pathway in vivo. What is potentially most interesting about Dscam is that it exists in multiple forms generated by alternative splicing. These different forms encode proteins with the same architecture but with sequence variations in three different Ig domains and in the transmembrane domain. Up to 38,000 isoforms of Dscam could potentially be generated. It is tempting to speculate that these isoforms might differ in their affinity for a specific ligand or recognize different ligands, or might act as co-receptors to modulate the activities of other guidance receptors. In either case, the complexity of this molecular system would allow for much fine tuning, which could, for example, provide the positional information required for retinotopic mapping.

7 Outlook

Although much exciting progress has been made in understanding the cellular and the molecular requirements for establishing neural connectivity in the *Drosophila* visual system, many important questions remain unsolved. The biggest unknown is how retinotopy is established in the lamina, in particular the molecular basis for sorting fibers along the dorsoventral axis. Similarly, the cues for targeting R axons to their appropriate layer in lamina and medulla remain to be identified. Additional challenges lie ahead. Little is known about the cellular assembly of the lamina cartridge (Huang et al. 1998), or about the rewiring of the R1–R6 axons in the pupa to achieve neural superposition (but see Clandinin and Zipursky 2000 for a pioneering study). Finally, the proximal neuropils of the optic lobe, medulla and lobula complex, with their complex cytoarchitecture are virtually unexplored. The full arsenal of recent advances in genetics and imaging technology will have to be brought to bear to investigate the development of these systems.

References

Anderson H (1978) Postembryonic development of the visual system of the locust, *Schistocerca gregaria*. II. An experimental investigation of the formation of the retina-lamina projection. J Embryol Exp Morphol 46:147–170

Braitenberg V (1967) Patterns of projection in the visual system of the fly. I. Retina- lamina projections. Exp Brain Res 3:271–298

Choi KW, Benzer S (1994) Migration of glia along photoreceptor axons in the developing *Drosophila* eye. Neuron 12:423–431

Clandinin TR, Zipursky SL (2000) Afferent growth cone interactions control synaptic specificity in the *Drosophila* visual system. Neuron 28:427–436

Cowan WM, Hunt RK (1985) The development of the retinotectal projection: an overview. In: Edelman GM, Gall WE, Cowan WM (eds) Molecular bases of neural development. Wiley, New York, pp 389–428

Cutforth T, Gaul U (1997) The genetics of visual system development in *Drosophila*: specification, connectivity and asymmetry. Curr Opin Neurobiol 7:48–54

Dickson BJ (2001) Rho GTPases in growth cone guidance. Curr Opin Neurobiol 11:103–110

Drescher U, Bonhoeffer F, Muller BK (1997) The Eph family in retinal axon guidance. Curr Opin Neurobiol 7:75–80

Fischbach KF, Technau G (1984) Cell degeneration in the developing optic lobes of the sine oculis and small-optic-lobes mutants of *Drosophila melanogaster*. Dev Biol 104:219–239

Flanagan JG, Vanderhaeghen P (1998) The ephrins and Eph receptors in neural development. Annu Rev Neurosci 21:309–345

Frost JA, Khokhlatchev A, Stippec S, White MA, Cobb MH (1998) Differential effects of PAK1-activating mutations reveal activity- dependent and -independent effects on cytoskeletal regulation. J Biol Chem 273:28191–28198

Garrity PA, Rao Y, Salecker I, McGlade J, Pawson T, Zipursky SL (1996) *Drosophila* photoreceptor axon guidance and targeting requires the dreadlocks SH2/SH3 adapter protein. Cell 85:639–650

Garrity PA, Lee CH, Salecker I, Robertson HC, Desai CJ, Zinn K, Zipursky SL (1999) Retinal axon target selection in *Drosophila* is regulated by a receptor protein tyrosine phosphatase. Neuron 22:707–717

Golic HG (1991) Site-specific recombination between homologous chromosomes in *Drosophila*. Science 252:958–961

Gong Q, Rangarajan R, Seeger M, Gaul U (1999) The netrin receptor frazzled is required in the target for establishment of retinal projections in the *Drosophila* visual system. Development 126:1451–1456

Hall A (1998) Rho GTPases and the actin cytoskeleton. Science 279:509–514

Hing H, Xiao J, Harden N, Lim L, Zipursky SL (1999) Pak functions downstream of Dock to regulate photoreceptor axon guidance in *Drosophila*. Cell 97:853–863

Hofbauer A, Campos-Ortega JA (1990) Proliferation pattern and early differentiation of the optic lobes in *Drosophila melanogaster*. Roux's Arch Dev Biol 198:264–274

Huang Z, Kunes S (1996) Hedgehog, transmitted along retinal axons, triggers neurogenesis in the developing visual centers of the *Drosophila* brain. Cell 86:411–422

Huang Z, Kunes S (1998) Signals transmitted along retinal axons in *Drosophila*: Hedgehog signal reception and the cell circuitry of lamina cartridge assembly. Development 125:3753–3764

Huang Z, Shilo BZ, Kunes S (1998) A retinal axon fascicle uses spitz, an EGF receptor ligand, to construct a synaptic cartridge in the brain of *Drosophila*. Cell 95:693–703

Kunes S, Steller H (1993) Topography in the *Drosophila* visual system. Curr Opin Neurobiol 3: 53–59

Kunes S, Wilson C, Steller H (1993) Independent guidance of retinal axons in the developing visual system of *Drosophila*. J Neurosci 13:752–767

Macagno ER (1978) Mechanism for the formation of synaptic projections in the arthropod visual system. Nature 275:318–320

Macagno ER (1979) Cellular interactions and pattern formation in the development of the visual system of *Daphnia magna* (Crustacea, Branchiopoda). I. Interactions between embryonic retinular fibers and laminar neurons. Dev Biol 73:206–238

Martin KA, Poeck B, Roth H, Ebens AJ, Ballard LC, Zipursky SL (1995) Mutations disrupting neuronal connectivity in the *Drosophila* visual system. Neuron 14:229–240

Meili R, Ellsworth C, Lee S, Reddy TB, Ma H, Firtel RA (1999) Chemoattractant-mediated transient activation and membrane localization of Akt/PKB is required for efficient chemotaxis to cAMP in *Dictyostelium*. EMBO J 18:2092–2105

Meinertzhagen I, Hanson I (1993) The development of the optic lobe. In: Martinez Arias A, Bates M (eds) The development of *Drosophila melanogaster*. Cold Spring Harbor Laboratory Press, Cold Spring Harbor, NY, pp 1363–1490

Meyerowitz EM, Kankel DR (1978) A genetic analysis of visual system development in *Drosophila melanogaster*. Dev Biol 62:112–142

Newsome TP, Asling B, Dickson BJ (2000a) Analysis of *Drosophila* photoreceptor axon guidance in eye-specific mosaics. Development 127:851–860

Newsome TP, Schmidt S, Dietzl G, Keleman K, Asling B, Debant A, Dickson BJ (2000b) Trio combines with dock to regulate Pak activity during photoreceptor axon pathfinding in *Drosophila.* Cell 101:283–294

Perez SE, Steller H (1996) Migration of glial cells into retinal axon target field in *Drosophila melanogaster.* J Neurobiol 30:359–373

Poeck B, Fischer S, Gunning D, Zipursky SL, Salecker I (2001) Glial cells mediate target layer selection of retinal axons in the developing visual system of *Drosophila.* Neuron 29:99–113

Power ME (1943) The effect of reduction in numbers of ommatidia upon the brain of *Drosophila melanogaster.* J Exp Zool 34:33–71

Rangarajan R, Gong Q, Gaul U (1999) Migration and function of glia in the developing *Drosophila* eye. Development 126:3285–3292

Rangarajan R, Courvoisier H, Gaul U (2001) Dpp and Hedgehog mediate neuron-glia interactions in *Drosophila* eye development by promoting the proliferation and motility of subretinal glia. Mech Dev 108:93–103

Rao Y, Zipursky SL (1998) Domain requirements for the Dock adapter protein in growth-cone signaling. Proc Natl Acad Sci USA 95:2077–2082

Rao Y, Pang P, Ruan W, Gunning D, Zipursky SL (2000) brakeless is required for photoreceptor growth-cone targeting in *Drosophila.* Proc Natl Acad Sci USA 97:5966–5971

Ready DF, Hanson TE, Benzer S (1976) Development of the *Drosophila* retina, a neurocrystalline lattice. Dev Biol 53:217–240

Schmucker D, Clemens JC, Shu H, Worby CA, Xiao J, Muda M, Dixon JE, Zipursky SL (2000) *Drosophila* Dscam is an axon guidance receptor exhibiting extraordinary molecular diversity. Cell 101:671–684

Scully AL, McKeown M, Thomas JB (1999) Isolation and characterization of Dek, a *Drosophila* eph receptor protein tyrosine kinase. Mol Cell Neurosci 13:337–347

Selleck SB, Steller H (1991) The influence of retinal innervation on neurogenesis in the first optic ganglion of *Drosophila.* Neuron 6:83–99

Selleck SB, Gonzalez C, Glover DM, White K (1992) Regulation of the G1-S transition in postembryonic neuronal precursors by axon ingrowth. Nature 355:253–255

Senti K, Keleman K, Eisenhaber F, Dickson BJ (2000) brakeless is required for lamina targeting of R1-R6 axons in the *Drosophila* visual system. Development 127:2291–2301

Servant G, Weiner OD, Herzmark P, Balla T, Sedat JW, Bourne HR (2000) Polarization of chemoattractant receptor signaling during neutrophil chemotaxis. Science 287:1037–1040

Stowers RS, Schwarz TL (1999) A genetic method for generating *Drosophila* eyes composed exclusively of mitotic clones of a single genotype. Genetics 152:1631–1639

White K, Kankel DR (1978) Patterns of cell division and cell movement in the formation of the imaginal nervous system in *Drosophila melanogaster.* Dev Biol 65:296–321

Winberg ML, Perez SE, Steller H (1992) Generation and early differentiation of glial cells in the first optic ganglion of *Drosophila melanogaster.* Development 115:903–911

Wolff T, Martin KA, Rubin GM, Zipursky SL (1997) The development of the *Drosophila* visual system. In: Cowan WM, Jessell TM, Zipursky SL (eds) Molecular and cellular approaches to neural development. Oxford University Press, Oxford, pp 474–508

Zhao ZS, Manser E, Chen XQ, Chong C, Leung T, Lim L (1998) A conserved negative regulatory region in alphaPAK: inhibition of PAK kinases reveals their morphological roles downstream of Cdc42 and Rac1. Mol Cell Biol 18:2153–2163

Xu T, Rubin (GM (1993) Analysis of genetic mosaics in developing and adult *Drosophila* tissues. Development 117:1223–1237

Homologies Between Vertebrate and Invertebrate Eyes

Volker Hartenstein[1] and Thomas A. Reh[2]

"The sudden appearance of the vertebrate eye in evolution is like the birth of Athena, full grown and fully armed from the brow of Zeus."
Gordon Wall, 1942

1 Eye Field Topology

1.1 Definition of the Anterior Brain/Eye Anlage

The eye of vertebrates derives from the eye field, an unpaired anlage that is located in the anterior part of the neural plate (Adelmann 1936; Li et al. 1997; Bernier et al. 2000). Together with the anlagen of the dorsal forebrain, midbrain, olfactory system and pituitary, the eye field forms the anterior neural plate, a neurectodermal domain that differs in its molecular properties from the posterior neural plate that gives rise to hindbrain and spinal cord. The anterior neural plate with its above-listed derivatives will be referred to as the anterior brain/eye anlage in the following. Molecularly, this anlage is characterized by the overlapping expression of several regulatory proteins, including *Otx1/2* (Simeone et al. 1993; Kablar et al. 1996) and *Tlx* (Yu et al. 1994; Hollemann et al. 1998). Hox genes, which provide the posterior part of the neural primordium with specific antero-posterior "identities", are not expressed in the anterior brain/eye anlage (reviewed in Holland and Graham 1995). Another characteristic that sets the anterior brain/eye anlage apart from the caudal neural plate is its reliance on different signals during neural induction. In mouse, for example, signals (including *cerberus* and *dickkopf*) emanate from the anterior endoderm and induce anterior brain/eye structures, whereas the posterior neural plate is induced by *noggin*, *chordin* and *follistatin* that originate from the dorsal mesoderm (reviewed in Brewster and Dahmane 1999).

[1] Department of Biology, University of California, Los Angeles, California, USA
[2] Department of Biological Structure, University of Washington, School of Medicine, SM-20, Seattle, Washington 98195, USA

Results and Problems in Cell Differentiation, Vol. 37
K. Moses (Ed.): Drosophila Eye Development

1.2 Topology of the Anterior Brain/Eye Anlage

Fate maps that are based on dye injection into single or small groups of cells at the neural plate stage have been worked out for representatives of all vertebrate classes. These fate maps agree in the major aspects of how the brain topology is projected on the neural plate (Fig. 1A; reviewed in Rubenstein et al. 1998). The anlage of the retina occupies a central position. The anlage of the ventral forebrain (septum, hypothalamus, and optic stalk) separates the retina anlagen in the midline. Anterior and lateral to the retinae are the presumptive olfactory bulb and cerebral cortex; further anterior is the domain that gives rise to the anterior pituitary, olfactory epithelium, and lens. The dorsal diencephalon and mesencephalon map posterior and lateral to the retinae. The retinae and the intervening domain that normally give rise to hypothalamus and optic stalk form one unpaired morphogenetic field of equipotential cells, known as the eye field. Cell-cell interactions, ultimately triggered by a signal derived from the prechordal plate (in mouse, the extraembryonic endoderm), are required to partition the eye field into its different domains (Adelmann 1937; Pera and Kessel 1997; Thomas and Beddington 1996; Li et al. 1997).

In the *Drosophila* embryonic head, a dorsal ectodermal domain that gives rise to the anterior brain and visual system is set apart from the posterior neurectoderm by similar criteria such as the anterior brain/eye anlage of vertebrates. These criteria include the absence of Hox genes and the overlapping expression of the *Drosophila otd* and *tll* genes (Hirth et al. 1995; Younossi-Hartenstein et al. 1997). The anlagen of various *Drosophila* head structures are laid out in a manner that bears strong resemblance to the topology of the anterior brain/eye anlage in vertebrates (summarized in Fig. 1B,C). To appreciate

Fig. 1. Comparison of eye development in vertebrates (*left*) and *Drosophila* (*right*) **A, B, C** Topology of the anterior brain/eye anlage (*green*) and eye field (*magenta*) in **A** vertebrates and **B, C** *Drosophila*. **A** and **B** represent schematic diagrams that show the fate map of the head structures prior to neurulation (dorsal view). Map positions of main neural structures of the head are indicated. **C** depicts the progenitors of the *Drosophila* brain and visual system at a later stage when the visual primordium has split into larval and adult eye, and inner/outer optic lobe. **D, E** Comparison of early eye and brain morphogenesis in **D** vertebrates and **E** *Drosophila*. In vertebrates, the anterior brain/eye anlage has invaginated and forms the neural tube. Structures located in the dorsal midline of the neural plate (*orange*) occupy ventromedial positions in the neural tube. The eye field evaginates as the optic cup. In *Drosophila*, neurulation proceeds in a different manner. Neuroblasts of the CNS, including the optic lobe, segregate from the surface ectoderm. Precursor cells located in the dorsal midline (neuroendocrine pars intercerebralis, *orange*) of the early fly embryo remain where they are, i.e., mid-dorsally. Precursors of the photoreceptors that form the larval and adult remain in the surface ectoderm. **G, H** BMP/Dpp and Shh/Hh signaling operate in vertebrates and *Drosophila* to partition the anterior brain/eye anlage. In vertebrates, high levels of Shh emanating from the prechordal plate and neural midline are required to set up ventral fates, including hypothalamus and optic stalk. BMPs control dorsal fates, including choroid plexus and dorsal retina. In *Drosophila*, Dpp emanating from the dorsal midline is required for both head epidermis and visual system. Secondarily, Hh acts on the eye field and triggers the differentiation of larval and (at a later stage) adult eye

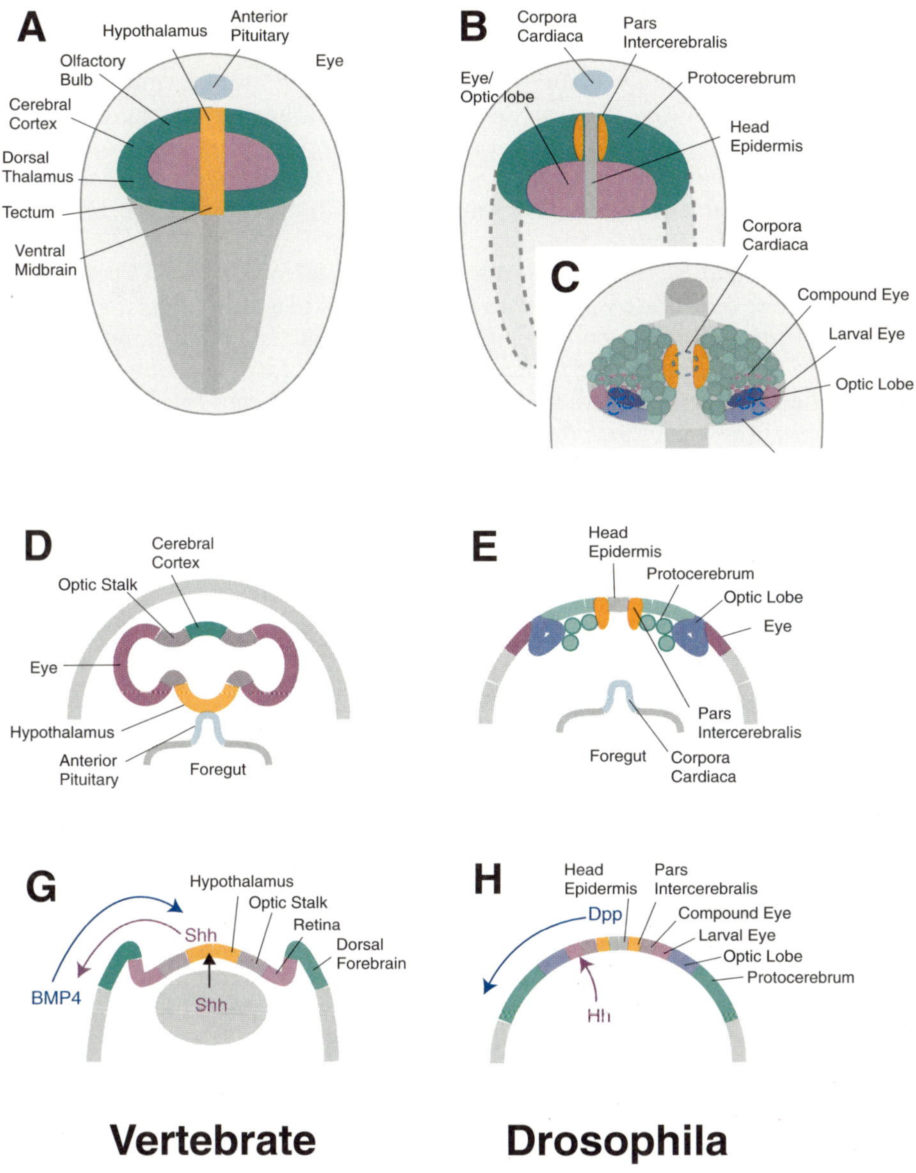
A
Hypothalamus
Anterior Pituitary
Eye
Olfactory Bulb
Cerebral Cortex
Dorsal Thalamus
Tectum
Ventral Midbrain
B
Corpora Cardiaca
Pars Intercerebralis
Eye/ Optic lobe
Protocerebrum
Head Epidermis
C
Corpora Cardiaca
Compound Eye
Larval Eye
Optic Lobe
D
Cerebral Cortex
Optic Stalk
Eye
Hypothalamus
Anterior Pituitary
Foregut
E
Head Epidermis
Protocerebrum
Optic Lobe
Eye
Pars Intercerebralis
Foregut
Corpora Cardiaca
G
Hypothalamus
Optic Stalk
Retina
Dorsal Forebrain
Shh
BMP4
Shh
H
Head Epidermis
Pars Intercerebralis
Dpp
Compound Eye
Larval Eye
Optic Lobe
Protocerebrum
Hh
Vertebrate
Drosophila

this similarity one needs to keep in mind that the neurectoderm of insects does not invaginate (Fig. 1E). As a result, early embryonic tissues located in the dorsal midline of the fly embryo remain where they are, i.e., mid-dorsally, whereas in vertebrates, they form the ventral midline of the neural tube (Fig. 1D). When taking into account this inverse topology, numerous similarities in part explain why dorsomedial structures in *Drosophila* share a number of functional and molecular similarities with the ventral forebrain in vertebrates.

A narrow strip of cells in the dorsal midline forms the head epidermis. Flanking these cells there is a domain that gives rise to neuroendocrine centers (the pars intercerebralis). As in vertebrates, cells that start out as an epithelial placode in the foregut anlage anterior to the eye field form neurohemal structures, called the corpora cardiaca (Copenhaver and Taghert 1991; DeVelasco et al., in prep.). These cells become innervated by the neuroendocrine neurons of the pars intercerebralis. The visual system maps to a dorso-lateral position in the eye field (Green et al. 1993; Dumstrei et al. 1998; Namba and Minden 1999). It includes the anlage of the compound eye, as well as the minute larval eye (Bolwig's organ; Fig. 1C). Besides these photoreceptors, the visual primordium gives rise to the so-called optic lobe, a part of the adult brain which receives the input from the photoreceptors. Thus, as further detailed below, the layered arrangement of target neurons of photoreceptors which, in vertebrates, form an integral part of the retina, are integrated in the protocerebrum in insects. The eye field domain surrounding the visual primordium anteriorly and laterally gives rise to the central protocerebrum.

2 The Expression Pattern of Regulatory Genes

When comparing the expression pattern of conserved regulatory genes required in anterior brain and eye development, one is struck by a number of similarities, but also some significant differences. *Tll*, *otd* and their vertebrate homologues are expressed in the fore/midbrain in vertebrates (Fig. 2M,N), and the protocerebrum in flies (Fig. 2O, P; Simeone et al. 1993; Yu et al. 1994; Hirth et al. 1995; Kablar et al. 1996; Younossi-Hartenstein et al. 1997; Hollemann et al. 1998). Strikingly similar is also the expression of *so* and its vertebrate counterpart, *six3/six6*: in both *Xenopus* and *Drosophila*, the gene at an early stage marks the unpaired eye field, that then splits up into the bilaterally symmetric eye primordium (Fig. 2A–D; Cheyette et al. 1994; Bernier et al. 2000; Zhou et al. 2000). In addition, *so* is expressed in the primordium of the foregut in a median patch that will give rise to the pituitary in vertebrates. In *Drosophila*, *so* is expressed at a similar position in cells that will form the stomatogastric nervous system and the corpora cardiaca. Other determinants of pituitary development, notably *lim-3* (reviewed in Parks et al. 1997), have *Drosophila* counterparts expressed in the stomatogastric nervous system and corpora cardiaca (Thor et al. 1999). A gene that reciprocally interacts with *six/so* in the eye in both vertebrates and *Drosophila* is eyes absent (*eya*; Bonini et al. 1993).

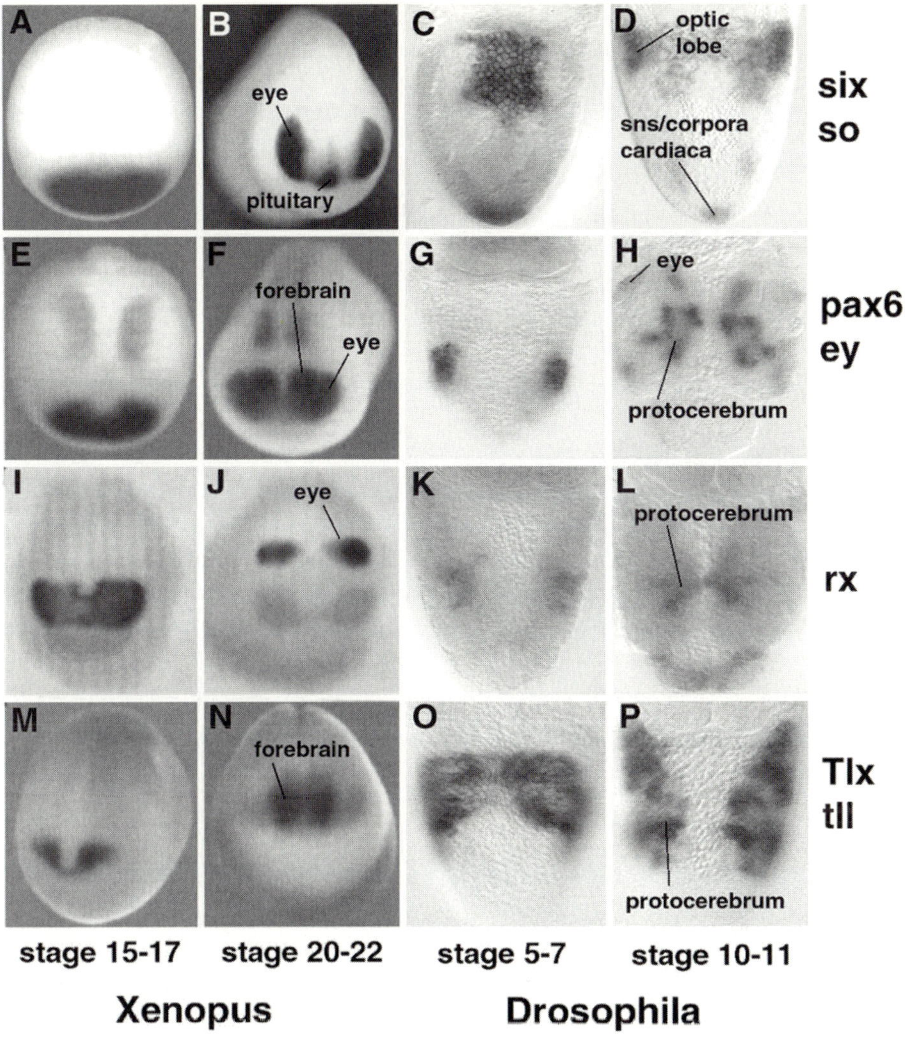

Fig. 2. Expression of regulatory genes six3/so, pax6/ey, rx, and Tlx/tll in the anterior brain/eye anlage of *Xenopus* and *Drosophila*, visualized by in situ hybridization to whole-mounted embryos. For both systems, an early stage (neural fold = *stage 15–17* for *Xenopus*; early gastrulation = *stage 5–7* for *Drosophila*) and a later stage (neural tube = *stage 20–22* for *Xenopus*, late extended germband = *stage 10–11* for *Drosophila*) are shown. *Xenopus* embryos are shown in anterior view, dorsal up; *Drosophila* embryos are in dorsal view, anterior down. [**A** and **B** from Mathers et al. 1997 (with permission); **E**, **F** and **M**, **N** from Hollemann et al. 1998 (with permission)]

In addition to the eye, *eya* is widely expressed in the nervous system and mesoderm of the trunk in vertebrates and *Drosophila*.

Other genes, notably *Rx*, *Pax6* and *Pax2* and their *Drosophila* counterparts, *rx*, *ey*, and *sparkling* are expressed quite differently in the two systems. In vertebrates, *Pax 6* and *Rx* are turned on in the eye field similar to *Six 3*

(Fig. 2E,F,I,J; Hirsch and Harris 1997; Li et al. 1997; Mathers et al. 1997). *Pax2* appears in the eye stalk, and mutations in *Pax2* result in a condition called coloboma, where the eye stalk and ventral retina fail to fuse (Nornes et al. 1990; Sanyanusin et al. 1995; Torres et al. 1996). In *Drosophila*, neither *rx* (Eggert et al. 1998) nor *ey* (Quiring et al. 1994; Daniel et al. 1999) nor *sparkling/pax2* (Fu and Noll 1998) appear in the eye field of the early embryo (Fig. 2G,H,K,L). *Rx* and *ey* are expressed in a subset of brain neuroblasts and their lineages, notably the mushroom body (Noveen et al. 2000; V. Hartenstein, unpubl.). Only at a late embryonic stage do these genes turn on in the primordium of the adult eye. *Sparkling* is expressed in sensory neurons in the embryo, and later in cone cells of the adult eye (Fu and Noll 1998). It is not expressed at all in the embryonic eye field.

3 Signaling Pathways That Partition the Eye Field

In both vertebrates and *Drosophila*, some of the signaling pathways controlling the partitioning and morphogenesis of the anterior brain/eye anlage have been identified, but many details are still elusive. In vertebrates, signals of the bone morphogenic protein (*BMP*) family are initially expressed in the ectoderm and inhibit the formation of the neural plate (Wilson and Hemmati-Brivanlou 1995). Signals derived from the organizer, among them Chordin, *Noggin*, *Cerberus*, and *Shh*, relieve this inhibition and, at the same time, begin to partition the emerging eye field (Fig. 1D; Piccolo et al. 1996; Zimmermann et al. 1996; Li et al. 1997). *BMPs*, in what may be considered a second phase of action, are released from the dorsal neural tube and are required for dorsal cell fates in the spinal cord, brain, and eye (Figs. 1F, 5B; Liem et al. 1995; Dudley et al. 1995; Furuta et al. 1997). Several *BMPs* are expressed in the developing eye at very early stages of optic vesicle formation. *BMP7* is expressed throughout the optic vesicle in mouse, and although the penetrance is variable, animals with homozygous deletion of this gene can have total loss of the eyes (Dudley et al. 1995; Luo et al. 1995). *BMP4* and *BMP2* are also expressed in the developing eye tissues, and although the BMP4 knockout mice die as embryos, Furuta and Hogan (1998) have found that in those animals that survive to late embryonic stages, the lens fails to develop. The lens can be rescued in explant cultures from the *BMP4*-deficient mice with beads soaked in *BMP4*. Overall, the results of these experiments, as well as those from experimental overexpression of *BMPs*, indicate that this family of factors has important roles in eye development; however, the precise functions are not yet clear.

Signals of the *Hedgehog* family, in particular *Sonic hedgehog* (*Shh*), form a gradient opposite to that of the BMPs. In all vertebrate systems, Shh is expressed in the notochord and prechordal plate, and later the ventral neural tube. Graded activity of *Shh* is crucially involved in specifying ventral neural fates along the entire length of the neural primordium (Fig. 5B). In the anterior brain/eye anlage, *Shh* induces hypothalamus and optic stalk (Fig. 1G).

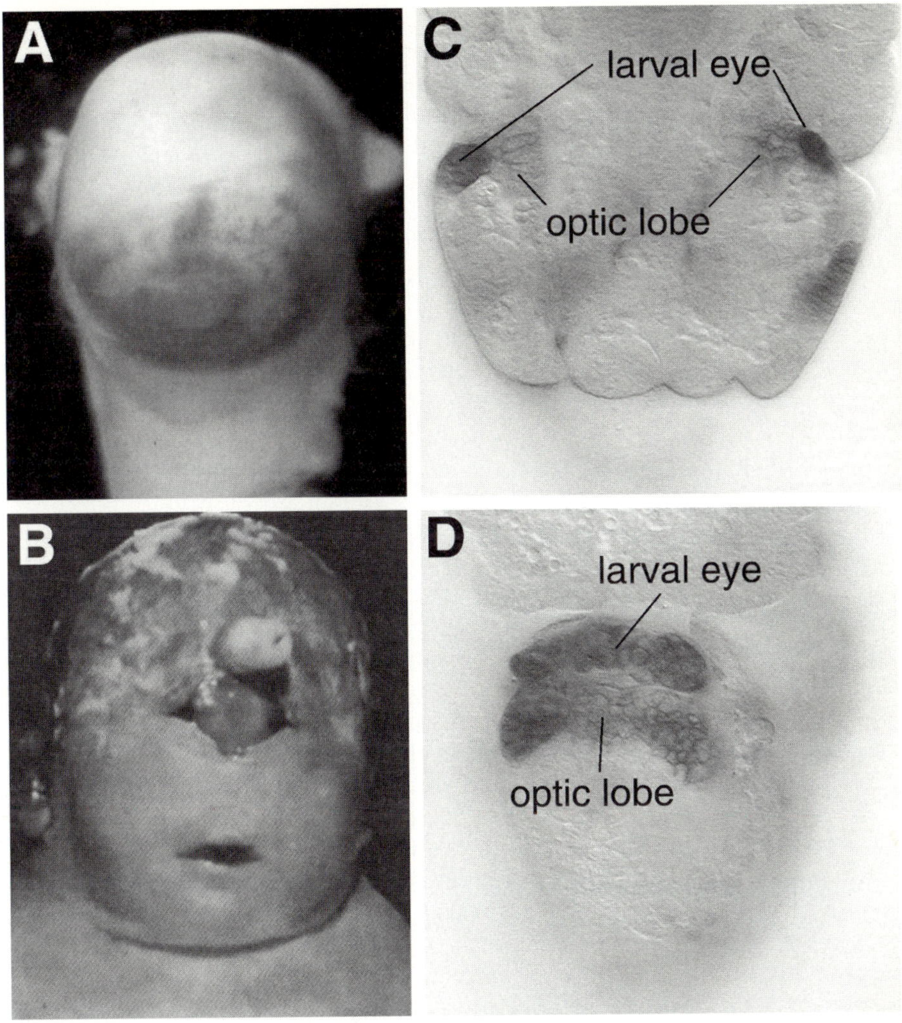

Fig. 3. Following treatment with cyclopamine, Cyclopia in **A** chick, **B** human and **D** *Drosophila*. **C** shows *Drosophila* wild-type embryo in dorsal view. Bilaterally symmetric *larval eye* and *optic lobe* are visualized by specific antibodies. In embryo with reduced Dpp level (**D**), dorsomedial tissue is transformed into larval eye and optic lobe. (**A** thanks to Henk Roelink)

Absence of *Shh* results in the loss of these medial forebrain structures, manifesting itself as cyclopia and holoprosencephaly (Fig. 3A; Chiang et al. 1996; reviewed in Goodrich and Scott 1998). Specific regulatory genes, such as *Pax2* (expressed in the optic stalk) and *Pax6* (in the retina), are under control of *Shh* (MacDonald et al. 1995). *Pax2* is positively regulated in cells close to the *Shh* source, whereas *Pax6* is repressed.

BMPs and Hh proteins play an important role in partitioning the anterior brain/eye anlage in the *Drosophila* embryo as well (Chang et al. 2001). The

BMP2/4 homologue Decapentaplegic (*Dpp*) is expressed at the blastoderm stage in the dorsal half of the trunk and head of the embryo (Fig. 1G). As gastrulation and neurulation proceeds, *Dpp* becomes restricted to a narrow posterior domain bordering the anterior brain/eye anlage. In line with this pattern of expression, genetic studies of loss and overexpression of *Dpp* support a model in which a dorso-ventrally directed *Dpp* gradient subdivides the anterior brain/eye anlage into its three main domains: the head midline ectoderm (requiring high *Dpp* levels), the visual primordium (moderate *Dpp* levels), and the protocerebral neurectoderm (low *Dpp*;). In embryos where the level of *Dpp* is reduced, head epidermis is "transformed" into visual primordium, resulting in a cyclopic visual system (Fig. 3B). By contrast, head epidermis is expanded if *dpp* is expressed at a higher level. Total loss of function of *dpp* leads to the absence of the head epidermis, as well as the visual system, accompanied by an expansion of the protocerebrum.

As previously described, vertebrate *Shh* is expressed in the notochord/prechordal plate and the floor plate where it plays a key role in promoting ventral fates in the brain and spinal cord (reviewed in Goodrich and Scott 1998). In the *Drosophila* embryo, *hh* is expressed quite differently from its vertebrate counterpart *Shh.* Initially activated by pairrule and gap genes, and then maintained by a paracrine Wingless signal, *hh* appears in a narrow transverse stripe in each segment (reviewed in Ingham 1995). It acts in a concentration-dependent manner to specify different epidermal (and possibly neural) fates along the ap axis of each segment. A function in controlling ventral neural cell fates, akin to the function established in the vertebrate neural tube, has not been reported. However, *Hh* is required in the anterior brain/eye anlage. The anterior-most (pre-antennal) stripe of *Hh* expression overlaps with the posterior boundary of the anterior brain/eye anlage. Loss of *hh* results in a strong reduction of the head epidermis, a reduction in the size of the brain and optic lobe, and the total absence of the larval and adult eye primordium. Heat shock-induced overexpression of *hh*, as well as loss of the *Hh* inhibitor *ptc*, causes an increase in larval eye neurons and optic lobe precursors. These cells also appear in the dorsal midline, resulting in a cyclops akin to that one caused by reduction in *Dpp* function (Fig. 3B; see above).

Genetic studies have begun to reveal the epistatic relationships between *Dpp*, *Hh*, and the various regulatory genes expressed in the anterior brain/eye anlage of the *Drosophila* embryo (Fig. 4B). Dpp activates *so* and *eya* in the visual primordium, as well as the Hox gene *zerknuellt* (*zen*; homologue of) in the dorsal midline epidermis. *Zen* in turn inhibits *so* and *eya*, as well as other genes that are initially expressed in an unpaired domain straddling the midline, including *otd* and *tll* (Chang et al. 2001). We see here a mechanism of "bilaterization" of the anterior brain/eye field that is quite different from the one that exists in vertebrates. Whereas in the latter, *Shh* signaling from the mesodermally derived prechordal plate is required to form a paired eye, separated by hypothalamus and eye stalk, neither mesodermal derivatives, nor *Hh* as a signal, play such a role in the fly embryo. Instead, graded activity of *Dpp*,

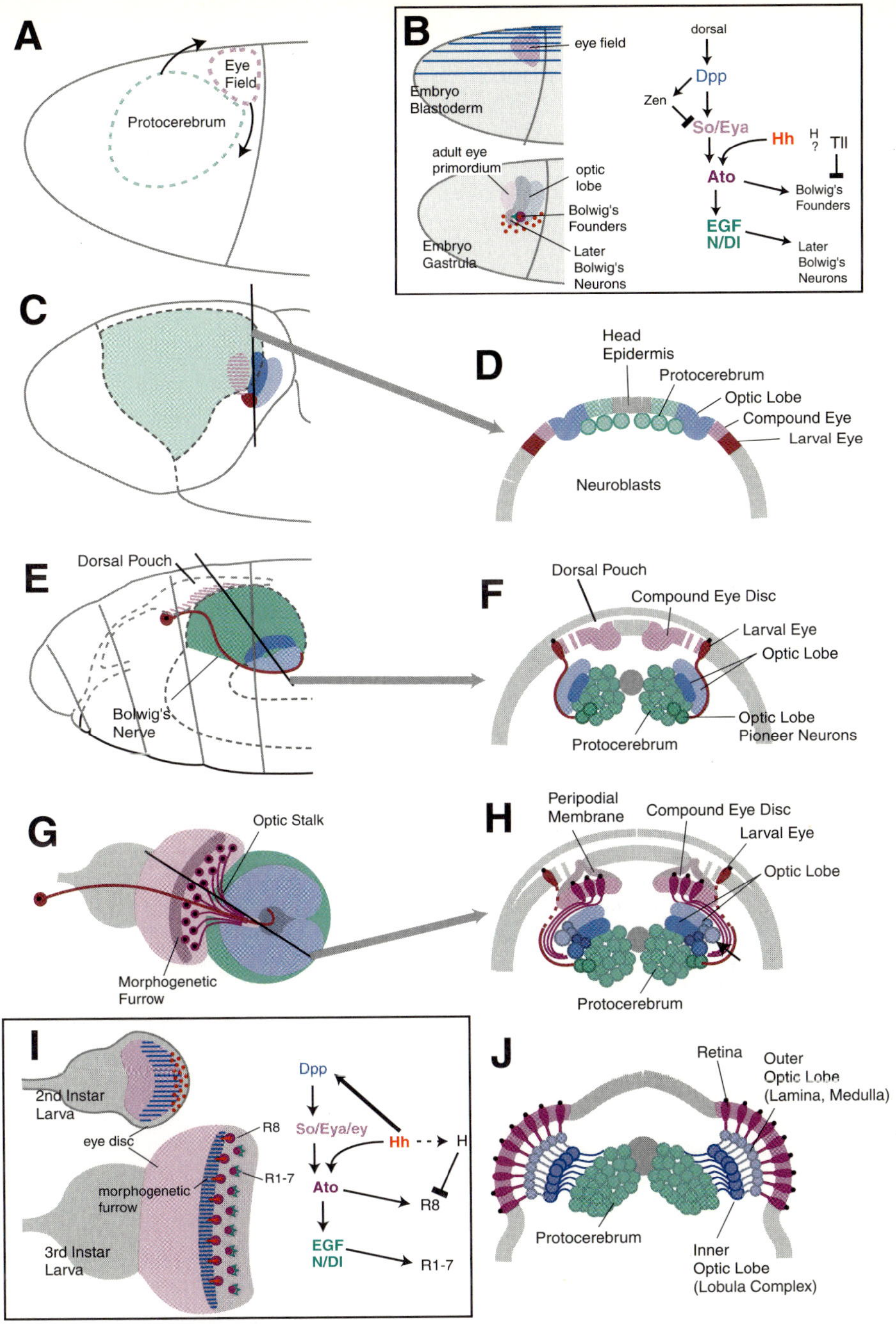

Fig. 4. Morphogenesis of the *Drosophila* visual system. *Left column* (**A**, **C**, **E**, **G**) shows schematic sketches of head of embryo (**A** blastoderm stage 5; **C** gastrulation, stage 7; **E** mature embryo, stage 17) and larva (**G** only brain/eye disc complex is shown) in lateral view. *Right column* (**D**, **F**, **H**) shows schematic cross sections of corresponding stages; plane of sectioning is indicated by line in side view next to section. **J** Cross section of adult brain and eye. Different derivatives of the anterior brain/eye anlage are indicated by different colors. **B** and **I** summarize gene interactions controlling the development of the visual system in the embryo (**B**) and compound eye (**I**). For details see text

via the dorsomedial transcriptional regulator *Zen*, split the early unpaired anlage into bilateral halves. The different role of *Hh* in *Drosophila* eye field patterning, when compared to its *Shh* counterpart in vertebrates, is also reflected in its regulation and its effect on target genes. As previously discussed, vertebrate *Shh* stimulates *Pax2* and inhibits *Pax6*. *Pax2* is not expressed in the *Drosophila* embryonic head, and the effect of Hh on *ey/Pax6* is the opposite to that in vertebrates, i.e., activating (Chang et al. 2001).

4 Eye Morphogenesis

In the preceding sections we attempted to emphasize the parallels in topology, gene expression, and signaling pathways between the vertebrate and *Drosophila* visual system. These parallels get increasingly obscured as one considers later, morphogenetic events that shape the inverted, (single lens) eye of vertebrates and the everted, (compound) eye of arthropods. These structural differences notwithstanding, there are a number of highly conserved molecular mechanisms controlling eye morphogenesis and differentiation in both systems. It will be the objective of Section 5 of this Chapter to attempt a rational hypothesis that explains the paradox, i.e., conserved genes controlling different morphogenetic processes.

4.1 *Drosophila* Visual System: Embryonic and Early Larval Development

Following the delineation of the anterior brain/eye anlage that takes place in the early embryo, the individual parts of the visual system separate from each other and follow different morphogenetic pathways. First, the stem cell-like progenitors (neuroblasts) of the protocerebrum (along with other neuroblasts that form around the same stage in the ventral neurectoderm) delaminate from the ectoderm and form an inner, proliferating cell layer (Fig. 4C,D). The eye field, giving rise to the visual primordium, remains part of the surface epithelium until later in embryogenesis. From its dorso-medial origin (Fig. 4A), it carries out a conspicuous lateral migration before it splits up into four domains: the primordia of the larval and adult eye, and the inner and outer optic lobe (Fig. 4C,D; Green et al. 1993; Cheyette et al. 1994; Daniel et al. 1999). The optic lobe primordium forms a triangular placode that invaginates. The posterior lip of this invagination will give rise to the lamina and medulla ("outer optic lobe"; Fig. 4J); the anterior lip gives rise to the lobula complex ("inner optic lobe"). Following its invagination, the optic lobe primordium forms an epithelial structure attached to the protocerebrum (Fig. 4E,F). During the larval period the optic lobe primordia expand. They form characteristic, U-shaped epithelial plates covering the lateral aspect of the larval brain (Fig. 4G,H). Eventually, starting at the edges of these plates, epithelial cells convert into stem cell-like neuroblasts that bud off lineages of neurons. Retinal axons impinging upon the developing optic lobe are required for both proliferation and differentiation of this structure. Interestingly, signals released from the retinal axons and organizing optic lobe structure include *Hh* and the *TGF* family member, *spi* (Kunes 2000), the same molecules that organize the progression of ommatidial cell differentiation in the eye disc (see below).

The larval eye, or Bolwig's organ, arises from the ventral-most portion of the visual primordium (Fig. 4C). As the optic lobe primordium invaginates, the larval eye remains in the head epidermis until late in embryogenesis when it moves in conjunction with head involution to reach its final position alongside the pharynx (Fig. 4E). The larval eye is composed of 12 photoreceptors that, unlike adult photoreceptors, do not form rhabdomeres; instead, they produce multiple, branched processes which carry the photopigment. Lens or pigment-forming cells are absent from the larval eye (Green et al. 1993).

Many genes expressed in and required for the morphogenesis of the visual system have been identified, but their regulatory interactions and exact developmental role remain largely elusive. Absence or strong defects of optic lobe and larval eye can be observed in embryos lacking the head gap genes *otd* and *tll*, the early eye genes *so* and *eya*, the photoreceptor-specific determinants *ato* and *glass*, as well as the *Hh*, *Dpp*, *EGFR*, and *Notch/Delta1* (*N/Dl*) signaling pathways. Recent studies have shed some light upon the interplay of intrinsic and extrinsic factors that selects the photoreceptor cells constituting the larval eye from within the visual primordium (Daniel et al. 1999; Suzuki and Saigo 2000; Chang et al. 2001). The specification of the larval eye bears a strong similarity to the mechanism controlling ommatidial development in the compound eye. The first step in larval eye formation is the expression of the proneural gene *ato* in a small cluster of two to three cells at the posterior boundary of the visual primordium. *Ato* expression is triggered by the *Hh* signal and requires the expression of *eya* and *so* in the visual primordium. In the compound eye disc, *ato* is expressed in R8, and is triggered by *Hh*. The *ato*-positive cells act as "larval eye founder cells": they secrete the signal *Spi* that acts on neighboring cells of the visual primordium and recruits them (via the *EGFR* signaling pathway) to express larval eye fate (in the compound eye disc, Spi secreted by R8 is crucial for the specification of photoreceptors R2–7). Two mechanisms counteract the larval eye promoting signal. One is the lateral inhibition pathway mediated by *N* and *Dl*; the other one is the expression of *tll* in most cells of the visual primordium. Only cells negative for *tll* develop as larval photoreceptors; the *tll*-positive cells form the optic lobe. Loss of *tll* function results in a largely increased larval eye and absence of optic lobe. In the compound eye, the *N/Dl* pathway is involved in restricting the number of R8 cells, as well as (at a later stage) the number of R7 cells (Cagan and Ready 1989).

The adult eye develops from a small population of cells that first appear at the time when larval photoreceptors begin to differentiate. At this stage the adult eye progenitors form a placode that is anteriorly adjacent to the larval eye (Fig. 4E,F). Molecularly, several genes also found later in the eye disc are first expressed at this placodal stage, among them *ey/Pax6* (interestingly, *ey/Pax6* is not expressed or required for the larval eye!). Most other factors known to promote ommatidial development, including *Hh* and *Dpp*, are not yet expressed in the embryonic eye disc primordium, but turn on later in the third instar larva. The eye disc primordium is folded inside the body during the course of head involution. It comes to lie in a fold called the dorsal pouch (Fig. 4E,F). During early larval development, the eye disc primordium invaginates from that fold to form the characteristic oval eye discs that grow continuously by cell proliferation. Beside the progenitors of the adult eye, other groups of cells destined to form the antenna and head capsule are included in the eye disc. The disc is composed of two layers: an inner, high cylindrical epithelial layer that includes the antennal and eye primordium, and a thin epithelium, the peripodial membrane that gives rise to the head capsule (Fig. 4H). Differentiation sets in during the third larval instar in response to hormonal stimuli (see later section for ommatidial differentiation).

An interesting case in cross-phyletic comparison is the optic stalk. In vertebrates the optic stalk is the structure that connects the retina with the ventral forebrain. It is formed by a folded neuroepithelium along which retinal ganglion cell axons grow towards the hypothalamus and dorsal midbrain. The situation is reversed in *Drosophila*, where the optic stalk originates from the eye disc (which, strictly speaking, is a specialized placode within the epidermis) and grows centrally towards the optic lobe (CNS!). In the embryo, the optic stalk is prefigured by the bundle of larval photoreceptor axons (Bolwig's nerve; Fig. 4E). On their path towards the optic lobe, these axons grow alongside the eye disc primordium. During the second larval instar, a thin epithelial tube grows out from the eye disc towards the brain (Fig. 4G). This optic stalk is later joined by retinal axons emanating from the differentiating ommatidia towards the optic lobe, as well as by glial cells that migrate in the opposite direction, from the optic lobe into the eye disc.

4.2 *Drosophila* Visual System: Adult Eye Differentiation

The compound eye is built of approximately 800 identical photoreceptive units, called ommatidia (see other chapters in this Volume). Each ommatidium contains eight photoreceptor cells (R1–R8) which are surrounded by two layers of pigment cells and a quartet of lens-forming cone cells. All cell types of the compound eye arise in a single epithelial layer. The mechanism by which the regular mosaic pattern of ommatidia is generated is dominated by the stereotyped temporal sequence in which the different cells are generated, as well as by cell–cell interactions among these cells. Ommatidial differentiation starts during the mid-third larval instar with the formation of one row of evenly spaced R8 cells which are born at the posterior edge of the eye disc. This row corresponds to the posterior-most row of ommatidia of the mature eye. Cells are committed to the R8 fate by expressing the proneural gene *ato* (for recent review, see Brennan and Moses 2000, and N.E. Baker, this Vol.). The mechanism that results in the even spacing of the first row of R8 cells involves *N/Dl* signaling (lateral inhibition), as well as more widely acting activators and repressors (H, Emc; Fig. 4I; see below). The nascent R8 cells send out signals that carry out two different functions. First, there are signals that act on the neighboring, undifferentiated tissue and induce the next row of R8 cells. The second type of signal emitted by the R8 cells recruits their immediate neighbors into the ommatidia. The first cells which join each R8 cell shortly after its determination are R2, 3, 4, and 5. The next cells to join the ommatidial clusters are R1 and R6, R7 and four cone cells (non-neuronal cells which will form the lens). Among the signals that in specific temporal and spatial combinations commit cells to these different fates are the EGFR ligand *Spi*, as well as *Dl*. EGFR and *N/Dl* signaling trigger specific combinations of intrinsic determinants of different ommatidial fate (see other chapters in this Volume).

The initiation of the first ommatidia and the progression of the morphogenetic wave that sequentially recruits all cells of the compound eye field is controlled by a network of signaling interactions which is virtually identical to those summarized above for the larval eye (compare diagrams in Fig. 4B,I). It is therefore justified to view the formation of the larval and adult eye as a continuum, controlled by the same molecular switches, and simply interrupted in late embryogenesis when the adult eye disc primordium is “put on ice”, allowing no further differentiation and promoting further expansion (proliferation) instead.

Starting in the early third instar larva, Hh is expressed at the posterior tip of the compound eye field (this would be a position close to the former larval eye; Fig. 4I; Greenwood and Struhl 1999; Curtiss and Mlodzik 2000). Hedgehog then triggers the expression of *Dpp* in a broad band that at this early stage covers the entire compound eye field (note: in the embryo, the expression of *Dpp* is independent of *Hh*; it is under the control of the maternal dorsal gradient). The effect of *Dpp* on ommatidial differentiation resembles a trigger mechanism that is required for the first batch/row of ommatidial precursors to appear. The subsequent induction and differentiation of additional ommatidia do not require *Dpp*. Additional experiments employing elimination or ectopic expression of *Dpp* and its receptor, *Tkv*, indicate that *Dpp* signaling turns on the expression of the early eye genes, *so* and *eya*, similar to its function in the embryonic eye field. Other molecules, such as the inhibitors of proneural genes, *H* and *Emc*, are also expressed in response to *Dpp*. (Greenwood and Struhl 1999, characterize cells in which this cocktail of genes is expressed, as having attained a “pre-proneural state”).

From a pre-proneural state, ommatidial precursors are guided towards a proneural state by other signaling events that ultimately are also under the control of *Hh*. Thus, as in the embryo, *Hh* promotes the expression of the proneural gene atonal, as well as inhibitors of ommatidial differentiation, including *Hairy* and *Emc*. This effect of *Hh* is indirect, given that clones lacking the *Hh* receptor Smoothened are still able to express *ato* and *H*. *Hh* must act via a secondary, short range signal that has not yet been identified, although it is known that *Raf* acts as one of the transducers.

Two additional widely conserved signaling pathways play a crucial role in the initiation and progression of ommatidial differentiation: *Wnt/wg* and *N/Dl*. The role of *N/Dl* signaling in compound eye development is complex. *N* activation is initially required to upregulate *ato* and thereby promote R8 formation. This role is mediated by *Dl*, which is expressed in a band coinciding with the morphogenetic furrow. At a slightly later stage, *Dl/N* mediated lateral inhibition restricts the number of *ato* positive photoreceptors. Both Dl and other signals, such as *Sca* and *Kuz*, are required for this inhibitory signaling step. These signals are formed by R8s and others, in particular the row of ommatidia that had previously formed. By this mechanism, the almost crystalline mosaic of ommatidia is assembled, with one well-formed row of nascent ommatidia controlling the spacing of the younger precursors of the next row.

4.3 Vertebrate Eye Morphogenesis

The vertebrate eye is a composite structure that is formed by derivatives of the forebrain vesicle (optic stalk, pigment epithelium, neural retina), the epidermal ectoderm (lens), and the ocular mesenchyme or neural crest (corneal epithelium; iris). Following the invagination of the neural tube and its separation from the epidermal ectoderm, the neural eye primordia evaginates from the ventrolateral wall of the forebrain vesicle (Fig. 5A). The paired evaginations are known as the optic vesicles, and they are in close contact at their distal part with the presumptive lens ectoderm (Fig. 5A,C). As the more distal part of the vesicle undergoes a considerable amount of cell proliferation, it thickens and begins to invaginate to form the optic cup (Fig. 5D). The optic cup is thus a two layered structure, with the apical surfaces of the two epithelia facing one another. The presumptive pigmented epithelium is only a single layered epithelium, while the distal part of the optic cup, that which will give rise to the neural retina, is already multilayered. At these early stages, the optic cup is primarily a dorsal expansion of the vesicle, and the attachment of the cup to the rest of the diencephalon is maintained through the optic stalk at its ventral side. As the cup expands, the ventral retina develops around the stalk, eventually leading to a more central location for the resulting optic nerve. Although the two sides of the ventral retina eventually fuse, the choroidal fissure is a discontinuity that represents the remnant of the unique aspect of ventral retinal development. At this same time, the lens placode first appears as a thickening in the ectoderm overlying the optic vesicle (Fig. 5D). The placode then invaginates in much the same way as the optic cup, and the lens vesicle is formed as it pinches off from the ectoderm.

The further development of the optic cup is characterized by extensive proliferation with the presumptive neural retina (Fig. 5 E), and a much more reduced proliferation in the optic stalk and presumptive pigmented

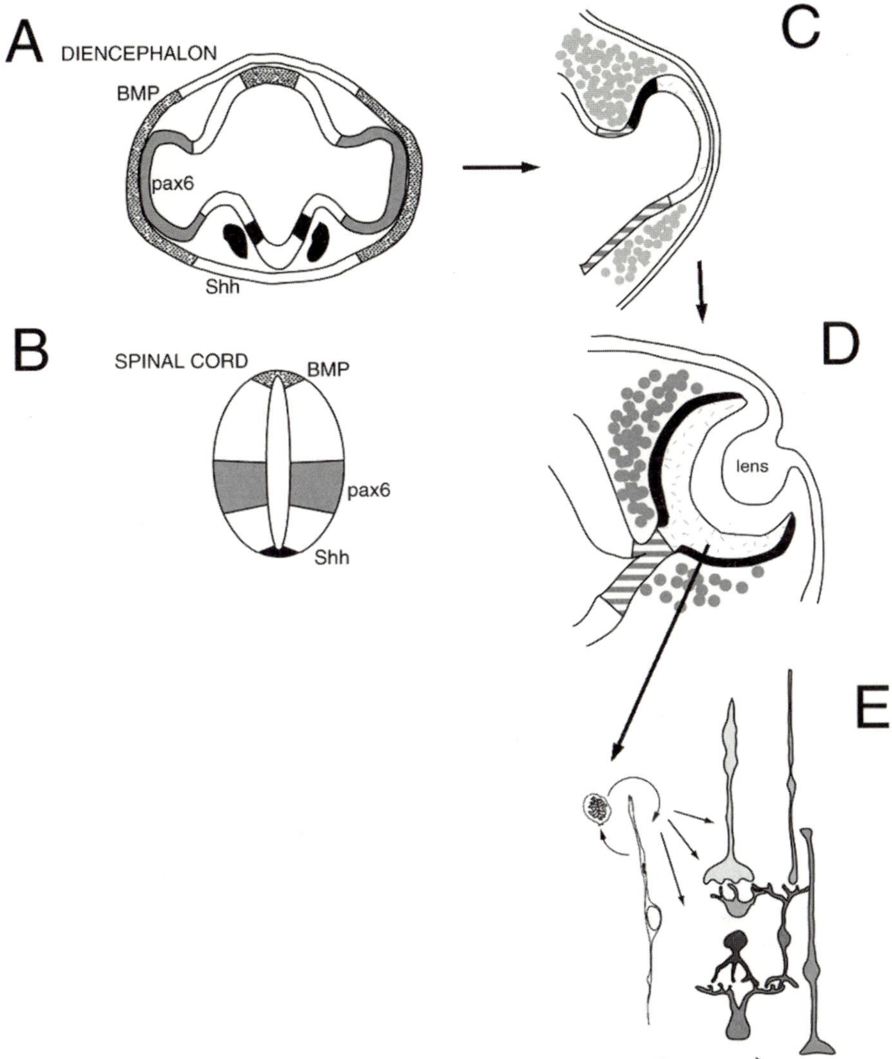

Fig. 5. The optic vesicles evaginate from the ventrolateral wall of the forebrain vesicle (**A**) and are in close contact at their distal part with the presumptive lens ectoderm (**C**). In **D** the vesicle has invaginated to form the optic cup and the *lens* has begun to form. In **E** a retinal progenitor is shown, as well as the differentiated retinal neurons that can be generated from the progenitor through successive cell divisions. **B** Section through the neural tube to show the relationships between sonic hedgehog in the ventral floor plate, and BMP in the dorsal roof plate. The pax6 in the spinal cord is expressed in a middle zone where the mutually antagonist signaling molecules SHH and BMP have somewhat balanced each other out. A similar balance between these two signaling molecules may be responsible for patterning the domain of pax6 expression in the diencephalon that will give rise to the optic vesicles

epithelium. The pigmented epithelium then begins to acquire pigmentation. The lens, initially a hollow sphere, is soon filled by the expansion of the cells closest to the optic cup. These cells elongate as they differentiate and fill the lumen of the lens vesicle. The continued expansion of the lens occurs by proliferation and differentiation of the adjacent lens epithelial cells. The formation of the other ocular structures, such as the iris, is quite complex and beyond the scope of this review.

Soon after the formation of the optic cup, the first retinal neurons become postmitotic and rapidly begin their differentiation (Fig. 5E). The first neurons to be generated by the progenitor cells are the retinal ganglion cells (see McCabe et al. 1999 for references), and the mechanisms that trigger their differentiation will be discussed in more detail below. Other retinal neurons are then generated in a sequence that is generally conserved throughout all vertebrates. After the ganglion cells, the cone photoreceptors and horizontal cells become postmitotic, then amacrine cells, rod photoreceptors, bipolar cells and finally Muller glia are produced in turn (see Reichenbach et al 1995 for review). However, it should be noted that even in those species with extremely protracted developmental programs (like monkeys) there is a considerable amount of overlap among the periods of genesis of these different neuronal types (LaVail et al. 1991). There are no strict lineage relationships among the cells, and clones of retinal progenitor cells typically form radial columns with combinations of multiple cell types (see Cepko 1993 for review).

The coordinated development of these various ocular tissues requires many interactions among them, and the nature of these interactions have been the subject of experimental embryological investigation for over 100 years. One of the first examples of inductive interactions between tissues was that which occurs between the retina and the lens (Spemann 1938; Grainger 1992). In addition, the lens is known to be critical for the growth of the neural retina(Ballard 1939). The pigmented epithelium is necessary for neural retinal growth (Raymond and Jackson 1995), and if the retina is removed early in development, the pigmented epithelium can transdifferentiate into neural retina (Coulombre and Coulombre 1965). In addition, the extraocular mesenchyme is critical for the development of the pigmented epithelial fate (Fuhrmann et al. 2000) from the cells of the optic vesicle, and the neural retinal fate requires an interaction with the lens ectoderm. These complex interactions among these tissues are quite different from those thought to pattern the *Drosophila* eye antennal disc, though ultimately these interactions may function through the regulation of homologous eye determination genes. For example, although pax6 is initially expressed throughout the optic vesicle, the presumptive pigmented epithelium loses the expression of this gene as it acquires the pigmentation distinctive of its differentiated fate. The downregulation of pax6 in the proximal (most dorsal) aspect of the optic vesicle is dependent on a signal from the extraocular mesenchyme related to activin (Fuhrmann et al. 2000), a homologue to Dpp, which, as noted above, is critical to restrict the eye field from the dorsal midline in *Drosophila*. It is also possible that, as in *Drosophila*, wingless related proteins play some role in restricting the domain of the neural retina, since wnt2b and wnt 5a are both expressed in the pigmented epithelium (Jasoni et al. 1999; Fuhrmann et al. 2000).

5 Retinal Differentiation

5.1 Homology of Cell Types in the Vertebrate Retina to the *Drosophila* Eye

Before describing any potential molecular homology that may exist in the next stages of eye development between the *Drosophila* and vertebrates, it is worth considering whether the cell types in the two very different types of eyes are in any way homologous. The basic sensory epithelium design found in vertebrates and most invertebrate groups alike is shown at the top of Fig. 6. Sensory cells (S) are surrounded by accessory cells (A) in many specialized sensory epithelia from both invertebrates and vertebrates. Examples include the olfactory epithelium and the auditory epithelium of vertebrates. This basic pattern has been modified extensively in different sensory epithelia and in different species. In the ommatidia of the retinal epithelium of the *Drosophila*, the basic sensory cell morphology is maintained and each group of eight photoreceptors is associated with several accessory cells. The accessory cells can be divided into two basic types: the pigment cells and the cone cells. The pigment cells can be subdivided on the basis of their position into the primary, secondary and tertiary pigment cells. The other basic type of accessory cell, the cone cell, produces the crystalline lens. Morphologically the *Drosophila* retina has two basic types of supporting cells and one basic type of neuron.

The vertebrate retina is an outgrowth of the central nervous system and differs in many aspects from the insect compound eye. First and foremost, it contains, beside photoreceptors, two layers of neuronal cells that are the targets of photoreceptor axons. In insects (and invertebrates in general; see Sect. 5), visual target neurons are located in the brain. Secondly, pigment cells and lens-forming cells are spatially segregated from the photoreceptors. Pigment cells form a homogenous epithelium that develops from the outer layer of the optic cup. Lens forming cells do not form part of the retina, but are induced from the epidermal ectoderm by the optic cup. Only a single type of accessory cell, the Muller glial cell, develops in the neural retina itself.

One should note that in vertebrates, more primitive light-sensing organs with less neuronal diversity exist, such as the unpaired medians "eyes", also known as either the pineal, frontal, or parietal eyes. In fact, nearly all of the neuronal cells in these primitive retinas fall into two classes, either photoreceptors or ganglion cells (Kappers 1965). In the pineal eye of the lamprey, some of the photoreceptor cells actually have axons that project similar to the photoreceptors found in invertebrate eyes. In the more complex paired lateral

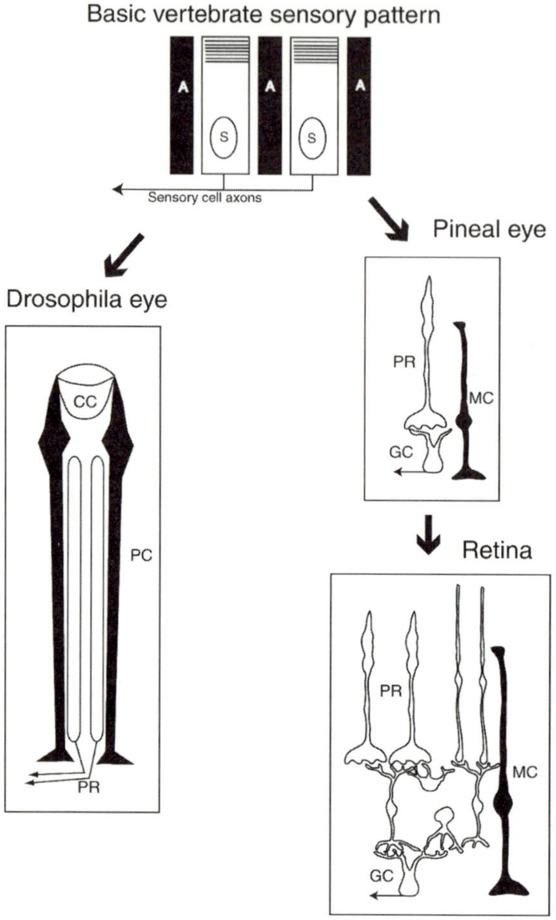

Fig. 6. The basic sensory epithelium design found in vertebrates and most invertebrate groups. Sensory cells (*S*) are surrounded by accessory cells (*A*) in many specialized sensory epithelia, such as the olfactory epithelium and the auditory epithelium of vertebrates. In the ommatidia of the *Drosophila* eye, each group of eight photoreceptors is associated with several different types of accessory cells, including the pigment cells and the cone cells. More primitive vertebrate light-sensing organs have increased neuronal diversity over that observed in the *Drosophila* ommatidia, but in these simpler vertebrate "eyes" there are only two types of retinal neurons, either photoreceptors or ganglion cells. The retina contains, beside photoreceptors, two layers of neuronal cells that are the targets of photoreceptor axons. Only a single type of accessory cell, the Muller glial cell, develops in the neural retina itself. However, it is interesting that in the more primitive pineal or parietal midline eye, the Muller cells may also express pigment or lens proteins, somewhat akin to the *Drosophila* ommatidium

eyes of vertebrates, the retina has further increased the diversity of cell types, and now there are many different classes of interneurons (Fig. 6).

If the above scheme represents an approximate picture of how sensory epithelia have become diversified from a common pattern, it begs the question as to which cell type in the vertebrate retina is homologous to the *Drosophila* photoreceptor cell, and which cell type in the *Drosophila* retina is homologous to the Muller cell. Several hypotheses can be put forward and two of these will be presented here. According to one hypothesis, the invertebrate photoreceptor cell is homologous to all the retinal neurons. In this way of thinking, a division of labor has occurred in the vertebrate retina to parcel out the functions of phototransduction and long distance signal transmission into separate cells. In addition, the vertebrate retina has many interneuronal cell types, such as the amacrine cells and the horizontal cells, that are critical for basic information processing (such as center-surround antagonism) within the retina. The second hypothesis would homologize photoreceptors of invertebrates with the rods and cones of vertebrates. The deeper layers of the retina containing the target neurons of rods and cones would have their homologues in the central neurons that make up the optic ganglia of the insect brain, i.e., the lamina, medulla, and lobula complex. According to this view, the visual system of the common bilaterian ancestor contained photoreceptors and target neurons processing visual input. In the evolutionary line leading to present day arthropods, both cell populations were separated during early development: precursors of visual target neurons segregated along with other neuroblasts and formed part of the central nervous system, and precursors of photoreceptors and accessory cells remained in the surface ectoderm that then gives rise to the compound eye. By contrast, in the line leading to chordates, precursors of both target neurons (ganglion cells, bipolars), photoreceptors and accessory cells stayed together, invaginated during neurulation, and then grew out from the neural tube as the optic cup.

Physiological, anatomical, and molecular evidence does not presently allow one to clearly favor one hypothesis over the other. Clearly, the vertebrate and *Drosophila* eye do perform homologous functions, and the photoreceptor cells of the vertebrate and invertebrate have much of their biochemistry and fundamental morphology in common (see D. Ready, this Vol., for review). The proteins involved in transduction and communication of light could well be regulated by homologous transcription factors in both vertebrate and invertebrate photoreceptors. For example, the chicken rhodopsin promoter contains a glass-like sequence that is homologous to that in *Drosophila* Rh1 (Sheshberadaran and Takahashi 1994). Therefore, at least two aspects of these cells could develop by homologous mechanisms: their patterned arrayed distribution, and their expression of genes relevant to their functions in light transduction and transmission of the signals to the visual centers.

One might hope that the expression of specific structural or regulatory genes might help to decide whether the *Drosophila* photoreceptors correspond to the vertebrate photoreceptors alone, or to all neuronal cell types found in the retina. However, molecules studied to date do not provide an unambiguous answer. Math5, the homologue of the proneural gene *atonal* that initiates R8 differentiation in flies, is expressed primarily in retinal ganglion cells in vertebrates (Kanekar et al. 1997), suggesting molecular affinities between these cell types. On the other hand, *atonal* is also expressed in the inner optic lobe of the *Drosophila* larval brain. Other genes, such as the Pit1/Oct1/Unc1 (POU) factor Brn-3 and its *Drosophila* homologue, *acj6*, are specifically expressed in ganglion cells of the vertebrate retina, and optic lobe in flies (Gan et al. 1996; Certel et al. 2000). One can hope that with the systematic study of many more homologous genes in the vertebrate and *Drosophila* visual system it will be possible to eventually determine with some confidence which cell types should be considered homologous.

In both vertebrates and invertebrates, the retinal photoreceptors must be patterned in arrays so that each part of the visual scene is processed in roughly the same way across its extent. Thus, the mechanisms that initiate and regulate the differentiation of the fly photoreceptors from the unpatterned epithelium could be conserved in the vertebrate retina. In principle, the same basic patterning mechanisms could work to organize all the types of vertebrate retinal neurons. The first neurons to become postmitotic in the vertebrate retina are not photoreceptors, but, rather, the retinal ganglion cells. Yet, as will be discussed more thoroughly below, these cells may well set up the initial pattern of the cells in the vertebrate retina using similar molecular mechanisms as those used in the fly eye. However, the problem is not as simple (!) for the vertebrate retina as it is in the fly. In *Drosophila* ommatidia, a founder cell organizes the other cells around it. In the vertebrate, it is not just the photoreceptors that are patterned in mosaic arrays; nearly all types of retinal neurons are distributed in orderly arrays (Cook and Chalupa 2000). More striking to consider is the fact that the arrays of the different types of retinal neurons are independent from one another in their distribution, and so cannot be organized around any one particular cell type. Therefore, if the vertebrate uses the same mechanisms as the *Drosophila* to select the first cell type from the undifferentiated epithelium, it may be that the vertebrate has had to re-use the patterning mechanisms over and over again for each type of neuron.

5.2 Ganglion Cells Differentiate in a Regular Array in the Developing Chick Retina

The retinal ganglion cells are the first neurons in the vertebrate to be generated from the mitotically active progenitor cells in the undifferentiated epithelium of the retina. Thus, they are in this way similar to the R8 photoreceptor in the *Drosophila* eye imaginal disc. In recent years, the pattern in the onset of ganglion cell differentiation has been characterized in fish and chicks. In general, the differentiation of neurons in the vertebrate retina begins in the center and proceeds outwards in a radial manner. The ganglion cells in the chick embryo retina appear to lead this general pattern, differentiating from the central to peripheral retina as a wave, much like an expanding circle. While not nearly as precise as the arrangement of the R8 photoreceptors, there does appear to be some degree of patterning in the arrangement of the newly forming retinal ganglion cells. Figure 7 shows a flat-mount preparation of the

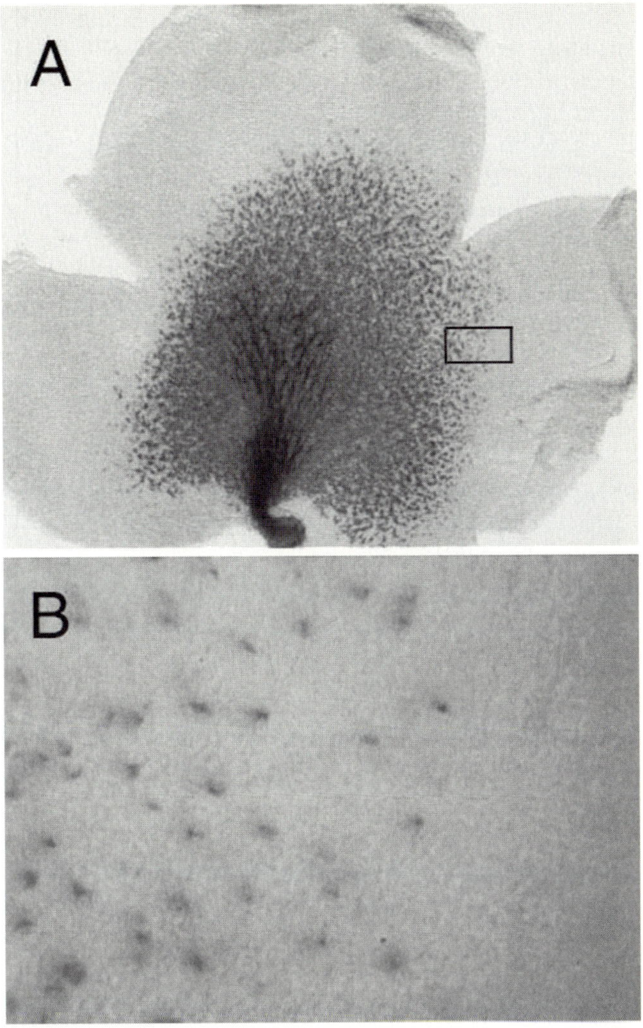

Fig. 7. **A** Low power and **B** higher power views of a flat-mount preparation of the chick retina to show the pattern of ganglion cells as they are being generated. The central to peripheral progression of labeled cells can be seen in **A**. In **B** the individual ganglion cells can be seen as they differentiate. The ganglion cells at the front of differentiation are typically not immediately adjacent to one another, but are spaced at somewhat regular intervals

chick retina at two different magnifications to show the pattern of neurofilament labeling, a marker for the ganglion cells at their final mitotic division and thereafter. The edge or front of ganglion cell differentiation is where the first neurons in the retina are differentiating. The ganglion cells at the front of differentiation are typically not immediately adjacent to one another but are spaced at somewhat regular intervals. All the cells between the ganglion cells are still undifferentiated progenitor cells, destined to give rise to later differentiating ganglion cells, as well as all other cell types, and as yet there are no layers to the epithelium – it is a pseudostratified epithelium with all the cells spanning the entire extent.

Ganglion cells are non-randomly distributed in the adult animals of most vertebrates that were analyzed. The regularity of the spacing between cells in mosaic arrangements in the retina is commonly analyzed by near-neighbor analysis. When the spacing among the cells at the front of differentiation is analyzed using a similar near-neighbor analysis, the cells are found to be non-randomly distributed, though not as regularly arrayed as they are in adults (McCabe et al. 1999). The ultimate regularity of the ganglion cell mosaics may be sharpened as the result of cell death and cell migrations later in development (Cook and Chalupa 2000). Nevertheless, some mechanism must exist to pattern the first ganglion cells into this non-random disposition.

The onset of ganglion cell differentiation has also been examined in the fish retina. In fish, there appear to be two somewhat distinct patterns of ganglion cell differentiation. Initially, the ganglion cells form in the ventronasal retina, and then proceed outward from the center "roughly like the hand on a clock sweeping the long way around from 6:01 to 5:59"(Hu and Easter 1999; Easter 2000; Masai et al. 2000). After this initial wave, the further expansion of the ganglion cell domain looks more like that observed in the chick retina, in that it proceeds radially (Fig. 8).

5.3 How Is the Wave of Ganglion Cells Initiated and Propagated?

A wave of differentiation, known as the morphogenetic furrow, passes across the unpatterned epithelium of the eye disc in *Drosophila*, as well as in more primitive insects (Friedrich and Benzer 2001). There is no clear morphological specialization analogous to the morphogenetic furrow at the front of ganglion cell differentiation in the vertebrate retina. In addition, there does not appear to be any clear cell cycle synchronization at the front of ganglion cell differentiation as there is at the morphogenetic furrow (McCabe et al. 1999). Despite these differences, there are many similarities between the morphogenetic furrow and the front of ganglion cell differentiation. Both represent the transition from undifferentiated progenitor cells to an array of differentiated neurons. Both appear to have a similar wave-like progression-although in *Drosophila* the furrow propagates in one direction across the eye disc, while in the vertebrate the front of differentiation can proceed either as an expanding circle, or as a radial sweep.

In the *Drosophila* eye, the current evidence indicates that the previously differentiated R8 photoreceptors secrete a diffusible factor, Hedgehog, that induces the differentiation of the adjacent undifferentiated cells. These new R8 photoreceptors in turn trigger the differentiation of the next row of cells, and the process continues across the eye until the entire epithelium has

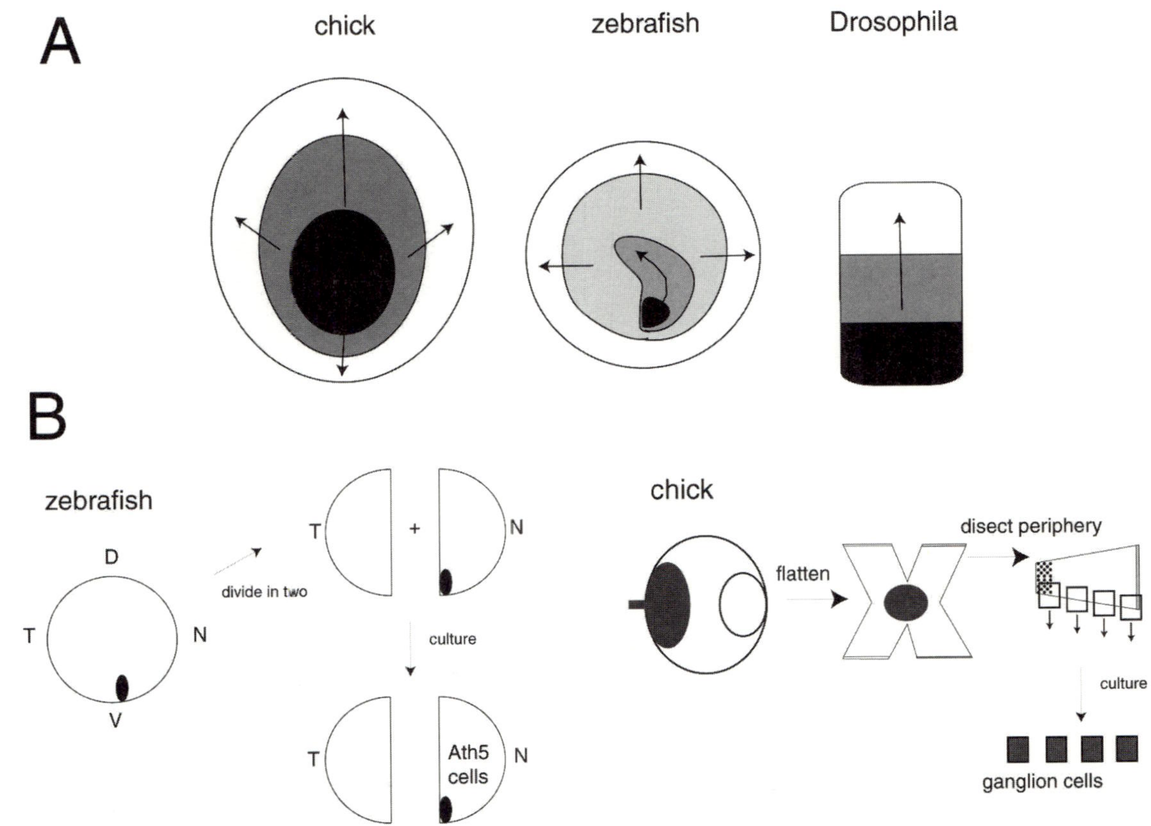

Fig. 8. The onset of ganglion cell differentiation has been examined in the chick and fish retinas. Both appear to have a wave-like progression of differentiation similar to that of the fly eye disc, although in *Drosophila* the furrow propagates in one direction across the eye disc, while in the vertebrate the front of differentiation can proceed either as an expanding circle, or as a radial sweep. Experiments have attempted to isolate the undifferentiated retina from the ganglion cells that have already developed to determine whether the previously developed ganglion cells are necessary for the development of ganglion cells in the undifferentiated regions of the retina. In the chick the answer appears to be "no" though the experiments in the fish indicate that some factor in the optic stalk may be needed to **initiate** ganglion cell development in the retina

differentiated (see Treisman and Heberlein 1998 for review). Are the previously differentiated ganglion cells in the central retina necessary for the progression of the front of ganglion cell differentiation?

This question has been addressed recently by two studies (McCabe et al. 1999 and Masai et al. 2000). In both studies, the source of the putative induction was isolated from the undifferentiated retina surgically. As noted above, there is no clear morphological feature, like the furrow, at the point where retinal ganglion cells begin their differentiation. McCabe et al (1999) defined the location of the onset of ganglion cell differentiation in the following way (Fig. 8): (1) the retinas were removed from the embryos and small peripheral pieces of retina were cut away; (2) the location of the onset of ganglion cell differentiation was determined by staining of the remaining central retina with a ganglion cell-specific antibody; (3) the small pieces of peripheral retina were cultured overnight, and labeled with the same antibody to assess whether retinal ganglion cells developed in the absence of the central retinal cells. The isolated pieces of peripheral retina (E3.5–4) that had been cultured overnight in the absence of the central retinal ganglion cells retained the ability to differentiate into ganglion cells in 100% of trials. Ganglion cell differentiation occurred in explants almost 500 μm from the normal front of differentiation. These results show that the previously differentiated ganglion cells in the central retina are not required to propagate the central-to-peripheral wave of ganglion cell differentiation; instead, new ganglion cells can develop in the absence of other ganglion cells. This result from the chick embryo experiment contrasts with that obtained in the zebrafish. Masai et al (2000) dissected the zebrafish retina into two fragments prior to the stage when any ganglion cells (as evidenced by *Ath5* expression) were expressed in the retina (Fig. 8). They then cultured the fragments for 24 h and found that the nasal fragments, containing the optic stalk, now expressed *Ath5* while the temporal fragments never expressed *Ath5*. This would suggest that the stalk contains a factor that is necessary for the onset of ganglion cell differentiation.

The two contrasting results from the fish and the chick are unlikely to be due to some fundamental difference in retinal structure, since the basic retinal architecture and development are very similar in the various vertebrate species. One possible explanation is that the experiment in the fish was done before any ganglion cells were present in the retina, and so addresses the issue of the initiation of ganglion cell differentiation. By contrast, the experiments in the chick embryo were done at a stage when there were already ganglion cells present in the central retina, and the results from chick might relate more to the issue of the propagation of the wave of differentiation, rather than its initiation. This is interesting in light of the fact that in *Drosophila* there is also evidence that the mechanisms for initiation of the furrow may be different from those involved in its progression. As we understand more about the molecular mechanisms involved in the differentiation of ganglion cells, we will get a clearer idea as to how these two processes relate to one another in the vertebrate.

5.4 Molecular Mechanisms of Ganglion Cell Differentiation

Since the front of ganglion cell differentiation is in some ways analogous to the fly eye disc furrow, it is tempting to speculate that common molecular mechanisms may be involved in the patterning of the cells in the two structures. While a great deal is known about the molecular events involved in the *Drosophila* eye disc furrow progression, much less in known about the vertebrate. Nevertheless, studies of the distribution of many key signaling molecules during vertebrate retinal development have shown some intriguing similarities.

In *Drosophila*, as described in great detail throughout this Volume, the morphogenetic furrow is propagated by an interaction between *Hedgehog* and *Dpp*. In the vertebrate, both of these molecules are known to be present in the developing retina. A source of *Sonic hedgehog* is located in the ventral diencephalon, adjacent to the optic stalk (Fig. 5A). *Shh* diffusing through the optic stalk could then be the critical factor that Masai et al (2000) found associated with the optic stalk that was required for *Ath5* expression. Once *Shh* from the stalk has initiated the first ganglion cells in the ventronasal retina, these new ganglion cells begin to express *Shh* as they differentiate, and they in turn induce new ganglion cells to form from neighboring progenitor cells. As the next group of ganglion cells differentiate, the process is repeated, thereby spreading the initial induction to progressively more peripheral regions of retina. There is evidence for such a mechanism in the early stages of fish retinal development. Although immunohistochemistry and in situ hybridization have not detected the presence of *Shh* in the ganglion cells at the stages when their wave of differentiation is proceeding across the retina (Stenkamp et al 2000), a fish strain with green fluorescent protein (GFP) driven by the *Shh* promoter/enhancers shows a progression of activity that closely parallels the wave of ganglion cell development (Neumann and Nuesslein-Volhard 2000). Since both *tiggywinkle hedgehog* and *Shh* are expressed by the ganglion cells at later stages of development, it may be that the wave of ganglion cell development is driven by levels of hedgehog proteins that are too low to be detected. Alternatively, it is possible that some other factor, such as *nodal* for example, is responsible for activating the *Shh* promoter, (Muller et al. 1999; Masai et al. 2000) in the fish ganglion cells. Recent data from the chick embryo also complicate a simple model of hedgehog activity in ganglion cell differentiation; Zhang and Yang (2001) have found that rather than promoting retinal ganglion cell differentiation, overexpression of *Shh* with a retrovirus inhibits ganglion cell development, and conversely, inhibition of *Shh* activity with anti-*Shh* antibodies leads to an increased differentiation of ganglion cells. The authors conclude that *Shh* derived from ganglion cells serves as a negative regulator of this cell fate from the progenitors. Despite these seemingly contradictory results, there does seem to be some critical function for *Shh* in retinal ganglion cell development. However, since *Shh* also appears to be important for the development of the rod photoreceptors (Levine et al. 1997; Stenkamp et al.

2000), and perhaps other cell types as well (see below), the results of any particular experiment may be difficult to unravel with the current techniques.

There are other similarities between the patterning of the ommatidia in *Drosophila* and the differentiation of ganglion cells in the vertebrate that are also noteworthy. First, many studies of the vertebrate retina have found that fibroblast growth factors (FGFs) are important for ganglion cell development (Pittack et al. 1991; Guillemont and Cepko 1992; Zhao and Barnstable 1996; Pittack et al. 1997; Desire et al. 1998; McCabe et al. 1999). Addition of FGF to frog, chick, or mammalian retina at early stages of development increases the differentiation of ganglion cells from the progenitors, while inhibition of FGF signaling, either through antisense, antibodies or pharmacological inhibitors of the FGF receptor, inhibit the differentiation of ganglion cells. Consistent with the possibility that a receptor tyrosine kinase activation is critical for ganglion cell formation, Neumann and Nuesslein-Volhard (2000) also reported that in zebrafish retina a wave of phospho-extracellular signal regulated kinase (ERK) activity overlaps with that of the *Shh*GFP expression. This is analogous to a similar wave of phospho-ERK activation at the morphogenetic furrow, that is thought to be downstream from the activation of the raf pathway (Green and Struhl 1999). While no report has described a role for FGF in the *Drosophila* ommatidial formation, the related receptor tyrosine kinase, EGFR, is well known to be important for this process. Therefore, it appears that at least part of this pathway has been conserved for the formation of the first retinal neurons to differentiate in both systems.

Several studies have also highlighted parallels between the functions of the *Notch/delta* pathway and basic helix-loop-helix (bHLH) genes in the *Drosophila* eye and the vertebrate retina. Using a temperature-sensitive allele of *Notch*, Cagan and Ready (1989) demonstrated that *Notch* is required for successive cell fate decisions by all the cells of the *Drosophila* eye. While early temperature shifts blocked the addition of photoreceptors and cones, later shifts affected bristles and pigment cells. Studies in the frog first established a similar role for the vertebrate *Notch* gene. Misexpression of an activated form of the *Notch* receptor in *Xenopus* retinal progenitor cells blocks their differentiation into retinal neurons (Dorsky et al. 1995). Widespread misexpression of the *Notch* ligand, *delta*, in either embryonic frog or chick retina also prevents differentiation of the retinal progenitor cells, presumably because of an activation of the *Notch* receptor. Expression of a dominant negative form of *delta* has the opposite effect, it causes a premature differentiation of the progenitor cells into the early generated types of retinal neurons (Dorsky et al. 1997; Henrique et al. 1997; see Rapaport and Dorsky 1998 for review).

Thus, *Notch* and *delta* mediate lateral inhibition, and presumably the relative numbers of the different types of neurons and glia in the vertebrate retina, in much the same way that they serve these functions in the *Drosophila* eye. The functions of *Notch* and *delta* in the regulation of cell fate are at least partly mediated through the basic helix-loop-helix transcription factors and their negative regulators, *hairy* and the *enhancer of split* complex. In

Drosophila, the *atonal* gene product is well established as a critical regulator of photoreceptor differentiation (Jarman et al. 1994). In vertebrates, homologues to the proneural class of transcription factors are also known to be expressed in the developing retina and appear to have similar functions (see Hassan and Bellen 2000 for review). While several vertebrate homologues to proneural bHLH genes that are expressed in the nervous system are also expressed in the developing retina (see Jasoni et al. 1994 or Jasoni and Reh 1996 for example), the gene known in various species as *Xath5* or *Math5* is the only proneural gene that is predominantly expressed in the retina (Kanekar et al. 1997; Brown et al. 1998; Liu et al. 2001; Wang et al. 2001). The gene is apparently the proneural gene for the retinal ganglion cells. It is expressed just prior to the stages when these cells begin their differentiation, and over-expression of the gene causes additional ganglion cells to be generated (Kanekar et al. 1997; Liu et al. 2001).Deletion of the gene in mice results in a profound reduction in the number of ganglion cells that develop (Wang et al. 2001). Thus, since this gene is expressed in the first retinal neurons to develop and is required for their differentiation, it can be considered the most homologous gene to the *Drosophila* atonal gene. Two clear differences, however, are (1) that the gene is not expressed in photoreceptors or required for their development, and (2) the rest of the retinal cells develop nearly normally despite the failure of ganglion cell formation in the knockout mice. While it may well be that other proneural genes, such as *NeuroD*, may have taken over the functions of *atonal* for the photoreceptors (Yan and Wang 1998), the development of the majority of retinal cells appears to occur relatively undisturbed, regardless of the absence of "founder cells" such as the ganglion cells (Wang et al. 2001).

5.5 Are the Mechanisms of Photoreceptor Differentiation Conserved?

In the *Drosophila* retina, the sequential recruitment of photoreceptors to the ommatidia is well known. The development of the R8 cell is critical for organizing the other cells into ommatidia, and as such acts as a founder cell for the other cells. As noted above, the diversity of neuronal type in the vertebrate makes direct comparison between cell types in the fly and the vertebrate difficult. In addition, since most of the different types of retinal neurons are arrayed in mosaics that appear to be independent in their distribution, it is unlikely that a single cell type "organizes" the remaining retinal cells around it. In addition, although there is good evidence that the phenotypes of retinal cells are regulated by factors in their microenvironment, to date there is not much evidence for a model of sequential cell-specific induction in the vertebrate retina, similar to that which is now well established for the *Drosophila* eye. Finally, as noted in the previous section, deletion of the retinal ganglion cells by *Math5* knockout, does not prevent other retinal cell types from differentiating in a relatively normal manner.

Nevertheless, the cone photoreceptors in many vertebrates have a very regular inter-relationship, consistent with a "founder cell" model like that in the *Drosophila* eye. In the fish retina, as in most vertebrates, the different types of cones, as defined both by their morphology and their color sensitivity, are in regular spatial relationships with respect to one another. The mosaic arrangement of the cones in the zebrafish retina is shown in Fig. 9. The mosaic consists of four different types of cones. A double cone, consisting of a red and green-sensitive pair is surrounded on one side by a blue-sensitive cone and on the other by a UV-sensitive cone. The basic pattern is repeated, but the blue and UV-sensitive cones alternate in rows between the double cone pairs (Fig. 9A). Several investigators have studied the development of the cones in fish, and have found that the opsin genes are expressed in a definite temporal sequence(Wickler and Rakic 1991; Bruhn and Cepko 1996; Stenkamp et al. 1996, 1997; Bumsted et al. 1997; Wickler et al. 1997). In their analysis of the sequence of differentiation of the cone types in the fish retina, Stenkamp et al (1997) have pointed out that the sequence of differentiation-red, green, blue, UV-is the same as the distance from the red cone of each of the different types of cones. Thus, it has been suggested that the red cones might act as founder cells for organizing the differentiation of the other cone types (Stenkamp et al. 1997; Wan and Stenkamp 2000). However, if the red cones act as the founder cells for organizing the mosaic in the fish, it is unlikely that this will be a general principle for all vertebrates, since the sequence of cone differentiation is not well conserved in other species. For example, in the mammalian retina, where there are only two basic types of cones, the S-opsin cones (blue) are the first cones to differentiate (Fig. 9B) and the M-opsin (green) cones differentiate later.

Is there any direct evidence to indicate that there is a "founder" cone that organizes the cone mosaic of vertebrates homologous to the R8 photoreceptor in flies? Nearly all the literature on cone mosaics is descriptive; however, recently there have been some experimental studies of cone development. Wan and Stenkamp (2000) tested the possibility that the rod photoreceptors were actually the organizers for the cone mosaic. While in most vertebrates, rod photoreceptors are generated later in development than the cones, in fish, the rods differentiate more rapidly than cones. However, when the generation of the rods was specifically inhibited, there was no clear effect on the organization of the cone mosaic, and the authors concluded that the rods were unlikely to be playing an organizer role. Recent data from work in mouse retina indicate that at least part of the mechanism by which developing cone fate is determined involves the thyroid hormone receptor. Newly generated cones express the TRbeta2 for a short time near their final mitotic division, and when the TRbeta2 gene is selectively deleted in mice, the cones all become S-opsin cones and there are no M-opsin cones in the retina (Fig. 9C). Thus, it appears that some type of signaling event is required to specify the identity of the different types of cones in the vertebrate, though we are just at the first stages in understanding the nature of this interaction. It is interesting that members of the

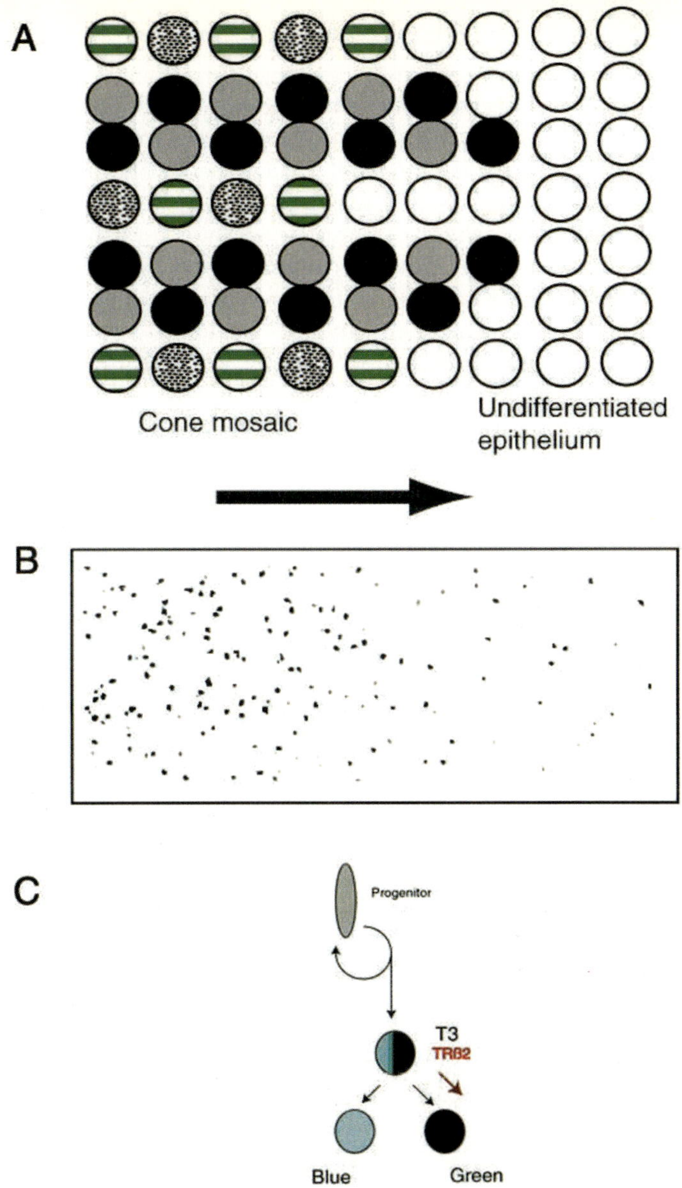

Fig. 9. **A** The origin of the mosaic arrangement of the cones in the zebrafish retina. The different types of cones, as defined both by their morphology and their color sensitivity, are in regular spatial relationships with respect to one another. The mosaic consists of four different types of cones. A double cone, consisting of a red and green-sensitive pair is surrounded on one side by a blue-sensitive cone and on the other by a UV-sensitive cone. The basic pattern is repeated, but the blue and UV-sensitive cones alternate in rows between the double cone pairs. The sequence of development of the different types of cones is shown as well. **B** The development of the blue cones in human retina. Although the cones in the human retina will eventually form relatively organized mosaics, note the lack of precise spatial organization as the cones first differentiate. **C** The green cone fate requires a specific isoform of the thyroid hormone receptor during development. In mice lacking this gene, all of the cones develop as blue cones

steroid/thyroid superfamily of transcription factors have also been shown to be required for normal furrow progression in the *Drosophila* eye imaginal disc and for proper photoreceptor morphogenesis.

The vertebrate has another basic type of photoreceptor, the rod photoreceptor, that responds to low levels of light. These are formed in a somewhat different process than the cones, since they are generally not as precisely patterned, and in mammals, are generated in such abundance that they make-up from 60–70% of all cells in the retina. Despite the lack of a precise spatial organization in these cells, there is evidence that several localized signaling molecules regulate their development. Once again, comparisons between the various vertebrate classes can be instructive. In zebrafish it has been known for some time that the rods develop from a ventral patch of retina and then they spread to other regions as described for the ganglion cells. The differentiation of these cells also appears to be dependent on a wave of Hedgehog, though this time, the Hedgehog is produced not by the other rod photoreceptors, but by adjacent cells in the retinal pigmented epithelium. A similar situation is true for the mammal, where *Indian hedgehog* expressed in the pigmented epithelium, along with *Shh* in the ganglion cells, essentially sandwich the progenitor cells to surround them with Hedgehog (Levine et al. 1997; Jensen and Wallace 1997). In both the fish and the mammalian retina, the addition of exogenous *Shh* protein causes an increase in the rate of rod photoreceptor differentiation. Thus, the data from these studies of the later role for Hedgehog indicate a conservation in the importance of this gene family for photoreceptor differentiation from *Drosophila* to vertebrates.

The other types of neurons in the vertebrate retina are also arranged in mosaics, and one might ask whether these cells are in any way aligned with the other cells they connect with. For example, there is a circuit formed by the S-cones, a specific type of bipolar cell, known as the blue cone bipolar, and a specific type of ganglion cell. Are the cells that make up this circuit directly aligned with one another? At this point there is no direct evidence that the cells that make up a circuit like this are all derived from a single progenitor by lineage, nor is there evidence that the blue-pathway ganglion cell directs the differentiation of the S-cone or blue cone bipolar cell. As noted above, some evidence would suggest the contrary, since cone photoreceptors are not lost when ganglion cells fail to form in the *Math5* knockout mice (Wang et al. 2001).

6 Evolution and Homologies

From the studies reviewed above, it is clear that there are many questions remaining about the evolutionary relationships among eyes from various organisms. In the past ten years, there has been a remarkable expansion in our knowledge of the expression and importance of several key factors that control eye development in both vertebrates and *Drosophila*. However, it is clear that this information does not all fit into a neat or simple model of eye evolution.

In this section, we highlight the similarities and differences and discuss some possible features of the eyes of the common ancestor between vertebrates and invertebrates.

As described in the first section, the transcription factors and signaling molecules that direct the cells of the embryo to develop as the eye fields are partly conserved from the initial stages of eye specification. The similarity in eye field topology between insects and vertebrates, and the conservation of a considerable number of signals and regulatory genes involved in eye field partitioning and later patterning of the retina, indicate that the bilaterian ancestor might have possessed a head in which photoreceptors, various brain structures and neuroendocrine cells were arranged in a manner that may have been similar to the one found in present day taxa. This does not imply the existence of a complex organ, such as presented by the present day eye, brain, or neuroendocrine glands. Rather, one might surmise that the bilaterian ancestor was a simple triploblastic animal in which small clusters of cells with the basic properties of photoreceptors, pigment cells, neuroendocrine cells, or central neurons were integrated in a neurectoderm. As a result of evolutionary change, these cell types further diversified and increased in number. Morphogenetic movements shaped the neurectoderm into more complex organs. For example, in regard to the eye, in one evolutionary line leading to the chordates, the neurectoderm invaginated to form a tube-like neural primordium that included all cells with the fate of photoreceptors, pigment cells, and target neurons. These cells then evaginated as the optic cup, induced the lens and other structures from the outer ectoderm and formed an eye. In another evolutionary line that resulted in the present day arthropods, cells with the fate of photoreceptors and pigment cells remained in the outer ectoderm and became organized into a compound eye. In simple invertebrates, including ctenophores and platyhelminths, photoreceptive organs consist of single or small clusters of photoreceptors typically associated with pigment cells and sometimes simple lens cells. Such "photo-sensilla" can be embedded in the epidermis (as in Ctenophores), or they are embedded in the brain (as in platyhelminths). In the phylogenetic line that led to chordates, photo-sensilla are internal structures that differentiate in the wall of the anterior tip of the neural tube. This is the case with the unpaired ocellus (Eakin and Kuda 1971; Katz 1983) in ascidian tadpoles, as well as the unpaired frontal eye in the cephalochordate Amphioxus (Lacalli et al. 1994). It is interesting to note that in both of these "primitive" representatives of the chordate phylum the eye is an unpaired structure, which may be relevant in interpreting the fact that the bilateral vertebrate eye derives from an eye field that initially also is an unpaired domain within the anterior neurectoderm. The photoreceptors in ascidians and cephalochordates send axons into the neuropile of the anterior brain (prosencephalon). Given the small number of photoreceptors, it is not surprising that the postsynaptic targets in these simple chordate brains do not form elaborate layered arrays, as in the retina and tectum of vertebrates. With the evolution of vertebrates, the eye developed into an image-forming organ with a greatly increased number of photoreceptors. Accessory cells (including pigment cells, lens forming cells, muscles, and others) also increased and became organized into a complex organ. The growing array of photoreceptors and their target neurons formed the neural retina which evaginated from the prosencephalon (reviewed in Lacalli et al. 1994) to form an optic cup.

With the expansion of the cell number in the eyes of both *Drosophila* and more visually sophisticated vertebrates, like the teleosts, it is clear that there is considerable advantage to patterning the light receptive elements into an organized array. The arrays of cells in the vertebrate retina and the *Drosophila* compound eye appear to be patterned in part by somewhat similar molecular mechanisms. However, while much can be learned from comparative approaches to the study of the patterning mechanisms for photoreceptors, it may be that these are much more general mechanisms for patterning neural (or even non-neural) elements for array-like information processing. Therefore, it might not be too surprising to find that a wave of patterning occurs in other sensory system maps in the brain of vertebrates, for example. The more specific patterning events may relate to control of neurogenic gene expression and defining the localized expression of the neurogenic genes.

Despite these interesting and compelling similarities in the development of vertebrate and invertebrate eyes, there are some significant differences. For example, it is not clear whether the vertebrate optic stalk has a *Drosophila* counterpart. As described above, the vertebrate optic stalk is the structure that connects the retina with the ventral forebrain. It is formed by a folded neuroepithelium along which retinal ganglion cell axons grow towards the hypothalamus and dorsal midbrain. In *Drosophila*, the peripheral axonal tract of the larva that is formed by axons of retinal photoreceptors projecting towards the optic lobe is commonly called optic stalk. This tract is pioneered by the thin embryonic nerve connecting the larval eye with the brain, and besides axons and glial cells, contains no other elements (e.g., neuroepithelial cells). This comparison reveals that, both in terms of composition (ganglion cells versus photoreceptor axons) and morphogenesis (folded outgrowth of neuroepithelium versus centripetal axon tract) the optic tract in vertebrates and flies is radically different, and it may be misleading to search for structural and molecular homologies. However, there is another structure in fly embryos that may correspond to the vertebrate optic stalk: the so-called optic lobe pioneer tract. The optic lobe pioneers are a group of neurons that receive input from the larval photoreceptors (Tix et al. 1989; Campos et al. 1995). Later they become incorporated into the optic lobe. What makes these cells interesting as candidates for a fly optic stalk is their projection; their axons are all crossed (like vertebrate retinal axons, forming one of the first commissural tracts of the embryonic brain (Nassif et al. 1998).

An issue related to possible homologies of the optic stalk is the question of what vertebrate structure may correspond to the fly optic lobe. This structure receives topographically ordered input from retinal axons, and consists of three ganglia: the lamina, medulla, and lobula complex. In vertebrates, photoreceptor axons terminate on bipolar cells, which in turn synapse with ganglion cells. Both of these populations of neurons form part of the retina. Should one homologize the deep layers of the retina (bipolar and ganglion cell layer) with the optic lobe in arthropods? Alternatively, has a division of labor given rise to separate cell classes in the vertebrate retina that have either light-receptive proteins or long axon projections, but not both as in the *Drosophila*? An even more radical model would homologize the optic lobe of the fly to the retina of the vertebrate; it should be recalled that all of the vertebrate retina derives from the brain, and none from the epidermal placode. In this way of thinking, the lens of the vertebrate would be more akin to the *Drosophila* eye, a rather heretical view first articulated by Sharp in 1885. He proposed that the lens was once a sensory organ, like the other cranial placodes (olfactory placode or otic placode), and that the present retina was its ganglion. Eventually the ganglion took over the sensory function, and neural development was suppressed in the lens placode.

While the fascinating story of the evolution of the eye is not nearly complete, the current pace of characterization of the genes involved in formation of both vertebrate and invertebrate eyes should facilitate our ability to draw clear, meaningful relationships between them. In addition, we should be mindful of the fact that nearly all the information on which we have based this review has been derived from only a very few species. Comparative biology of the extensive variation in visual organs requires the examination of a much wider diversity of animals if we are to have a sound basis for our analysis.

References

Adelmann HB (1936) The problem of cyclopia. Part II. Q Rev Biol 11:284–304

Ballard WW (1939) Mutual regulation between eye-ball and lens in *Amblystoma* studied by means of heterotopic transplantation. J Exp Zool 81:261–285

Bernier G, Panitz F, Zhou X, Hollemann T, Gruss P, Pieler T (2000) Expanded retina territory by midbrain transformation upon overexpression of Six6 (Optx2) in *Xenopus* embryos. Mech Dev 93:59–69

Bonini NM, Leisserson WM, Benzer S (1993) The *eyes absent* gene: genetic control of cell survival and differentiation in the developing *Drosophila* eye. Cell 72:379–395

Brennan CA, Moses K (2000) Determination of *Drosophila* photoreceptors: timing is everything. Cell Mol Life Sci 57:195–214

Brewster R, Dahmane N (1999) Getting a-head of the organizer: anterior-posterior patterning of the brain. Bioessays 21:631–636

Brown NL, Kanekar S, Vetter ML, Tucker PK, Gemza DL, Glaser T (1998) Math5 encodes a murine basic helix-loop-helix transcription factor expressed during early stages of retinal neurogenesis. Development 125:4821–4833

Bruhn SL, Cepko CL (1996) Development of the pattern of photoreceptors in the chick retina. J Neurosci 16:1430–1439

Bumsted K, Jasoni C, Szel A, Hendrickson A (1997) Spatial and temporal expression of cone opsins during monkey retinal development. J Comp Neurol 378:117–134

Cagan RL, Ready DF (1989) *Notch* is required for successive cell decisions in the developing *Drosophila* retina. Genes Dev 3:1099–1112

Campos AR, Lee KJ, Steller H (1995) Establishment of neuronal connectivity during development of the *Drosophila* larval visual system. J Neurobiol 28:313–329
Cepko C (1993) Lineage versus environment in the embryonic retina. Trends Neurosci 16:96–97
Certel SJ, Clyne PJ, Carlson JR, Johnson WA (2000) Regulation of central neuron synaptic targeting by the *Drosophila* POU protein, Acj6. Development 127:2395–2405
Chang T, Mazotta J, Dumstrei K, Dumitrescu A, Hartenstein V (2001) Dpp and Hh signaling in the *Drosophila* embryonic eye field (submitted)
Cheyette BNR, Green PJ, Martin K, Garren H, Hartenstein V, Zipursky SL (1994) The *Drosophila sine oculis* locus encodes a homeobox gene required for the development of the entire visual system. Neuron 12:977–996
Chiang C, Litingtung Y, Lee E, Young KE, Corden JL et al. (1996) Cyclopia and defective axial patterning in mice lacking Sonic hedgehog gene function. Nature 383:407–413
Cook JE, Chalupa LM (2000) Retinal mosaics: new insights into an old concept. Trends Neurosci 23(1):26–34
Copenhaver PF, Taghert PH (1991) Origins of the insect enteric nervous system: differentiation of the enteric ganglia from a neurogenic epithelium. Development 113:1115–1132
Coulombre JL, Coulombre AJ (1965) Regeneration of neural retina from the pigmented epithelium in the chick embryo. Dev Biol 12:79–92
Curtiss J, Mlodzik M (2000) Morphogenetic furrow initiation and progression during eye development in *Drosophila*: the roles of decapentaplegic, hedgehog and eyes absent. Development 127:1325–1336
Daniel A, Dumstrei K, Lengyel J, Hartenstein V (1999) *tailless* and *atonal* control cell fate in the embryonic visual system. Development 126:2945–2954
Desire L, Courtois Y, Jeanny JC (1998) Suppression of fibroblast growth factors 1 and 2 by antisense oligonucleotides in embryonic chick retinal cells in vitro inhibits neuronal differentiation and survival. Exp Cell Res 241:210–221
Dorsky RI, Chang WS, Rapaport DH, Harris WA (1997) Regulation of neuronal diversity in the *Xenopus* retina by Delta signalling. Nature 385:67–70
Dorsky RI, Rapaport DH, Harris WA (1995) Xotch inhibits cell differentiation in the *Xenopus* retina. Neuron 14:487–496
Dudley AT, Lyons KM, Robertson EJ (1995) A requirement for bone morphogenetic protein-7 during the development of the mammalian kidney and eye. Genes Dev 9:2795–2807
Dumstrei K, Nassif C, Abboud G, Aryai A, Aryai AR, Hartenstein V (1998) EGFR signaling is required for the differentiation and maintenance of neural progenitors along the dorsal midline of the *Drosophila* embryonic head. Development 125:3417–3426
Eakin RM, Kuda A (1971) Ultrastructure of sensory receptors in Ascidian tadpoles. Z Zellforsch Mikrosk Anat 112:287–312
Easter Jr. SS (2000) Let there be sight. Neuron 27:193–195
Eggert T, Hauck B, Hildebrandt N, Gehring WJ, Walldorf U (1998) Isolation of a *Drosophila* homolog of the vertebrate homeobox gene *Rx* and its possible role in brain and eye development. Proc Natl Acad Sci USA 95:2343–2348
Friedrich M, Benzer S (2001) Divergent decapentaplegic expression patterns in compound eye development and the evolution of insect metamorphosis J Exp Zool 288:39–55
Fu W, Noll M (1997) The Pax2 homolog *sparkling* is required for development of cone and pigment cells in the *Drosophila* eye. Genes Dev 11:2066–2078
Fuhrmann S, Levine EM, Reh TA (2000) Extraocular mesenchyme patterns the optic vesicle during early eye development in the embryonic chick. Development 127:4599–4609
Furuta Y, Hogan BLM (1998) BMP4 is essential for lens induction in the mouse embryo. Genes Dev 12:3764–3775
Furuta Y, Piston DW, Hogan BLM (1997) Bone morphogenetic proteins (BMPs) as regulators of dorsal forebrain development. Development 124:2203–2212
Gan L, Xiang M, Zhou L, Wagner DS, Klein WH, Nathans J (1996) POU domain factor Brn-3b is required for the development of a large set of retinal ganglion cells. Proc Natl Acad Sci USA 93:3920–3925

Goodrich LV, Scott MP (1998) Hedgehog and patched in neural development and disease. Neuron 21:1243–1257

Grainger RM (1992) Embryonic lens induction: shedding light on vertebrate tissue determination. Trends Genet 8:349–355

Greenwood S, Struhl G (1999) Progression of the morphogenetic furrow in the *Drosophila* eye: the roles of Hedgehog, Decapentaplegic and the Raf pathway. Development 126(24):5795–5808

Guillemot F, Cepko CL (1992) Retinal fate and ganglion cell differentiation are potentiated by acidic FGF in an in vitro assay of early retinal development. Development 114:743–754

Green P, Hartenstein AY, Hartenstein V (1993) Embryonic development of the *Drosophila* visual system. Cell Tissue Res 273:583–598

Greenwood S, Struhl G (1999) Progression of the morphogenetic furrow in the *Drosophila* eye: the roles of Hedgehog, Decapentaplegic and the Raf pathway. Development 126:5795–5808

Hassan BA, Bellen HJ (2000) Doing the MATH: is the mouse a good model for fly development? Genes Dev 14:1852–1865

Henrique D, Hirsinger E, Adam J, Le Roux I, Pourquie O, Ish-Horowicz D, Lewis J (1997) Maintenance of neuroepithelial progenitor cells by Delta-Notch signalling in the embryonic chick retina. Curr Biol 7:661–670

Hirth F, Therianos S, Loop T, Gehring WJ, Reichert H, Furukubo-Tokunaga K (1995) Developmental defects in brain segmentation caused by mutations of the homeobox genes *orthodenticle* and *empty spiracles* in *Drosophila*. Neuron 15:769–778

Hirsch N, Harris WA (1997) *Xenopus Pax*-6 and retinal development. J Neurobiol 32:45–61

Holland PW, Graham A (1995) Evolution of regional identity in the vertebrate nervous system. Perspect Dev Neurobiol 3:17–27

Hollemann T, Bellefroid E, Pieler T (1998) The *Xenopus* homologue of the *Drosophila* gene *tailless* has a function in early eye development. Development 125:2425–2432

Hu M, Easter SS (1999) Retinal neurogenesis: the formation of the intial central patch of postmitotic cells. Dev Biol 207:309–321

Ingham PW (1995) Signalling by Hedgehog family proteins in *Drosophila* and vertebrate development. Curr Opin Gen Dev 5:492–498

Jarman AP, Grell EH, Ackerman L, Jan LY, Jan YN (1994) Atonal is the proneural gene for *Drosophila* photoreceptors. Nature 369:398–400

Jasoni CL, Reh TA (1996) Temporal and spatial pattern of MASH-1 expression in the developing rat retina demonstrates progenitor cell heterogeneity. J Comp Neurol 369:319–327

Jasoni CL, Walker MB, Morris MD, Reh TA (1994) A chicken achaete-scute homolog (CASH-1) is expressed in a temporally and spatially discrete manner in the developing nervous system. Development 120:769–783

Jensen AM, Wallace VA (1997) Expression of Sonic hedgehog and its putative role as a precursor cell mitogen in the developing mouse retina. Development 124:363–371

Kablar B, Vignali R, Menotti L, Pannese M, Andreazzoli M, Polo C, Giribaldi MG, Boncinelli E, Barsacchi G (1996) *Xotx* genes in the developing brain of *Xenopus laevis*. Mech Dev 55: 145–158

Kanekar S, Perron M, Dorsky R, Harris WA, Jan LY, Jan YN, Vetter ML (1997) Xath5 participates in a network of bHLH genes in the developing *Xenopus* retina. Neuron 19:981–984

Kunes S (2000) Axonal signals in the assembly of neural circuitry. Curr Opin Neurobiol 10:58–62

Katz MJ (1983) Comparative anatomy of the tunicate tadpole *Ciona intestinalis*. Biol Bull Mar Biol Lab, Woods Hole 164:1–17

Lacalli TC, Holland ND, West JE (1994) Landmarks in the anterior central nervous system of amphioxus larvae. Philos Trans R Soc Lond B 344:165–185

LaVail MM, Faktorovich EG, Hepler JM, Pearson KL, Yasumura D, Matthes MT, Steinberg RH (1991) Basic fibroblast growth factor protects photoreceptors from light-induced degeneration in albino rats. Ann N Y Acad Sci 638:341–347

Levine EM, Roelink H, Turner J, Reh TA (1997) Sonic hedgehog promotes rod photoreceptor differentiation in mammalian retinal cells in vitro. J Neurosci 17:6277–6288

Li HS, Tierney C, Wen L, Wu JY, Rao Y (1997) A single morphogenetic field gives rise to two retina primordia under the influence of the prechordal plate. Development 124:603–615

Liem KF, Tremml G, Roelink H, Jessell TM (1995) Dorsal differentiation of neural plate cells induced by BMP-mediated signals from epidermal ectoderm. Cell 82:969–979

Liu W, Mo Z, Xiang M (2001) the Ath5 proneural genes function upstream of Brn3 POU domain transcription factor genes to promote retinal ganglion cell development. Proc Natl Acad Sci USA 98:1649–1654

Luo G, Hofmann C, Bronckers AL, Sohocki M, Bradley A, Karsenty G (1995) BMP-7 is an inducer of nephrogenesis, and is also required for eye development and skeletal patterning. Genes Dev 9:2808–2820

Macdonald R, Barth KA, Xu Q, Holder N, Mikkola I, Wilson S (1995) Midline signaling is required for *Pax* gene regulation and patterning of the eyes. Development 121:3267–3278

Masai I, Stemple DL, Okamoto H, Wilson SW (2000) Midline signals regulate retinal neurogenesis in zebrafish. Neuron 27:251–263

McCabe KL, Gunther EC, Reh TA (1999) The development of the pattern of retinal ganglion cells in the chick retina: mechanisms that control differentiation. Development 126:5713–5724

Mathers PH, Grinberg A, Mahon KA, Jamrich M (1997) The *Rx* homeobox gene is essential for vertebrate eye development. Nature 387:603–607

Namba R, Minden JS (1999) Fate mapping of *Drosophila* embryonic mitotic domain 20 reveals that the larval visual system is derived from a subdomain of a few cells. Dev Biol 212:465–476

Nassif C, Noveen A, Hartenstein V (1998) Embryonic development of the *Drosophila* brain I. The pattern of pioneer tracts. J Comp Neurol 402:10–31

Neumann CJ, Nuesslein-Volhard C (2000) Patterning of the zebrafish retina by a wave of sonic hedgehog activity. Science 289:2137–2139

Nornes HO, Dressler GR, Knapik EW, Deutsch U, Gruss P (1990) Spatially and temporally restricted expression of Pax2 during murine neurogenesis. Development 109:797–809

Noveen A, Daniel A, Hartenstein V (2000) Early development of the *Drosophila* mushroom body: the roles of *eyeless* and *dachshund*. Development 127:3475–3488

Parks JS, Adess ME, Brown MR (1997) Genes regulating hypothalamic and pituitary development. Acta Paediatrica Suppl 423:28–32

Pera EM, Kessel M (1997) Patterning of the chick forebrain anlage by the prechordal plate. Development 124:4153–4162

Piccolo S, Sasai Y, Lu B, De Robertis E (1996) Dorsoventral patterning in *Xenopus*: inhibition of ventral signals by direct binding of chordin to BMP-4 Cell 86:589–598

Pittack C, Grunwald GB, Reh TA (1997) Fibroblast growth factors are necessary for neural retina but not pigmented epithelium differentiation in chick embryos. Development 124:805–816

Pittack C, Jones M, Reh TA (1991) Basic fibroblast growth factor induces retinal pigment epithelium to generate neural retina in vitro. Development 113:577–588

Quiring R, Walldorf U, Kloter U, Gehring WJ (1994) Homology of the *eyeless* gene of *Drosophila* to the *Small eye* gene in mice and *Aniridia* in humans. Science 265:785–789

Rapaport DH, Dorsky RI (1998) Inductive competence, its significance in retinal cell fate determination and a role for Delta-Notch signaling. Semin Cell Dev Biol 9:241–247

Raymond SM, Jackson IJ (1995) the retinal pigmented epithelium is required for development and maintenance of the mouse neural retina. Curr Biol 5:1286–1295

Reichenbach A, Siegel A, Rickmann M, Wolff JR, Noone D, Robinson SR (1995) Distribution of Bergmann glial somata and processes: implications for function. J Hirnforsch 36:509–517

Rubenstein JL, Shimamura K, Martinez S, Puelles L (1998) Regionalization of the prosencephalic neural plate. Annu Rev Neurosci 21:445–477

Sanyanusin P, Schimmenti LA, McNoe LA, Ward TA, Pierpont MEM, Sullivan MJ, Dobyns WB et al. (1995) Mutation of the *PAX* 2 gene in a family with optic nerve colobomas, renal anomalies and vesicoureteral reflux. Nat Genet 9:358–363

Sharp (1885) Cited in Walls GL (ed) The vertebrate eye and its adaptive radiation. Cranbrook Press, Bloomfield Hills, Michigan, 785 pp

Sheshberadaran H, Takahashi JS (1994) Characterization of the chicken rhodopsin promoter: identification of retina-specific and glass-like protein binding domains. Mol Cell Neurosci 5:309–318

Simeone A, Acampora D, Mallamaci A, Stornaiuolu A, D'Apice MR, Nigro V, Boncinelli E (1993) A vertebrate gene related to orthodenticle contains a homeodomain of the bicoid class and demarcates anterior neurectoderm in the gastrulating mouse embryo. EMBO J 12:2735–2745

Stenkamp DL, Barthel LK, Raymond PA (1997) Spatiotemporal coordination of rod and cone photoreceptor differentiation in goldfish retina. J Comp Neurol 382:272–284

Stenkamp DL, Frey RA, Prabhudesai SN, Raymond PA (2000) Function for Hedgehog genes in zebrafish retinal development. Dev Biol 220:238–252

Stenkamp DL, Hisatomi O, Berthel LK, Tokunaga F, Raymond PA (1996) Temporal expression of rod and cone opsins in embryonic goldfish retina predicts the spatial organization of the cone mosaic. Invest Ophthalmol Vis Sci 37:363–376

Suzuki T, Saigo K (2000) Transcriptional regulation of atonal required for *Drosophila* larval eye development by concerted action of eyes absent, sine oculis and hedgehog signaling independent of fused kinase and cubitus interruptus. Development 127:1531–1540

Thomas P, Beddington R (1996) Anterior primitive endoderm maybe responsible for patterning the anterior neural plate in the mouse embryo. Curr Biol 11:1487–1496

Thor S, Andersson SG, Tomlinson A, Thomas JB (1999) A LIM-homeodomain combinatorial code for motor-neuron pathway selection. Nature 397:76–80

Tix S, Minden JS, Technau GM (1989) Pre-existing neuronal pathways in the developing optic lobes of *Drosophila*. Development 105:739–746

Torres M, Gomez-Pardo E, Gruss P (1996) Pax 2 contributes to inner ear patterning and optic nerve trajectory. Development 122:3381–3391

Treisman JE, Heberlein U (1998) Eye development in *Drosophila*: formation of the eye field and control of differentiation. Curr Top Dev Biol 39:119–158

Wan J, Stenkamp DL (2000) Cone mosaic development in the goldfish retina is independent of rod neurogenesis and differentiation. J Comp Neurol 423:227–242

Wang SW, Kim BS, Ding K, Wang H, Sun D, Johnson RL, Klein WH, Gan L (2001) Requirement for math5 in the development of retinal ganglion cells. Genes Dev 15:24–29

Wickler KC, Rakic P (1991) Relation of an array of early-differentiating cones to the photoreceptor mosaic in the primate retina. Nature 351:397–400

Wickler KC, Rakic P, Bhattacharyya N, Macleish PR (1997) Early emergence of photoreceptor mosaicism in the primate retina revealed by a novel cone-specific monoclonal antibody. J Comp Neurol 377:500–508

Wilson PA, Hemmati-Brivanlou A (1995) Induction of epidermis and inhibition of neural fate by Bmp-4 Nature 376:331–333

Yan RT; Wang SZ (1998) NeuroD induces photoreceptor cell overproduction in vivo and de novo generation in vitro. J Neurobiol 36:485–496

Younoissi-Hartenstein A, Green P, Liaw G, Rudolph K, Lengyel J, Hartenstein V (1997) Control of early neurogenesis of the *Drosophila* brain by the head gap genes *tll*, *otd*, *ems*, and *btd*. Dev Biol 182:270–283

Yu RT, McKeown M, Evans RM, Umesona K (1994) Relationship between *Drosophila* gene tailless and a vertebrate nuclear receptor Tlx. Nature 370:375–379

Zhang XM, Yang XJ (2001) Regulation of retinal ganglion cell production by Sonic hedgehog. Development 128:943–957

Zhao S, Barnstable CJ (1996) Differential effects of bFGF on development of the rat retina. Brain Res 723:169–176

Zhou X, Hollemann T, Pieler T, Gruss P (2000) Cloning and expression of xSix3, the *Xenopus* homologue of murine Six3. Mech Dev 91:327–30

Zimmermann LB, De Jesus-Escobar J, Harland RM (1996) The Spemann organizer signal noggin binds and inactivates bone morphogenetic protein-4 Cell 86:599–606

Applications of the *Drosophila* Retina to Human Disease Modeling

Nancy M. Bonini[1] and Mark E. Fortini[2]

1 Introduction

The renowned Dutch naturalist Anton van Leeuwenhoek, an early inventor of the microscope, was fascinated by the new scientific perspectives afforded by his new apparatus. Writing just over three centuries ago, Leeuwenhoek marveled at the exquisite detail of the fly retina revealed by the microscope, and he commented upon its scientific implications:

If we now see that provident Nature works in all Creatures, from the biggest to the smallest, almost in one and the same way, and if we remember that each round protuberance of the Eye is composed of many superposed scale-like parts, as I have said about the eye of the Dragon-Fly, and still is provided with it perfect roundness, which most ingenious structure exceeds anything we see on the Earth with the naked Eye, we must say again: Away with the opinion of Aristotle and all those who still follow him and want to maintain that flying Creatures or any other living Creature is generated from rotten matter and who are trying to obscure the Truth with their writings. (Leeuwenhoek 1695)

Leeuwenhoek would doubtless be pleased to know that the theory of spontaneous generation has long since been discredited, but he might be surprised by the considerable attention currently devoted by scientists to the beautifully patterned insect compound eye. Over the past 20 years or so, biologists have discovered that the retina of one insect in particular, the fruit fly *Drosophila melanogaster*, is very amenable to studies on the development of multicellular organisms. These studies have sought to answer questions that are relevant to the development of many different animals, including humans. How is a complex array of differentiated cells generated from an initially homogeneous reservoir of undifferentiated cells? Are cell fates determined primarily by intrinsic cellular factors or by extrinsic cues? What are the molecular mechanisms that underlie pattern formation and cell fate acquisition during development? In their search for answers to these and other questions, biologists

[1] Department of Biology, Howard Hughes Medical Institute, University of Pennsylvania, 415 S. University Avenue, Philadelphia, Pennsylvania 19104-6018, USA
[2] Department of Genetics, University of Pennsylvania School of Medicine, Philadelphia, Pennsylvania 19104, USA

Results and Problems in Cell Differentiation, Vol. 37
K. Moses (Ed.): Drosophila Eye Development

have come to realize that many of the molecular mechanisms that control the formation and function of the fly eye are conserved in higher organisms, and that in many cases, malfunction of these mechanisms plays a key role in human disease.

In this chapter, we discuss how both traditional "forward genetic" screens and more molecularly based "reverse genetic" approaches (Fig. 1) using the *Drosophila* retina have elucidated genetic networks and biochemical mechanisms with clear applications to human disease biology. Decades of classical genetic screens have uncovered hundreds of mutants with informative phenotypes in eye development, tissue patterning, and visual phototransduction, revealing distinct molecular pathways and identifying their protein components. More recently, the realization that flies and humans share many structurally and functionally related gene families has spurred interest in using *Drosophila* to investigate known culprits of human diseases, particularly neurodegenerative diseases such as Huntington's disease, Parkinson's disease, and Alzheimer's disease. In this "reverse genetic" strategy, fly genes with similarity to specific human disease genes are functionally characterized by mutational inactivation, and normal and mutant forms of the fly proteins or their mammalian counterparts are overexpressed in transgenic *Drosophila* to assess their effects on neuronal differentiation and survival. Both the forward and reverse genetic approaches will no doubt benefit greatly from the recently completed sequencing of the entire *Drosophila* genome (Adams et al. 2000), techniques for expressing foreign proteins in flies (Brand and Perrimon 1993), as well as targeted alteration of fly genes by homologous recombination (Rong and Golic 2000).

2 Classic Forward Genetics in the *Drosophila* Retina

In the decades before recombinant DNA technology became available, *Drosophila* workers used classical mutagenesis and screening methods to isolate genetic mutants with interesting phenotypes, many of which exhibited abnormal retinal phenotypes. Over the past two decades, the realization that the fly eye is well-suited to studies of specific processes, such as cell patterning, apoptosis, and phototransduction, led to systematic genetic screens for mutations affecting these processes. Molecular characterization of the genes uncovered in these screens has by no means been completed, but such studies have identified key proteins and underlying mechanisms in eye development and function, many of which have parallels among human disease genes.

2.1 Eye Specification Genes

The *Drosophila* retina has been used to elucidate a network of genes involved in eye development. These gene interactions are now being extended to verte-

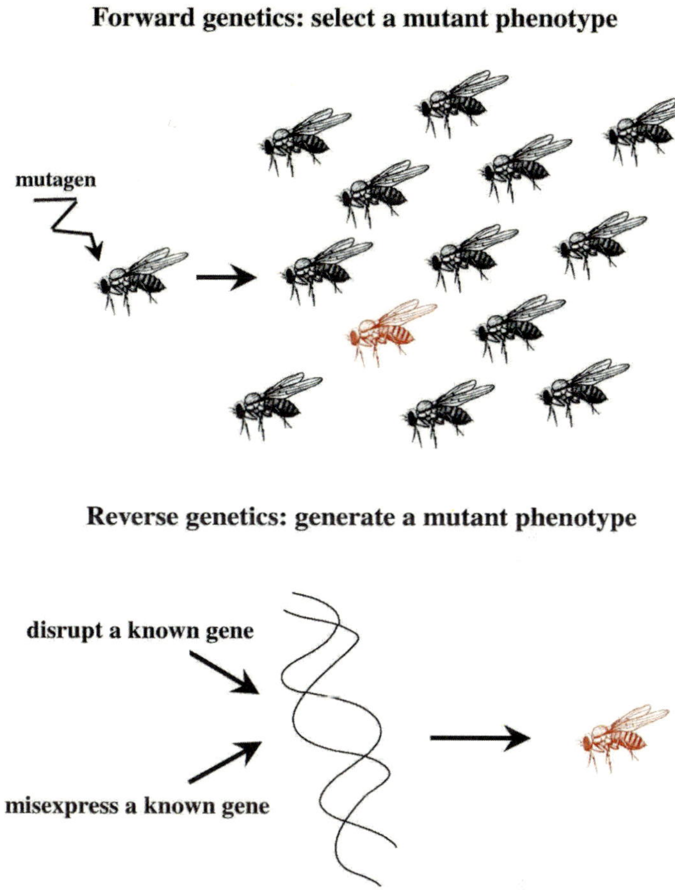

Fig. 1. Approaches for using *Drosophila* in genetic manipulations. *Top* The classical, forward genetic approach of mutagenizing flies, then selecting for a specific phenotype in the progeny. By this approach, genes are initially defined based on their function as assessed by mutant phenotype, and subsequently cloned to determine their molecular nature. *Bottom* The reverse genetic approach of manipulating a specific gene of interest, either by introducing the gene as a transgene for misexpression or by targeting a specific gene for mutagenesis. The potential function of the already known gene is then analyzed in the progeny flies that are mutant for that gene or that express the specific gene in a particular pattern

brate systems to ask whether the same genes are involved in vertebrate eye development, and to examine the relationships among them. A fundamental similarity between eye development in *Drosophila* and in humans was established by the remarkable finding that a homologous gene is involved in early events of both fly and vertebrate eye development. Upon cloning various *Drosophila* homeobox genes, a paired domain- and homeodomain-containing gene homologous to the vertebrate *Pax-6* gene was isolated and shown to correspond to the *Drosophila eyeless* mutation, which causes a dramatic reduced eye phenotype (Quiring et al. 1994). In humans, mutation of the *Pax-6* gene is responsible for the eye abnormality termed *Aniridia*, which ranges from mild phenotypes such as cataracts to severe phenotypes including loss of the iris (Hanson and van Heyningen 1995). Similarly, mutation of the *Pax-6* gene in mice is responsible for the *Small eye* mutant, which also exhibits developmental defects in eye formation (Hanson and van Heyningen 1995). The fact that mutation of homologous genes in flies, mice, and humans results in a similar disruption of eye development indicates that key aspects of early eye formation in vertebrates and flies are evolutionarily related, despite the dramatically different eye structures in the different organisms.

A central role of the *eyeless/Pax-6* gene in eye formation was subsequently established by transgenic expression of the gene in *Drosophila* – targeted misexpression of *eyeless* directs the formation of ectopic eyes, leading to flies with structurally normal eyes on their legs, wings, and other body parts (Halder et al. 1995). Subsequently, it has been observed that many other genes involved in fly eye specification have similar overexpression effects and mutant phenotypes, and research has focused on establishing the relationships among them and the roles of their vertebrate homologues. Much of this work has focused on the *eya* (*eyes absent*), *so* (*sine oculis*) and *dac* (*dachshund*) genes. These genes are involved in normal eye development in *Drosophila*, with primary roles early in eye formation as reflected by mutant phenotypes resembling that of *eyeless* mutants (Bonini et al. 1993; Cheyette et al. 1994; Mardon et al. 1994; Serikaku and O'Tousa 1994). Targeted over-expression of *eya* and *dac* also directs ectopic eye formation (Bonini et al. 1997; Shen and Mardon 1997); moreover, coexpression of either *dac* or *so* together with *eya* dramatically potentiates ectopic eye formation (Chen et al. 1997; Pignoni et al. 1997). The Eya protein binds in vitro and in yeast two-hybrid studies to the So and Dac proteins (Chen et al. 1997; Pignoni et al. 1997; Bui et al. 2000), raising the possibility that a large protein complex exits that directs eye formation. Whereas these genes were initially assumed to function downstream of *eyeless*, the regulatory pathways are complex as the proteins appear to have roles both downstream and upstream of *eyeless*. This observation suggests that the eye formation pathway – perhaps like most pathways – is not a simple linear pathway, but is comprised of complex regulatory loops.

With the identification of vertebrate homologues of these genes, their potential roles in vertebrate eye formation and other developmental events are being explored. This approach does not always yield simple answers. There are typically numerous vertebrate homologues of any one *Drosophila* gene, and the vertebrate genes generally have functions in many tissues, or they have redundant functions. A role for *eya* homologues in vertebrate eye development has been revealed (Xu et al. 1997; Azuma et al. 2000), but the *Eya1* gene also has a critical role in kidney and ear formation, with mutations in the gene being a leading cause of human hereditary deafness (Abdelhak et al. 1997; Xu et al. 1999). Moreover, some mutations mapping to the mammalian *Six* genes, counterparts of the fly *so* gene, result in holoprosencephaly (Wallis et al. 1999) whereas others are associated with anopthalmia (Gallardo et al. 1999). In both *Drosophila* and vertebrates, this same genetic network, or parts of it, appears to be involved in functions other than eye formation, such as development of specific brain centers in flies (Kurusu et al. 2000; Noveen et al. 2000) and muscle in the vertebrate limb (Heanue et al. 1999). These findings suggest that this genetic network might code, in an evolutionary sense, for a function more general than eye formation, and that it has been co-opted for use in multiple developmental contexts. Thus, the fly eye can provide a basis for defining the genes that comprise such networks, and place their activities relative to those of other genes involved in the same process.

2.2 Developmental Signaling Pathway Mutants

A number of conserved signaling pathways play major roles in the development of organisms ranging from the nematode *C. elegans* to man. Among these are the Ras/MAPK pathway, the Wnt pathway, the Hedgehog pathway, the Notch pathway, and the Transforming Growth Factor-β (TGF-β) pathway. In humans, mutations in components of several of these pathways contribute to disease. For instance, mutations in human *ras* genes are found in most colorectal cancers (Barbacid 1990), and human Hedgehog gene mutations cause basal nevus syndrome (Hammerschmidt et al. 1997). Malfunction of the human Wnt pathway has been implicated in mammary tumors and other cancers (Bienz and Clevers 2000). Similarly, mutations in genes encoding human Notch cell-surface receptors or one of their ligands, Jagged1, cause T-cell leukemia, an adult onset neurological disorder, and the developmental disorder Alagille syndrome (Artavanis-Tsakonas et al. 1999). Since all of these pathways also operate in *Drosophila*, they can be studied in a well-defined developmental context, as well as used in modifier screens to define additional components of the pathways.

In the *Drosophila* retina, the Hedgehog, TGF-β, and Wnt pathways are all important in regulating the progression of a wave of cell differentiation that patterns the immature eye tissue into the ordered ommatidial array of the adult eye (see Lee and Treisman, this Vol.). A Wnt pathway involving the *frizzled* gene is also involved in the establishment of cell polarity in ommatidial clusters (see Mlodzik, this Vol.). The Ras/MAPK pathway, which functions downstream of receptor tyrosine kinases such as Sevenless and the Epidermal Growth Factor Receptor (EGFR), transmits signals that induce numerous cell types of the fly retina (see Kumar, this Vol.). The Ras/MAPK core pathway appears to act both in remarkably specific cell-fate inductive interactions between specific ommatidial cells (the Sevenless pathway) as well as in more widespread signaling events needed for cell proliferation and survival (the EGFR pathway). Notch signaling activity is needed for the allocation of progenitor cells into evenly spaced clusters of ommatidial precursors within the morphogenetic furrow (see Baker, this Vol.), subsequent inductive events that specify the photoreceptors and other specific cell types (Cagan and Ready 1989), and cell death signals that eliminate superfluous cells to create the highly ordered lattice of the fully differentiated adult eye (Miller and Cagan 1998). In addition, recent studies have shown that Notch is involved in the establishment of a large-scale territorial border, the equator, that separates dorsal and ventral compartments and prevents cell intermingling across the compartment boundary (Cho and Choi 1998; Papayannopoulos et al. 1998), as well as the rotation and asymmetry of photoreceptor arrays within individual ommatidial clusters (Cooper and Bray 1999; Fanto and Mlodzik 1999). Extensive analyses of these pathways in *Drosophila* retinal development have contributed important insights into their developmental functions and molecular mechanisms, which are likely to be relevant to their involvement in human disease.

The *Drosophila* retina has proven especially useful for large-scale genetic modifier screens to identify additional molecular components of these highly conserved signaling pathways. In a pioneering screen of this type, a temperature-sensitive engineered form of the Sevenless receptor tyrosine kinase was used to provide just enough pathway activity to permit normal R7 photoreceptor cell development in most ommatidia across the *Drosophila* eye. Dominant mutations were then isolated that reduced signaling efficiency enough to significantly impair R7 specification (Simon et al. 1991). Among the genes defined by this "sensitized" screen were four that were critical for EGFR as well as Sevenless signaling, including Ras1, Son of Sevenless, Corkscrew, and Drk, confirming biochemical evidence from mammalian studies that receptor tyrosine kinase signals are transmitted through a Ras/MAPK kinase cascade, and identifying new components of the pathway that have since been found to act in the corresponding mammalian pathways. Additional screens for modifiers of activated or dominant negative components of this Ras/MAPK pathway, again performed using the fly eye, have continued to expand the list of genes and putative components functioning in this complex developmental pathway (Dickson et al. 1996; Karim et al. 1996; Therrien et al. 2000). Similar screens in the *Drosophila* retina have been applied to dissection of other pathways, including the Notch signaling pathway (Fig. 2; Brand and Campos-Ortega 1990; Fortini and Artavanis-Tsakonas 1994; Verheyen et al. 1996).

3 Phototransduction Cascade Mutants

A particular type of neurodegeneration seen in *Drosophila* is that caused by malfunction of the phototransduction cascade, analogous to the human neurodegenerative condition retinitis pigmentosa (RP). Mutant variants of several *Drosophila* photoreceptor cell-specific proteins, including rhodopsin (*ninaE* gene product), structural proteins and other factors needed for rhabdomere integrity or rhodopsin trafficking to the rhabdomeres (*chaoptic*, *rab6*, *ninaA*, and *ninaC* gene products), and retinal phospholipase C (*norpA* gene product)

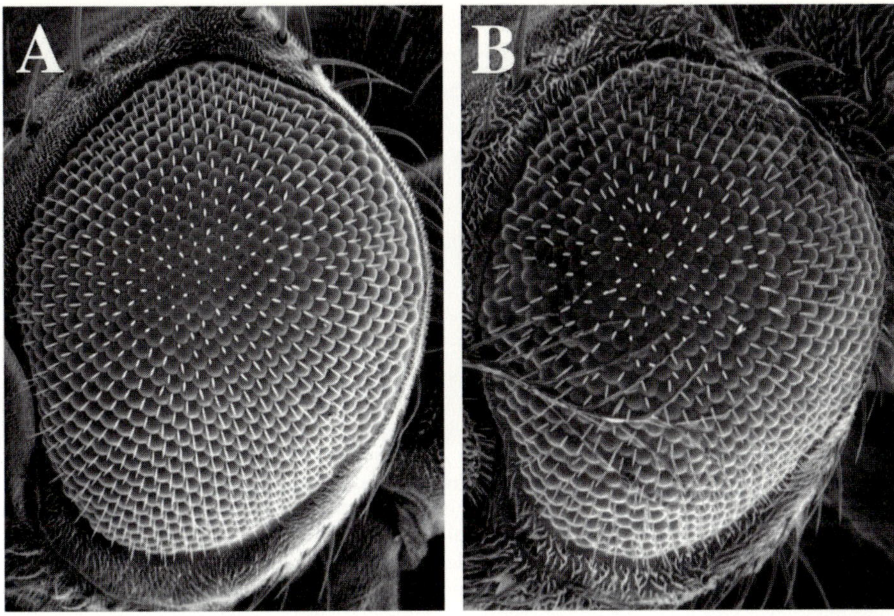

Fig. 2. Example of a dominant modifier mutation that influences the Notch phenotype in the eye. **A**) Eye phenotype of a fly bearing the *Notch* transheterozygous genotype N^{ts1}/fa^{g2}, the starting genotype for a temperature-sensitized modifier screen. The eye is normal in structure, although Notch signaling is reduced compared to wild-type. **B** N^{ts1}/fa^{g2} fly eye carrying a dominant enhancer modifier, in this case, a new mutation of the gene *mastermind*. The eye now shows a disordered surface structure, which is easily detected by examination under a dissecting microscope. (Fortini and Artavanis-Tsakonas 1994)

all cause gradual, light-independent degeneration, indicating that photoreceptor survival requires active phototransduction in flies (Montell and Rubin 1988; Colley et al. 1995; Kurada and O'Tousa 1995; Shetty et al. 1998; Van Vactor et al. 1998). Conversely, severely reduced Arrestin2 activity leads to a very rapid, light-dependent photoreceptor degeneration that can be counteracted by loss of phospholipase C function (Dolph et al. 1993). As Arrestin2 normally inactivates the phosphorylated form of rhodopsin by blocking its interaction with transducin, these observations imply that sustained phototransduction cascade activity can produce immediate and irreversible neuronal damage. Relatively rapid *Drosophila* photoreceptor degeneration is also seen in the *retinal degeneration* (*rdg*) mutants, such as *rdgB* and *rdgC*, which alter a putative rhabdomeric protein transporter and a calcium-dependent rhodopsin serine/threonine phosphatase, respectively (Steele and O'Tousa 1990; Steele et al. 1992; Vithelic et al. 1993).

The neuronal pathophysiologies exhibited by these *Drosophila* phototransduction mutants might be relevant to the complex genetic heterogeneity of RP

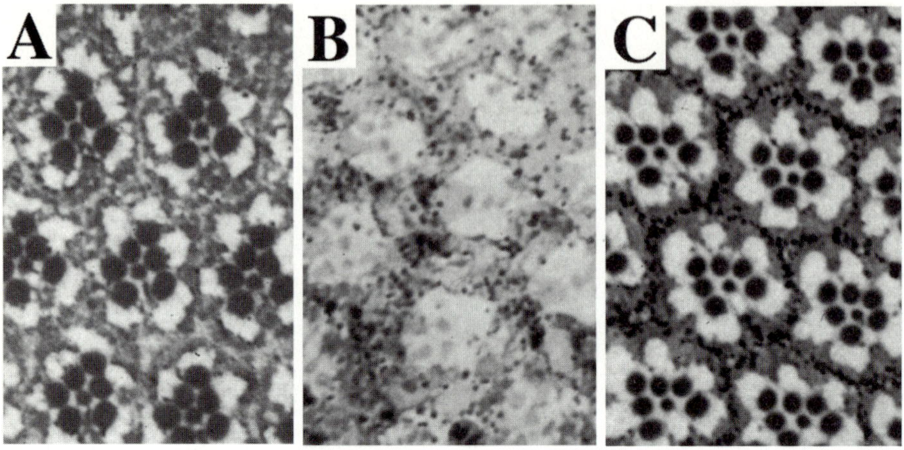

Fig. 3. Prevention of retinal degeneration induced by a rhodopsin mutation upon co-expression of an anti-cell death gene. **A** The *ninaE* rhodopsin mutant shows no degeneration when flies are raised in the dark, and the eye structure is normal. **B** The *ninaE* mutant eye exhibits severe degeneration when flies are raised in the light. **C** Flies that co-express the baculoviral P35 protein with the *ninaE* mutation and that are raised in the light show a normal eye structure. (Photomicrographs courtesy of F. Davidson; see Davidson and Steller 1998)

in human populations. RP can be caused by dominant, recessive, and mitochondrial mutations, and some pedigrees display digenic inheritance patterns implicating collaborating mutations at two different loci (Dryja and Berson 1995). Over 70 different dominant rhodopsin mutations have been detected in RP, and other mutations map to genes encoding peripherin and the β-subunit of rod cGMP-phosphodiesterase. Despite key differences between vertebrate and invertebrate phototransduction mechanisms, such as the active depolarization of *Drosophila* photoreceptors as opposed to the active hyperpolarization of their mammalian counterparts, identification and phenotypic analysis of retinal degeneration mutants in the fly might uncover promising candidate genes and biochemical events contributing to the etiology of RP and associated disorders, such as age-related macular degeneration. Furthermore, the *Drosophila* mutants prove useful for exploring therapeutic treatments for the analogous human conditions. Neurodegeneration seen in *rdgB* mutants can be delayed or prevented by administration of calcium-channel blockers to the retina (Sahly et al. 1992). Furthermore, late-stage inhibition of apoptosis by baculoviral P35 survival factor protects photoreceptors from degeneration in flies bearing a dominant rhodopsin gene mutation equivalent to a severe human RP mutation (Fig. 3; Davidson and Steller 1998). Rescued photoreceptors retain visual function in these flies, raising the possibility that modulating select phototransduction parameters or inhibiting neuronal apoptosis could be of therapeutic value in human RP syndromes.

4 Creating Directed Models for Human Disease

A complementary approach to classical genetics is to create a model for a specific disease in *Drosophila* using transgenic and reverse genetic technologies. This approach may include directed expression in flies of a mutant form of a human disease gene, or applying reverse genetics to inactivate an endogenous *Drosophila* gene that is related to a known human disease gene. In flies, it is reasonably straightforward to generate transgenic animals that express a human gene, and target that expression to specific cell or tissue types. The eye is particularly amenable to such an approach because of its well-characterized developmental origins and because it is dispensable for viability and fertility. Mutant gene expression and abnormal phenotypes that are thus restricted to the eye are advantageous for the critical subsequent step: application of *Drosophila* genetics to search for modifier mutations that might define interacting genes. By this approach, it is possible to uncover pathways that involve the normal function of the gene, as well as those disrupted by the mutant disease protein. This strategy has met with notable success in the modeling of human neurodegenerative diseases in *Drosophila*, as illustrated by several recent examples considered below.

4.1 Polyglutamine Expanded Repeat Disease

4.1.1 Modeling the Disease in the *Drosophila* Retina

The directed approach has been used to model human polyglutamine repeat disease in *Drosophila*. Human polyglutamine disease has a well-defined molecular mechanism: expansion of a CAG repeat-encoding glutamine within the open reading frame of the respective disease proteins (Zoghbi and Orr 2000). The expanded polyglutamine domain confers a dominant, toxic property on the disease protein, leading to neuronal dysfunction and degeneration. The longer the repeat, the earlier the onset and the more severe the disease (Gusella and MacDonald 2000). The dominant, gain-of-function nature of these diseases is well-suited for the misexpression approach. A number of polyglutamine diseases have been modeled in *Drosophila*, including SCA1 (spinocerebellar ataxia type 1), SCA3 (also called Machado-Joseph disease MJD) and Huntington's disease.

The SCA3/MJD model was the first fly model for a human neurodegenerative disease (Fig. 4; Warrick et al. 1998), followed by a Huntington's disease fly model (Jackson et al. 1998). Both of these models employ truncated versions of the respective disease proteins. The control, non-disease form of the proteins have short polyglutamine repeats within the normal range, whereas the mutant, disease forms of the proteins have long polyglutamine repeats at the high end of the range found in human disease, typically associated with juvenile-onset disease. By directing the protein selectively to the fly eye, Warrick et

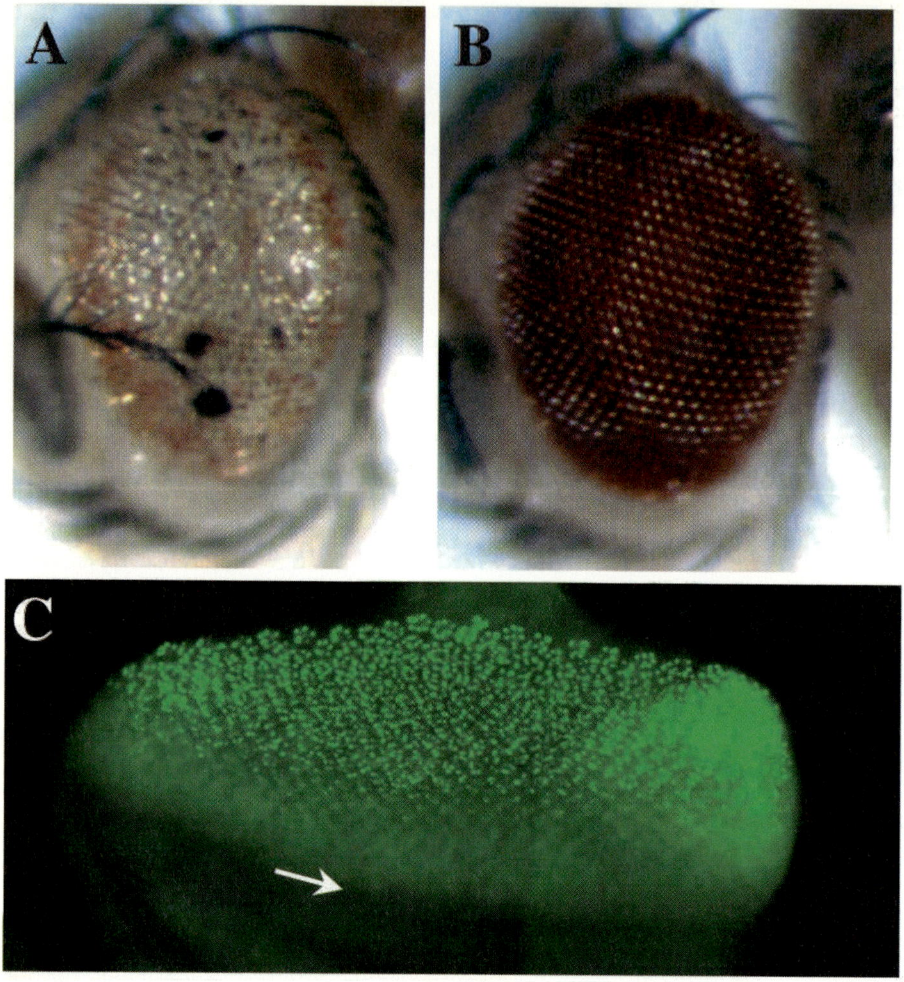

Fig. 4. Neurodegenerative phenotype caused by expression of pathologically active MJD protein in the *Drosophila* retina, and its suppression by the molecular chaperone Hsp70. **A** Fly expressing the mutant disease form of the MJD protein with an expanded polyglutamine repeat, exhibiting severe degeneration of the eye. **B** Co-expression of the molecular chaperone Hsp70 with the mutant disease protein results in an eye that appears phenotypically normal with the pigmentation restored. **C** Abnormal protein aggregates, termed nuclear inclusions, induced in the developing eye cells by the mutant MJD protein. Older cells (*top*) show larger, more prominent inclusions compared to the younger cells. Onset of expression is at the furrow (*arrow*). Co-expression of the chaperone does not alter the appearance of the abnormal protein aggregates as determined by immunocytochemistry (see Warrick et al. 1999)

al. (1998) showed that expression of a truncated MJD protein with a normal repeat of Q27 had no phenotype, whereas the protein with an expanded repeat of Q78 led to late-onset, progressive degeneration. Similarly, Jackson et al. (1998) found that expression of a truncated Huntingtin protein with an expanded repeat of Q120 resulted in adult-onset loss of integrity of the photoreceptor neurons, while a repeat length of Q75 resulted in less severe degeneration. The phenotypes of these expanded repeat genes in the eye are uniform for each construct and can be visualized using a low-power dissecting microscope or by observing the photoreceptor neurons under illumination of the retina. Thus, it is possible to apply genetic screens to define modifiers that have an effect on the phenotype due to alterations in gene dosage, as noted above for the analysis of developmental signaling pathways in the fly eye (see Fig. 2).

A fly model using the SCA1 protein has more recently been described (Fernandez-Funez et al. 2000). This model uses the full-length SCA1 protein with a Q30 repeat within the normal range, or alternatively with an expanded repeat of Q82. These proteins are identical to those used in mouse transgenic models, allowing subsequent comparison of the phenotypic features of the two model systems. In this case, both the normal control form of the protein, and the mutant expanded repeat form of the protein result in a rough eye phenotype in *Drosophila*. Although the phenotype is described as degenerative, the proteins may have an effect on development as revealed by the roughness of the eye, which is reminiscent of mutants affecting ommatidial development. As the Q30 phenotype appears to be a mild example of the Q82 phenotype, this result suggests that both forms of the protein can cause similar phenotypes, depending upon gene dosage and relative protein toxicity. This finding prompted a re-examination of the mouse transgenic phenotype, leading to the realization that increasing the dosage of the "normal" Q30 protein in mouse cerebellar cells induces degeneration, providing an example of how information derived from the fly eye model can lead to novel insights into aspects of human disease.

4.1.2 Defining Modifiers of Polyglutamine Disease Using *Drosophila* Genetics

A number of approaches have been taken to find novel genes that interact with polyglutamine disease proteins to modulate protein toxicity and/or neurodegeneration. These experiments illustrate the variety of applications of fly genetics to the study of human neurodegenerative disease, and demonstrate the utility of the fly eye for such modifier screens. The *Drosophila* polyglutamine models created to date are misexpression models, typically generated by using the GAL4-UAS transgenic system of directed expression (Brand and Perrimon 1993). The resulting eye phenotypes are sensitive to gene dosage, meaning that the systems are sensitized, such that dominant modifiers can be recovered in screens that detect phenotypic alterations due to reducing a gene's dosage by 50%, or that modify by directed coexpression of a second gene. One can also pursue other approaches, such as testing the role of suspected modifier genes that might be anticipated to affect the phenotype induced by abnormal polyglutamine protein. Through the candidate gene approach, the molecular chaperone Hsp70 was first shown to act as a modulator of neurodegenerative disease (Warrick et al. 1999). Coexpression of Hsp70 mitigates polyglutamine toxicity, restoring the external eye structure completely, and the internal retinal structure partially. Conversely, inhibition of normal levels of Hsp70 family members enhances degeneration, revealing a more severely degenerate eye. Biochemical studies reveal that Hsp70 modulates the aggregation of the protein, even though the protein aggregates appear unchanged when visualized by immunocytochemistry (Chan et al. 2000).

Approaches to identify novel genes that modulate polyglutamine toxicity include both new mutagenesis and testing existing collections of P element loss-of-function and gain-of-function overexpression lines for ability to modulate polyglutamine toxicity. By mobilizing a P-element containing UAS sequences in the fly genome, it is possible to generate lines that will drive the expression of a nearby gene upon insertion of the UAS element near the promoter region. This strategy allows the isolation of genes that modify the phenotype due to directed co-overexpression together with the polyglutamine protein. In this way, J-domain containing co-modulators of Hsp70, including Hsp40, have been recovered and shown to suppress polyglutamine phenotypes (Kazemi-Esfarjani and Benzer 2000). In a separate study, by crossing polyglutamine protein-expressing fly lines to existing collections of lethal mutants and overexpression fly lines, a large number of genes have been defined that display modest modifying effects on the eye phenotype of the SCA1 model (Fernandez-Funez et al. 2000). These genes include both components of the proteasome and chaperone pathways, as well as many RNA-binding proteins that are proposed to specifically modulate the activity of the SCA1 protein, which is itself thought to be an RNA-binding protein. The advantage of this approach is that it allows rapid identification of modifiers for which molecular data may already exist. The disadvantage is that the modifiers tend to be weak compared to *de novo*-induced mutations, the latter of which are selected due to their relatively strong modification of the transgenic phenotype.

4.2 Applications to Parkinson's Disease

The same approach – expressing a human disease gene in the fly in order to phenocopy the related human neurodegenerative disease – has been applied to Parkinson's disease, and illustrates the generality of the approach first demonstrated with human polyglutamine disease. Rare familial forms of dominant Parkinson's disease are associated with mutations in the protein alpha-synuclein (Polymeropoulos et al. 1997; Kruger et al. 1998). Alpha-synuclein is also a major component of the abnormal inclusions called Lewy bodies that occur in Parkinson's disease (Spillantini et al. 1997). Although alpha-synuclein itself may be highly divergent or absent in flies (Rubin et al. 2000), it is still feasible to misexpress the human gene in flies in order to generate a model for the disease to which genetic modification screens can be applied. Expression of alpha-synuclein in the fly, including wild-type and mutant forms, leads to selective loss of dopaminergic neurons accompanied by the formation of aggregated protein, reminiscent of Lewy bodies (Feany and Bender 2000). Whereas screens based on modification of dopaminergic cell loss are time-consuming and laborious, it has also been suggested that directed expression of alpha-synuclein in the eye disrupts the normal retinal structure, despite the fact that photoreceptor neurons are not dopaminergic neurons (Feany and Bender 2000). This eye phenotype provides a system in which one can define genes that can modify the phenotype due to directed alpha-synuclein expression, analogous to the expanded repeat protein models described above. Furthermore, Parkinson's disease has also been linked to loss-of-function mutations in additional genes, including *parkin* and *ubiquitin C-terminal hydrolase L1* (Kitada et al. 1998; Leroy et al. 1998), homologues of which have been detected in the *Drosophila* genome (Rubin et al. 2000). It might therefore prove possible to define the role of these genes in flies, and to create additional overexpression models. Environmental factors, such as pesticides, can also induce degeneration of dopaminergic neurons and phenocopy Parkinson's disease in a rat model (Betarbet et al. 2000), suggesting that a drug-insult approach may be applicable to flies and might facilitate the isolation of genes involved in the drug mechanism of action.

4.3 Alzheimer's Disease-Related Presenilin

A combination of traditional loss-of-function genetics and eye-directed expression in *Drosophila* has recently been applied to genes related to human Alzheimer's disease, including β-amyloid precursor protein (APP) and Presenilin (human PS-1 and PS-2). Mutations in these genes in humans and transgenic mice leads to an accelerated formation of extracellular amyloid plaques and intraneuronal filaments, the neuronal lesions characteristic of Alzheimer's disease (Selkoe 1999). APP undergoes complex proteolytic processing that generates short 40–42 amino acid amyloid peptides that are a major constituent of the amyloid plaques, and mutations in APP that cause disease lead to increased proteolysis and amyloid deposition. The Presenilin genes encode transmembrane proteins that are likely to function as novel aspartyl proteases that perform one of the proteolytic cleavages of APP (Wolfe et al. 1999). This cleavage event occurs within the transmembrane domain of APP, and mutant variants of human PS1 and PS2 favor production of the more neurotoxic 42-amino acid amyloid peptides over the shorter 40-amino acid peptides, both of which result from cleavage at slightly different sites within the transmembrane domain of APP (Selkoe 1999). Remarkably, Presenilins are also required for the transmembrane cleavage of the Notch receptor in response to ligand binding, releasing the intracellular domain of Notch which translocates to the nucleus and functions as a transcriptional co-activator (De Strooper et al. 1999; Struhl and Greenwald 1999; Kopan and Goate 2000). Finally, some studies have suggested that Presenilins are involved in the regulation of apoptosis, rendering cells more susceptible to apoptotic stimuli (Deng et al. 1996; Guo et al. 1996, 1997; Wolozin et al. 1996; Janicki and Monteiro 1997) or giving rise to C-terminal fragments with anti-apoptotic activities (Vito et al. 1996, 1997).

Drosophila has highly conserved homologues of these human disease genes, and using standard *Drosophila* genetics, loss-of-function mutations have been generated to analyze the normal functions of the genes in the nervous system. Homozygous deletion of the fly APP gene causes behavioral defects in visual phototaxis that can be reversed by providing the flies with a functional human APP gene (Luo et al. 1992). Loss of Presenilin in *Drosophila* causes a neurogenic phenotype – identical to characteristic phenotypes caused by loss of the canonical neurogenic gene *Notch* (Struhl and Greenwald 1999; Ye et al. 1999). Overexpression of Presenilin in the *Drosophila* eye causes cell death, although the levels of apoptosis are far lower than is seen with overexpression of "killer genes" that function directly in the apoptotic pathway in flies (Ye and Fortini 1999). Instead, overexpression of Presenilin at levels several fold higher than normal results in infrequent cell death across the fly retina, whereas mild overexpression of wild-type or Alzheimer's disease-associated Presenilins at levels approximately two fold above normal produces no apparent apoptotic phenotypes. The apoptotic effects of Presenilin in the fly retina are modulated by Notch pathway activity, and the high levels of Presenilin expression interfere with Notch receptor synthesis. Under these conditions, normal Presenilin-dependent processes, such as Notch signaling, may be impaired, resulting indirectly in the elimination of developmentally aberrant cells by apoptosis. In this case, the *Drosophila* retina was useful in assessing the apoptotic effects of Presenilin in a whole-tissue model, in contrast to earlier mammalian cell culture studies. Furthermore, the fly model provided the opportunity of connecting these effects to a known developmental signaling pathway – the Notch pathway – by taking advantage of *Drosophila* genetics and mutant phenotypes not readily available in cell culture systems.

5 Concluding Remarks

With the completion of the *Drosophila* genomic sequence, it is now clear that at least 60% of human disease genes identified to date have homologues in the fruit fly. Given the ease of manipulation with transgenic expression methods, the non-essential nature of the fly eye, and its well-described developmental history and technical approaches, the *Drosophila* retina will remain an experimental model of choice for deciphering functions and interactions of genes homologous to human disease genes. Through either endogenous mutations – for which technical advances are constantly forthcoming – or ectopic gene expression studies, the eye provides a critical tool for genetic approaches to be applied to complex human diseases. An astounding degree of functional homology has been found between fly and mammalian genes, such that mammalian counterparts can be used to replace endogenous fly genes, and expression of human disease genes yields phenotypes in the fly that reproduce the fundamental features of the respective diseases.

The true power of *Drosophila* as applied to the problems of human disease will be realized through its genetics. Whereas one can test specific ideas about disease mechanism in a complex nervous system in vivo with *Drosophila*, the greatest advantage provided by simple model organisms is the ability to screen for genes whose activities affect disease progression without any preconceived notions of which genes might be involved. Unbiased, systematic screening methods are a well-established approach in experimental organisms such as yeast, nematodes, and fruit flies, and the *Drosophila* retina is particularly well suited to genetic screens to identify modifiers of a human disease gene phenotype or to uncover mutations that produce phenotypes resembling the cellular pathologies seen in particular human diseases. Although it is unlikely that all features of a given human disease will be replicated in *Drosophila*, the fly can be used to apply genetics to those features that are conserved. In this regard, it is worth recalling again the words of Anton van Leeuwenhoek, who examined the insect compound eye under his simple microscope and concluded that "provident Nature works in all Creatures, from the biggest to the smallest, almost in one and the same way" (Leeuwenhoek 1695). The *Drosophila* retina has proved this point in recent years, and will continue to be one of the premier molecular genetic systems in which to gain insights into the complexities of human disease genes.

Acknowledgements. We thank F. Davidson for providing figure material. Research in N.M.B.'s lab is supported by grants from the David and Lucile Packard Foundation, the Huntington's Disease Society of America Coalition for the Cure, and the National Institutes of Health. Research in M.E.F.'s lab is supported by grants from the National Institutes of Health, the Alzheimer's Association, and the American Heart Association. N.M.B. is an Assistant Investigator of the Howard Hughes Medical Institute. We apologize to all of our colleagues whose work we were unable to cite due to space limitations.

References

Abdelhak S, Kalatzis V, Heilig R, Compain S, Samson D, Vincent C, Weil D, Cruau, C, Shaly I, Leibovici M, Bitner-Glindzicz M, Francis M, Lacomb, D, Vigneron J, Characho R, Boven K, Bedbeder P, Van Regemorter N, Weissenbach J, Petit C (1997) A human homologue of the *Drosophila* eyes absent gene underlies Branchio-Oto-Renal (BOR) syndrome and identifies a novel gene family. Nat Genet 15:157–164

Adams MD, Celniker SE, Holt RA et al. (2000) The genome sequence of *Drosophila melanogaster*. Science 287:2185–2195

Artavanis-Tsakonas S, Rand MD, Lake RJ (1999) Notch signaling: cell fate control and signal integration in development. Science 284:770–776

Azuma N, Hirakiyama A, Inoue T, Asaka A, Yamada M (2000) Mutations of a human homologue of the *Drosophila eyes absent gene (EYA1)* detected in patients with congenital cataracts and ocular anterior segment anomalies. Hum Mol Genet 9:363–366

Barbacid M (1990) ras oncogenes: their role in neoplasia. Eur J Clin Invest 20:225–235

Betabet R, Sherer TB, MacKenzie G, Garcia-Osuna M, Panov AV, Greenamyre JT (2000) Chronic systemic pesticide exposure reproduces features of Parkinson's disease. Nat Neurosci 3:1301–1306

Bienz M, Clevers H (2000) Linking colorectal cancer to Wnt signaling. Cell 103:311–320

Bonini NM, Leiserson WM, Benzer S (1993) The *Drosophila eyes absent* gene directs ectopic eye formation in a pathway conserved between flies and vertebrates. Cell 72:379–395

Bonini NM, Bui QT, Gray-Board GL, Warrick JM (1997) The *Drosophila eyes absent* gene directs ectopic eye formation in a pathway conserved between flies and vertebrates. Development 124:4819–4826

Brand AH, Perrimon N (1993) Targeted gene expression as a means of altering cell fates and generating dominant phenotypes. Development 118:401–415

Brand M, Campos-Ortega J (1990) Second-site modifiers of the *split* mutation of Notch define genes involved in neurogenesis in *Drosophila melanogaster*. Roux's Arch Dev Biol 198:275–285

Bui QT, Zimmerman JE, Liu H, Gray-Board GL, Bonini NM (2000) Functional analysis of an eye enhancer of the *Drosophila* eyes absent gene: differential regulation by eye specification genes. Genetics 155:709–720

Cagan RL, Ready DF (1989) *Notch* is required for successive cell decisions in the developing *Drosophila* retina. Genes Dev 3:1099–1112

Chan HYE, Warrick JM, Gray-Board GL, Paulson HL, Bonini NM (2000) Mechanisms of chaperone suppression of polyglutamine disease: selectivity, synergy and modulation of protein solubility in *Drosophila*. Hum Mol Gen 9:2811–2820

Chen R, Amoui M, Zhang Z, Mardon G (1997) Dachshund and Eyes absent proteins form a complex and function synergistically to induce ectopic eye development in *Drosophila*. Cell 81:893–904

Cheyette BNR, Green PJ, Martin K, Garren H, Hartenstein V, Zipursky SL (1994) The *Drosophila sine oculis* locus encodes a homeodomain-containing protein required for the development of the entire visual system. Neuron 12:977–996

Cho KO, Choi KW (1998) Fringe is essential for mirror symmetry and morphogenesis in the *Drosophila* eye. Nature 396:272–276

Colley NJ, Cassill JA, Baker EK, Zuker CS (1995) Defective intracellular transport is the molecular basis of rhodopsin-dependent dominant retinal degeneration. Proc Natl Acad Sci USA 92:3070–3074

Cooper MT, Bray SJ (1999) Frizzled regulation of Notch signalling polarizes cell fate in the *Drosophila* eye. Nature 397:526–530

Davidson FF, Steller H (1998) Blocking apoptosis prevents blindness in *Drosophila* retinal degeneration mutants. Nature 391:587–591

Deng G, Pike CJ, Cotman CW (1996) Alzheimer-associated presenilin-2 confers increased sensitivity to apoptosis in PC12 cells. FEBS Lett 397:50–54

De Strooper B, Annaert W, Cupers P, Saftig P, Craessaerts K, Mumm JS, Schroeter EH, Schrijvers V, Wolfe MS, Ray WJ, Goate A, Kopan R (1999) A presenilin-1-dependent γ-secretase-like protease mediates release of Notch intracellular domain. Nature 398:518–522

Dickson BJ, van der Straten A, Dominguez M, Hafen E (1996) Mutations modulating Raf signaling in *Drosophila* eye development. Genetics 142:163–171

Dolph PJ, Ranganathan R, Colley NJ, Hardy RW, Socolich M, Zuker CS (1993) Arrestin function in inactivation of G protein-coupled receptor rhodopsin in vivo. Science 260:1910–1916

Dryja TP, Berson EL (1995) Retinitis Pigmentosa and allied diseases. Invest Ophthamol Vis Sci 36:1197–1200

Fanto M, Mlodzik M (1999) Asymmetric Notch activation specifies photoreceptors R3 and R4 and planar polarity in the *Drosophila* eye. Nature 397:523–526

Feany MB, Bender WW (2000) A *Drosophila* model of Parkinson's disease. Nature 404:394–398

Fernandez-Funez P, Nino-Rosales ML, de Gouyon B, She W-C, Luchak JM, Martinez P, Turiegano E, Benito J, Capovilla M, Skinner PJ, McCall AE, Canal I, Orr HT, Zoghbi HY, Botas J (2000) Identification of genes that modify ataxin-1-induced neurodegeneration. Nature 408:101–106

Fortini ME, Artavanis-Tsakonas S (1994) The Suppressor of Hairless protein participates in Notch receptor signaling. Cell 79:273–282

Gallardo ME, Lopez-Rios J, Fernaud-Espinosa I, Granadino B, Sanz R, Ramos C, Ayuso C, Seller MJ, Brunner HG, Bovolenta P, Rodriguez de Cordoba S (1999) Genomic cloning and characterization of the human homeobox gene SIX6 reveals a cluster of SIX genes in chromosome 14 and associates SIX6 hemizygosity with bilateral anophthalmia and pituitary anomalies. Genomics 61:82–91

Guo Q, Furukawa K, Sopher BL, Pham DG, Xie J, Robinson N, Martin GM, Mattson MP (1996) Alzheimer's PS-1 mutation perturbs calcium homeostasis and sensitizes PC12 cells to death induced by amyloid β-peptide. Neuroreport 8:379–383

Guo Q, Sopher BL, Pham DG, Furukawa K, Robinson N, Martin GM, Mattson MP (1997) Alzheimer's presenilin mutation sensitizes neural cells to apoptosis induced by trophic factor withdrawal and amyloid β-peptide: involvement of calcium and oxyradicals. J Neurosci 17: 4212–4222

Gusella JF, MacDonald ME (2000) Molecular genetics: unmasking polyglutamine triggers in neurodegenerative disease. Nat Rev 1:109–115

Halder G, Callaerts P, Gehring WJ (1995) Induction of ectopic eyes by targeted expression of the *eyeless* gene of *Drosophila*. Science 267:1788–1792

Hammerschmidt M, Brook A, McMahon AP (1997) The world according to *hedgehog*. Trends Genet 13:14–21

Hanson I, van Heyningen V (1995) Pax6: more than meets the eye. Trends Genet 11:268–272

Heanue TA, Reshef R, Davis RJ, Mardon G, Oliver G, Tomarev S, Lassar AB, Tabin CJ (1999) Synergistic regulation of vertebrate muscle development by Dach2, Eya2, and six1, homologs of genes required for *Drosophila* eye formation. Genes Dev 13:3231–3243

Jackson GR, Salecker I, Dong X, Yao X, Arnheim N, Faber PW, MacDonald ME, Zipursky SL (1998) Polyglutamine-expanded human Huntingtin transgenes induce degeneration of *Drosophila* photoreceptor neurons. Neuron 21:633–642

Janicki S, Monteiro MJ (1997) Increased apoptosis arising from increased expression of the Alzheimer's disease-associated Presenilin-2 mutation (N141I). J Cell Biol 139:485–495

Karim FD, Chang HC, Therrien M, Wassarman DA, Laverty T, Rubin GM (1996) A screen for genes that function downstream of Ras1 during *Drosophila* eye development. Genetics 143:315–329

Kazemi-Esfarjani P, Benzer S (2000) Genetic suppression of polyglutamine toxicity in *Drosophila*. Science 287:1837–1840

Kitada T, Asakawa S, Hattori N, Matsumine H, Yamamura Y, Minoshima S, Yokochi M, Mizuno Y, Shimizu N (1998) Mutations in the *parkin* gene cause autosomal recessive juvenile parkinsonism. Nature 392:605–608

Kopan R, Goate A (2000) A common enzyme connects Notch signaling and Alzheimer's disease. Genes Dev 14:2799–2806

Kruger R, Kuhn W, Muller T, Woitalla D, Graeber M, Kosel S, Przuntek H, Epplen JT, Schols L, Riess O (1998) Ala30-to-pro mutation in the gene encoding alpha-synuclein in Parkinson's disease. Nat Genet 18:106–108

Kurada P, O'Tousa JE (1995) Retinal degeneration caused by dominant mutations in *Drosophila*. Neuron 14:571–579

Kurusu M, Nagao T, Walldorf U, Flister S, Gehring WJ, Furukubo-Tokunaga K (2000) Genetic control of development of the mushroom bodies, the associative learning centers in the *Drosophila* brain, by the *eyeless*, *twin of eyeless*, and *Dachshund* genes. Proc Natl Acad Sci USA 97:2140–2144

Leeuwenhoek A van (1695) Letter of 18 May 1695. In: Palm LC (ed) (1979) The collected letters of Antoni van Leeuwenhoek, vol 10. Swets and Zeitlinger, Lisse, p 251

Leroy E, Boyer R, Auburger G, Leube B, Ulm G, Mezey E, Harta G, Brownstein MJ, Jonnalagada S, Chernova T, Dehejia A, Lavedan C, Gasser T, Steinbach PJ, Wilkinson KD, Polymeropoulos MH (1998) The ubiquitin pathway in Parkinson's disease. Nature 395:451–452

Luo L, Tully T, White K (1992) Human amyloid precursor protein ameliorates behavioral deficit of flies deleted for *Appl* gene. Neuron 9:595–605

Mardon G, Solomon NM, Rubin GM (1994) *Dachshund* encodes a nuclear protein required for normal eye and leg development in *Drosophila*. Development 120:3473–3486

Miller DT, Cagan RL (1998) Local induction of patterning and programmed cell death in the developing *Drosophila* retina. Development 125:2327–2335

Montell C, Rubin GM (1988) The *Drosophila ninaC* locus encodes two photoreceptor cell specific proteins with domains homologous to protein kinases and myosin heavy chain head. Cell 52: 757–772

Noveen A, Daniel A, Hartenstein V (2000) Early development of the *Drosophila* mushroom body: the roles of eyeless and dachshund. Development 127:3475–3488

Papayannopoulos V, Tomlinson A, Panin VM, Rauskolb C, Irvine KD (1998) Dorsal-ventral signaling in the *Drosophila* eye. Science 281:2031–2034

Pignoni F, Hu B, Zavitz KH, Xiao J, Garrity PA, Zipursky SL (1997) The eye-specification proteins So and Eya form a complex and regulate multiple steps in *Drosophila* eye development. Cell 91:881–891

Polymeropoulos MH, Lavedan C, Leroy E, Ide SE, Dehejia A, Dutra A, Pike B, Root H, Rubenstein J, Boyer R, Stenroos ES, Chandrasekharappa S, Athanassiadou A, Papapetropoulos T, Johnso, WG, Lazzarini AM, Duvoisin RC, Di Iorio G, Golbe LI, Nussbaum RL (1997) Mutation in the alpha-synuclein gene identified in families with Parkinson's disease. Science 276:2045–2047

Quiring R, Walldorf U, Kloter U, Gehring WJ (1994) Homology of the *eyeless* gene of *Drosophila* to the *small eye* gene in mice and *aniridia* in humans. Science 265:785–789

Rong YS, Golic KG (2000) Gene targeting by homologous recombination in *Drosophila*. Science 288:2013–2018

Rubin GM, Yandell MD, Wortman JR et al. (2000) Comparative genomics of the eukaryotes. Science 287:2204–2215

Sahly I, Bar Nachum S, Suss-Toby E, Rom A, Peretz A, Kleiman J, Byk T, Selinger Z, Minke B (1992) Calcium channel blockers inhibit retinal degeneration in the retinal-degeneration-B mutant of *Drosophila*. Proc Natl Acad Sci USA 89:435–439

Selkoe DJ (1999) Translating cell biology into therapeutic advances in Alzheimer's disease. Nature 399 (Suppl):A23–31

Serikaku MA, O'Tousa JE (1994) *sine oculis* is a homeobox gene required for *Drosophila* visual system development. Genetics 138:1137–1150

Shen W, Mardon G (1997) Ectopic eye development in *Drosophila* induced by directed *dachshund* expression. Development 124:45–52

Shetty KM, Kurada P, O'Tousa JE (1998) Rab6 regulation of rhodopsin transport in *Drosophila*. J Biol Chem 273:20425–20430

Simon MA, Bowtell DDL, Dodson GS, Laverty TR, Rubin GM (1991) Ras1 and a putative guanine nucleotide exchange factor perform crucial steps in signaling by the Sevenless protein tyrosine kinase. Cell 67:701–716

Spillantini MG, Schmidt ML, Lee VM-Y, Trojanowski JQ, Jakes R, Goedert M (1997) alpha-synuclein in Lewy bodies. Nature 388:839–840

Steele F, O'Tousa JE (1990) Rhodopsin activation causes retinal degeneration in *Drosophila rdgC* mutant. Neuron 4:883–890

Steele FR, Washburn T, Rieger R, O'Tousa JE (1992) *Drosophila retinal degeneration C* (*rdgC*) encodes a novel serine/threonine protein phosphatase. Cell 69:669–676

Struhl G, Greenwald I (1999) Presenilin is required for activity and nuclear access of Notch in *Drosophila*. Nature 398:522–525

Therrien M, Morrison DK, Wong AM, Rubin GM (2000) A genetic screen for modifiers of a kinase suppressor of ras-dependent rough eye phenotype in *Drosophila*. Genetics 156:1231–1242

Van Vactor D Jr, Krantz DE, Reinke R, Zipursky SL (1988) Analysis of mutants in chaoptin, a photoreceptor cell-specific glycoprotein in *Drosophila*, reveals its role in cellular morphogenesis. Cell 52:281–290

Verheyen EM, Purcell KJ, Fortini ME, Artavanis-Tsakonas S (1996) Analysis of dominant enhancers and suppressors of activated Notch in *Drosophila*. Genetics 144:1127–1141

Vihtelic TS, Goebl M, Milligan S, O'Tousa JE, Hyde DR (1993) Localization of *Drosophila retinal degeneration B*, a membrane-associated phosphatidylinositol transfer protein. J Cell Biol 122: 1013–1022

Vito P, Lacanà E, D'Adamio L (1996) Interfering with apoptosis: Ca^{2+}-binding protein ALG-2 and Alzheimer's disease gene ALG-3. Science 271:521–525

Vito P, Ghayur T, D'Adamio L (1997) Generation of anti-apoptotic Presenilin-2 polypeptides by alternative transcription, proteolysis, and caspase-3 cleavage. J Biol Chem 272:28315–28320

Wallis DE, Roessler E, Hehr U, Nanni L, Wiltshire T, Richieri-Costa A, Gillessen-Kaesbach G, Zackai EH, Rommens J, Muenke M (1999) Mutations in the homeodomain of the human SIX3 gene cause holoprosencephaly. Nat Genet 22:196–198

Warrick JM, Paulson H, Gray-Board GL, Bui QT, Fischbeck K, Pittman RN, Bonini NM (1998) Expanded polyglutamine protein forms nuclear inclusions and causes neural degeneration in *Drosophila*. Cell 93: 939–949

Warrick JM, Chan HYE, Gray-Board GL, Chai Y, Paulson H, Bonini NM (1999) Suppression of polyglutamine-mediated neurodegeneration in *Drosophila* by the molecular chaperone HSP70. Nat Genet 23:425–428

Wolfe MS, Xia W, Ostaszewski BL, Diehl TS, Kimberly WT, Selkoe DJ (1999) Two transmembrane aspartates in presenilin-1 required for presenilin endoproteolysis and γ-secretase activity. Nature 398:513–517

Wolozin B, Iwasaki K, Vito P, Ganjei JK, Lacanà E, Sunderland T, Zhao B, Kusiak JW, Wasco W, D'Adamio L (1996) Participation of Presenilin 2 in apoptosis: enhanced basal activity conferred by an Alzheimer mutation. Science 274: 1710–1713

Xu PX, Woo I, Her H, Beier DR, Maas RL (1997) Mouse *Eya* homologues of the *Drosophila eyes absent* gene require *Pax6* for expression in cranial placodes. Development 124:219–231

Xu PX, Adams J, Peters H, Brown MC, Heaney S, Maas R (1999) Eya1-deficient mice lack ears and kidneys and show abnormal apoptosis of organ primordia. Nat Genet 23:113–117

Ye Y, Fortini ME (1999) Apoptotic activities of wild-type and Alzheimer's disease-related mutant Presenilins in *Drosophila melanogaster*. J Cell Biol 146:1351–1364

Ye Y, Lukinova N, Fortini ME (1999) Neurogenic phenotypes and altered Notch processing in *Drosophila Presenilin* mutants. Nature 398:525–529

Zoghbi HY, Orr HT (2000) Glutamine repeats and neurodegeneration. Annu Rev Neurosci 23: 214–247

Subject Index

Printing (Computer to Film): Saladruck, Berlin
Binding: Stürtz AG, Würzburg